Wearable Technology and Mobile Innovations for Next-Generation Education

Janet Holland
Emporia State University, USA

A volume in the Advances in Educational Technologies and Instructional Design (AETID) Book Series

Information Science REFERENCE
An Imprint of IGI Global

Published in the United States of America by
Information Science Reference (an imprint of IGI Global)
701 E. Chocolate Avenue
Hershey PA, USA 17033
Tel: 717-533-8845
Fax: 717-533-8661
E-mail: cust@igi-global.com
Web site: http://www.igi-global.com

Library of Congress Cataloging-in-Publication Data

Names: Holland, Janet, 1955- editor.
Title: Wearable technology and mobile innovations for next-generation education / Janet Holland, editor.
Description: Hershey, PA : Information Science Reference, 2016. | Includes bibliographical references and index.
Identifiers: LCCN 2015051295| ISBN 9781522500698 (hardcover) | ISBN 9781522500704 (ebook)
Subjects: LCSH: Educational technology. | Wearable technology.
Classification: LCC LB1028.3 .W398 2016 | DDC 371.33--dc23 LC record available at http://lccn.loc.gov/2015051295

This book is published in the IGI Global book series Advances in Educational Technologies and Instructional Design (AETID) (ISSN: 2326-8905; eISSN: 2326-8913)

British Cataloguing in Publication Data
A Cataloguing in Publication record for this book is available from the British Library.

For electronic access to this publication, please contact: eresources@igi-global.com.

Advances in Educational Technologies and Instructional Design (AETID) Book Series

Lawrence A. Tomei
Robert Morris University, USA

ISSN: 2326-8905
EISSN: 2326-8913

Mission

Education has undergone, and continues to undergo, immense changes in the way it is enacted and distributed to both child and adult learners. From distance education, Massive-Open-Online-Courses (MOOCs), and electronic tablets in the classroom, technology is now an integral part of the educational experience and is also affecting the way educators communicate information to students.

The **Advances in Educational Technologies & Instructional Design (AETID) Book Series** is a resource where researchers, students, administrators, and educators alike can find the most updated research and theories regarding technology's integration within education and its effect on teaching as a practice.

Coverage

- Bring-Your-Own-Device
- Curriculum Development
- Educational Telecommunications
- Game-Based Learning
- Web 2.0 and Education
- Adaptive Learning
- Social Media Effects on Education
- Virtual School Environments
- Instructional Design
- Hybrid Learning

IGI Global is currently accepting manuscripts for publication within this series. To submit a proposal for a volume in this series, please contact our Acquisition Editors at Acquisitions@igi-global.com or visit: http://www.igi-global.com/publish/.

The Advances in Educational Technologies and Instructional Design (AETID) Book Series (ISSN 2326-8905) is published by IGI Global, 701 E. Chocolate Avenue, Hershey, PA 17033-1240, USA, www.igi-global.com. This series is composed of titles available for purchase individually; each title is edited to be contextually exclusive from any other title within the series. For pricing and ordering information please visit http://www.igi-global.com/book-series/advances-educational-technologies-instructional-design/73678. Postmaster: Send all address changes to above address.

Titles in this Series

For a list of additional titles in this series, please visit: www.igi-global.com

Revolutionizing Education through Web-Based Instruction
Mahesh Raisinghani (Texas Woman's University, USA)
Information Science Reference • copyright 2016 • 391pp • H/C (ISBN: 9781466699328) • US $185.00 (our price)

Emerging Tools and Applications of Virtual Reality in Education
Dong Hwa Choi (Park University, USA) Amber Dailey-Hebert (Park University, USA) and Judi Simmons Estes (Park University, USA)
Information Science Reference • copyright 2016 • 360pp • H/C (ISBN: 9781466698376) • US $180.00 (our price)

Handbook of Research on Cloud-Based STEM Education for Improved Learning Outcomes
Lee Chao (University of Houston - Victoria, USA)
Information Science Reference • copyright 2016 • 481pp • H/C (ISBN: 9781466699243) • US $300.00 (our price)

User-Centered Design Strategies for Massive Open Online Courses (MOOCs)
Ricardo Mendoza-Gonzalez (Instituto Tecnologico de Aguascalientes, Mexico)
Information Science Reference • copyright 2016 • 323pp • H/C (ISBN: 9781466697430) • US $175.00 (our price)

Handbook of Research on Estimation and Control Techniques in E-Learning Systems
Vardan Mkrttchian (HHH University, Australia) Alexander Bershadsky (Penza State University, Russia) Alexander Bozhday (Penza State University, Russia) Mikhail Kataev (Tomsk State University of Control System and Radio Electronics, Russia & Yurga Institute of Technology (Branch) of National Research Tomsk Polytechnic University, Russia) and Sergey Kataev (Tomsk State Pedagogical University, Russia)
Information Science Reference • copyright 2016 • 679pp • H/C (ISBN: 9781466694897) • US $235.00 (our price)

Handbook of Research on Learning Outcomes and Opportunities in the Digital Age
Victor C.X. Wang (Florida Atlantic University, USA)
Information Science Reference • copyright 2016 • 851pp • H/C (ISBN: 9781466695771) • US $425.00 (our price)

Handbook of Research on Active Learning and the Flipped Classroom Model in the Digital Age
Jared Keengwe (University of North Dakota, USA) and Grace Onchwari (University of North Dakota, USA)
Information Science Reference • copyright 2016 • 450pp • H/C (ISBN: 9781466696808) • US $300.00 (our price)

Handbook of Research on Applied Learning Theory and Design in Modern Education
Elena Railean (European University of Moldova, Moldova) Gabriela Walker (University of South Dakota, USA) Atilla Elçi (Aksaray University, Turkey) and Liz Jackson (University of Hong Kong, Hong Kong)
Information Science Reference • copyright 2016 • 866pp • H/C (ISBN: 9781466696341) • US $400.00 (our price)

IGI GLOBAL
DISSEMINATOR OF KNOWLEDGE
www.igi-global.com

701 E. Chocolate Ave., Hershey, PA 17033
Order online at www.igi-global.com or call 717-533-8845 x100
To place a standing order for titles released in this series, contact: cust@igi-global.com
Mon-Fri 8:00 am - 5:00 pm (est) or fax 24 hours a day 717-533-8661

List of Reviewers

Susan Adam, *Emporia State University, USA*
Lisbeth Amhag, *Malmo University, Sweden*
Ibrahim Arpaci, *Gaziosmanpasa University, Turkey*
Bradley Barker, *University of Nebraska-Lincoln, USA*
Mike S. Butler, *Emporia State University, USA*
Alessio Drivet, *University of Turin, Italy*
Robert Gibson, *Emporia State University, USA*
Azizul Hassan, *The Cardiff Metropolitan University, UK*
Talitha Rowe Hudgins, *Utah Valley University, USA*
Jason Ribeiro, *University of Calgary, Canada*
Érika Bertozzi de Aquino Mattos, *Universidade Federal Fluminense, Brazil*
Mehtab Mehdi, *Northern Border University, Saudi Arabia*
Damara Goff Paris, *Emporia State University, USA*
Daniel Watanabe, *Olympic College, USA*
Thomas A. Woolman, *On Target Technologies, USA*

Table of Contents

Detailed Table of Contents

Very little research has examined the applications wearable technologies can have in education despite industry experts forecasting major growth in the global market for these innovative devices (Futuresource Consulting, 2015). Notwithstanding this profound challenge, in the next 3 to 5 years school districts will need to make multimillion-dollar investments into supporting these technologies as they begin to enter the learning environment. Given this gap in the research literature, this chapter explores the potential decision-making challenges school districts and their leaders will have to face when wearables become more commonplace in education. Using the Strategic Model for Technology Acquisition (Ribeiro, 2015), the author outlines both the innovative opportunities and potential problems wearables pose for both school leaders and stakeholders in the near future. Moreover, this chapter is meant to serve as an informative resource for education leaders and presents strategic and research-validated approaches for procuring and supporting wearable technologies.

Health and fitness wearable technology has been around for about 45 years. The most common wearables initially included heart rate monitors, pedometers, and GPS systems. While most of these were initially available as stand-alone devices, many are now available in combination devices such as fitness trackers and smart watches. Health and fitness wearables are used in a wide variety of situations including sports, athletics, personal health, fitness, education, and medical settings. A brief history of each wearable is provided, basic functions are explained, selected research is presented, and potential future developments are discussed. Some information is provided regarding current models and features.

Digital badges provide credentials to demonstrate successful student acquisition of knowledge, skills, and experience. The digital badges can especially excel at demonstrating new and emerging technology experiences through the use of wearable devices tailored to prepare students for the modern work place. Digital badges become virtual tracking systems making it possible to show students' specific learning, skills, and ultimately their employability. Not only do they record data on the individual student, they can track the institution and issuer each student has interacted with. With each successive use of such badges, a virtual narrative can be recorded reflecting how the badge was earned, what criteria was used, what standards were met, the rigor, and the different levels of learning.

Wearable Technologies represent an emerging theme. Probably the next emerging market where companies will focus. These devices are replacing entire categories of electronic objects in everyday life and affect the way we live, work and socialize. Among the many applications available, the author limits attention to the field of the smart video cameras. This chapter examines some of the most interesting applications of wearable cameras, with special reference to the Italian situation. In particular, the text traces a summary of the main applications in sports, spying, police, army, education, health, disabilities, and lifelogging. A part is devoted to "wearable extensions" and the concept of augmented reality.

Wearable technologies and appliances are making inroads into a variety of consumer and commercial applications, including leisure and entertainment, health care, and gaming. A natural evolution of the technology is in academe where faculty and students have begun exploring the possibilities of the technology in a variety of settings, most visibly in libraries where the process of seeking information using such devices holds significant promise. This chapter provides an inventory of the state of wearable technology, its challenges, its possibilities, and how it might be used in academe, including a study regarding the ability to access common library indexes using two wearable appliances: Google Glass and a smart watch.

The decision to implement new technologies such as smart glasses, smart phones, and tablets in an educational setting without determining optimal use scenarios is an evident universal problem as the adoption of such mobile platforms becomes widespread. Semi-structured interviews are conducted with two Science and Technology (S & T) and three Information Technology (IT) public school instructors to further investigate this significant problem. The constant comparative method was used to analyze the qualitative data resulting from these interviews. Preliminary results demonstrate the educational use of tablet computers has several advantages, along with a few limitations needing to be addressed. Specifically, one of the main limitations of these new instructional technologies is the lack of interactive content, embodied in the audio, video, and pictorial multimedia. The urgent need to address this limitation has motivated the development of multimedia software to work seamlessly on tablet computers.

Section 3
Mobile and Wearable Technology Applications

Assistive technology including wearables, have a positive impact towards changing the lives of people with disabilities. Advances in glass, electronics, sensors, and a wide range of digital devices are substantially improving how people with disabilities (PWD) navigate the world, and allow greater possibilities for PWD who are competing in today's job market. This chapter explores the following topics related to disabilities: Historical developments changing the lives of people with physical and sensory disabilities, laws pertaining to technology access, technological advances in wearables for individuals who are deaf, blind, paralyzed, or other mobility and speech impairments, socio-cultural and vocational implications of wearables for individuals who are deaf, deaf-blind or speech impaired, and the applications needed for wearables to meet the on-the-job demands.

The aim of this study is to analyze mobile technologies for student centered learning in a distance higher education program with a focus on mobile online webinars (web-based seminars or conferencing) using mobile applications such as laptops, smart phones, or tablets. As wearable technologies continue to grow it could very well extend to smart glasses, smart watches etc. These tools can provide face-to-face interactions, recording flipped classrooms and parallel chat communications. The data collection consists of observations of ten online face-to-face webinars with 22 students, six interviews, and two surveys. Theoretically, the study joins the research tradition of Computer-Supported Collaborative Learning with emphasis on collaboration, and Computer Self-Efficacy concerned with individuals' media and information literacy. Important conclusions from the study demonstrated mobile interactions increased student centered learning on theoretical concepts, assisted in the ability to review information critically, and provided experiences bridging professional teaching practices.

Google AdSense, a cost per click (CPC) advertising programme for those wishing to place ads on their website. Arguable viewed as promising in business, this literature review identifies key features of this promise when applied to education. This chapter connects business and education to identify Google AdSense as a valid tool and subject matter to be considered for use within the mobile and wearable technology curriculum. This chapter suggests considering positive outcomes generated using Google AdSense for incorporation into an education curriculum. Google AdSense should be applied to create more reasons to attract diverse user groups. This chapter also identifies the importance of Google needing a set of regulatory policies and procedures to avoid misuse of Google AdSense. In cases where monetary transactions occur, a possibility of violating the set terms and conditions exists. The chapter concludes with a clear call for inclusion of Google AdSense in an education curriculum.

This chapter explains the design and execution of a pre-service teacher training case study using Augmented Reality (AR), Quick Response (QR) codes, and social media mobile applications installed on iPads. The ScavengAR Hunt activity centered on a story narrative inspired by works of fine art using trigger images from the AR mobile application ARART® and incorporated elements of the board game Clue®. Pre-service teachers in the study were divided into groups of 4-6 and assigned specific, individual roles related to a mobile application used in the ScavengAR Hunt, and completed specific tasks while exploring the campus of a midwestern university. The research monitored the groups in real-time through reports submitted on Twitter and responses from QR code scans. The ScavengAR Hunt served as a model for designing a mobile learning activity incorporating multiple mobile applications.

In this chapter, we discuss the need for wearable technologies for earth science field research, a discipline that can be both extremely physically rigorous and even potentially dangerous at times. The field is also technically challenging because of environmental conditions as well as often being conducted in remote offline regions. Current trends in available wearable technology devices suitable for earth science applications are explored, in addition to describing possible modifications to existing wearable technologies in order to make them deployable, practical offline-capable geoscience systems. Offline Health and safety monitoring systems designed to operate on a low-powered personal body-area network system as well as offline (non-internet dependent) wearable communication technologies are particularly addressed. These types of wearable technologies represent some of the most complex aspects in this field as well being some of the most customizable systems with highly configurable components for remote back country fieldwork functionality.

This chapter describes the efforts and pilot study results of the Nebraska 4-H Wearable Technologies (WearTec) project, funded by the National Science Foundation's Innovative Technology Experiences for Students and Teachers (ITEST) program. The overall goal for the three-year WearTec project is to study a

Foreword

Google, Apple, Samsung, and many others have always been known for innovations in almost every area they can get their hands onto, and now have launched new, and might I mention amazing, technologies in the fields of wearables. These new forms of technology are being touted as the next big thing in portable technology. These wearable ideas are instant information units, allowing users hands free control of powerful computer devices in real time. These devices currently contain as much processing power as your middle line mobile cellular device, like the iPhone 4, Galaxy s4, and even the iPad first generation. The possibilities are endless among millions of users of wearables. Currently, these devices have the potential to bring exciting new possibilities to teachers and students alike.

Here are some exciting new ways Google and wearables are going to change and revolutionize the classroom. To begin, Google search alone has changed the way we look for things. Admit it, Google search is phenomenal, and this new technology of wearable devices will allow the student/teacher to stay connected to an interactive environment featuring Google Search and other online tools, all the time. Teachers and students can refer to topics related to their studies on the go. No messing with computers, iPads, or cell phones in the middle of class, and no distractions, just straight connection to that amazingly powerful service Google is known for. Google Glass and other wearables help students to create visually rich projects and presentations. Students can take pictures and videos with a wink, publish documents with a spoken command, and even connect with fellow colleagues from around the world at the blink of an eye. The same pictures and videos can be integrated into their presentations in PowerPoint or in Google Documents. Learning new languages and speaking them can now be done on the go with Glass and Google Translator. This allows for a breakthrough in English Language Learners (ELL) and English as a Second Language (ESOL) communication, students and teachers can now translate written text into their native tongue, opening new doors for ELL education. Distance learning is now a lot easier than before with all forms of wearable technology, including Google glass. Webinars, trainings, walkthroughs, and even field trips are live and on the go. Teachers can attend trainings anywhere in the world and see the entire training from the first person perspective. Students can attend field trips in other countries or even watch their teacher experience something that would not be able to be done in class. Flipping the classroom is made even easier. Teachers can record lessons from home, and then present them in class. This lesson can then be paired alongside the physical teacher in the classroom, which effectively makes the teacher his or her own co-teacher. Desktop control and PowerPoint presentations can be controlled with the nod of a head, wink, or vocal command. Providing hands free lessons allowing the teacher the ability to move around the room or the classroom and guide students all while teaching the lesson. Only time will tell whether smart glasses will be built and used effectively inside and outside of the classroom.

After investigating all the different functions and possibilities, just briefly, of Google Glass/smart glass technology and all forms of wearables, one can see the positive influence it will have on our education and world today. However, we are not the only company with the idea of educational development and applications of smart glasses in education. After data evaluation and research, we have found over 4,481 developers trying to profit from the application and development of smart glass and other wearable technologies in the education sector. This means that over 32,000 users of the current Google Glass brand, along slide millions of other wearable technologies leads to almost a 15% market share. From a business standpoint there will be extensive competition for the development of smart glass technology. Any teacher that has seen technology flourish over the past 20 years can see that if we start at this time in the market for smart glass and wearable technology, we will be the central hub for all things wearable in the classroom. We will be ready and prepared for new innovative ways to educate our future youth. In addition, if we are this prepared, we will be ahead of the game with training teachers who obtain this amazing technology and build more effectiveness among those who currently or possibly will use it in their classroom. Wearables are the future of education, and we need to be ready by educating ourselves of the possibilities of its success.

Adam Winkle
EduGlasses, USA

Preface

Wearable and mobile technologies are causing highly significant advances affecting everything from business, to education, personal health, and all of the sciences across a wide range of discipline areas. We are seeing everything wearable becoming smart, from glasses, to hearables, wristbands, watches, jewelry, rings, shoes, insoles, clothing, wrist drones, contact lenses, just about anything worn close to the body, on the body or even body implanted under the skin or swallowed using sensors to help us gather, use, and manipulate information. Smart phone and tablet based mobile technologies have traditionally been defined in three ways; 1) consisting of the functional components associated with wireless networking, 2) mobility, and 3) ubiquitous computing with on-demand power to access anytime anywhere. However, mobile and wearable technologies overlap due to the portability. With all the exciting, new, emerging mobile and wearable technologies we need to know what is working and why, so we can continue progress on how to design them better so they provide ever improving, natural, intuitive, quality user experiences.

When wearables affect our senses, especially sight and sound, they are augmenting our realities. The augmented reality needs to meet the objectives of the moment, communicate effectively, run dependably, and have sufficient human computer interaction design testing to ensure long term success. When the needs of the end users are met, the technology is more likely to flourish. When the technologies are ready for adoption, instructional designers can assist with effective and efficient training/ instruction within each unique business, or learning environment.

Wearable Technologies are changing the way we live, work, play, and interact. They provide users with a unique immersive form of augmented reality to enhance communications. Wearables provide users with more freedom by using miniaturized lightweight portable devices. The devices can include heads up displays, hands free controls, voice activation, touch screens/pads, or even body embedded sensors, with extensive Internet access and communications as we move through the day. No longer tethered, no longer weighted down, wearable designs are working to be seamlessly integrated parts of our enhanced day.

This book defines wearable technologies, their development, how they are currently being used, technical considerations, unique features, relevant examples in both business and education, an awareness of the effects of various interfaces, the implications for training and education, and sharing lessons learned and knowledge gained. At this time, there are no other publications like the one offered here. Since the field of wearables is so new it is important to share lessons learned and best practices to keep the technologies and applications focused on addressing consumer/learner needs with user-friendly designs to offer valuable resources and communications, often not possible with other media.

One of the limitations discovered in the development of this book is how new the technologies are. Sometimes it seems like a race to collect what we can of what is out there, how it is being applied, how well it is working as the technology races ahead.

The target audience for this book includes trainers, instructors, professors, researchers, policy makers affecting mobile and wearable technologies, instructional designers, technology professionals, as well as students, both undergraduate and postgraduate. For the most part, consisting of practitioners interested in the various areas of wearable and mobile technologies.

There are no current academic/ research based publications to inform us on wearables mobile technologies within the framework of teaching, training, and learning. This book offers readers one of the first opportunities to learn how to build upon appropriate models for the acquisition of wearable and mobile technologies for increased utility to facilitate effective applications.

My initial interest in Wearable Technologies stemmed from reading the NMC Horizon Report on a collaborative effort between NMC and Educause Learning Initiative (ELI). The report identifies emerging technologies likely to have an impact on learning, teaching, and creative inquiry over the next one to five years. It is designed to assist leaders and practitioners to develop a strategic technology plan. The report provides in-depth insight into how the trends are impacting the adoption of educational technology. Wearable and mobile technologies were found on their list of emerging technology trends.

The next step was researching books and articles on the topic of wearable technologies. In the process, I ran across the WT Wearable Technologies website and found some very exciting products and news items. I signed up for their summer 2014 San Francisco, CA conference. Attending the conference was so amazing as I was able to hear from professionals working in a wide variety of fields, meeting engineers building the products, personnel from so many start up companies, and ones already well established. At the end of the conference I left with a really good preview into the future of wearables for the next five years. Confirmed by my research and reading all of the current tech news I know there is no denying the current financial growth with the ultimate success of the wearable and mobile technologies is very promising indeed. When the market is growing so fast and so strongly in every industry it is only a matter of time until it reaches education.

I was one of the lucky ones to get a pair of Google Glass, Explorers Edition, to use for research purposes. I wore the glasses for one month and kept a journal of my trials and tribulations. To me it was so thrilling to be in on a new technology with the potential to literally change the face of computing. In this moment, hands free, temple touchpad, voice activated commands are all very appealing. Reading all of the bad press, was in contrast to my overall positive personal experiences. As a person who normally never wants excessive attention, I found it was actually a lot of fun to wear my smart glasses. It was so fun meeting people who were so curious about it. Wearing Glass sparked conversations with strangers, I let them try it out, and they took a lot of selfie photos to share on social media. Are improvements needed? Sure, like anything new it takes time to evolve, consider the needs of all of the target users, and lets be honest, to determine how to make a profit to support further development.

To give you a sample of some applications developed for using Glass; consider how individuals with autism can acquire assistance with facial interpretations, reminders to take medicine, describing the environment for blind individuals, medical uses, health records, hands free online data, hands free images while performing surgery, fire prevention and safety, airport personnel tasks, auto inspections, police work, electric fault detection, water fault detection, augmented reality, security, NASA research and development, and so many more. Just by the sheer number of new players found entering the smart glass market clearly demonstrates the anticipated growth. Some of the smart glass companies working on development include Huawei, Sony, Meta Pro, Vuzix, Optinvent, Recon Jet, GlassUp, Atheer Labs, Epson Moverio, Samsung, Epiphany, and Microsoft HoloLens. Each device has a different look, purpose, and features. To be successful in this arena it is important to know your audience, their perceptions, goals, needs, and values, while working within a cost attainable framework to be successful.

Looking at the bigger picture of wearables and mobile technologies, we are currently shifting from a niche market to more widespread adoption. The overall field is greatly expanding in all discipline areas. Consumers' desires for innovative applications using haptic technologies for extending the human experiences of touch, sight, hearing, and smell continue to grow. In the following paragraphs you will learn more about some of the wearables currently in development to get a preview of the trends and possible applications.

Haptics can provide a critical bridge for individuals with disabilities, or for those working at a distance, or needing enhanced user experiences by providing improved interactions. Considering approximately 45 million people are blind and 36 million have hearing loss, it is not surprising to see how individuals with disabilities are starting to find some of their needs being met with new wearable mobile devices. Currently in development is footwear with sensors to assist the visually impaired to navigate/map their world hands free without sight.

Bionic hearing implants and even a sensing glove are in development to translate sign language into voice. One glove in development will translate digital media using braille letter vibrations onto hand touch points. Already available are glasses with a smart camera attached to translate text to voice for items to be read. Or, there are glasses with voice directions for navigation. Another company offers a mobile imaging eye exam to detect diabetes induced blindness. There is even a smart watch with a braille screen. Clothing with haptics can provide tactile sensations simulating sport actions or even simulate a hug from a distant loved one. Haptics, referred to as a taptic engine, are used on the Apple smart watch for notifications to differentiate between different functions.

A wireless neuro headset can be used to manipulate virtual objects using only thoughts. This type of technology is especially beneficial to individuals with ALS or movement and communications disorders, to be able to use the device to interpret electrical signals to control an external device or share communications using only thoughts. Imagine a wheelchair or keyboard or game controlled hands free by thoughts alone. Headsets and glass attachments are being developed and improved to facilitate brain focus, concentration, and relaxation.

Since 1 in 10 adults now owns a fitness tracker, most people are well aware of the tremendous number of fitness tracker options on the market. Overall, we are beginning to see some merging of fitness tracker features and smart watches. The watch versions tend to offer additional notifications and apps. Fitness band features are even expanding into additional wearable devices such as shoes/ socks/ clips/ and bracelets embedded with sensors. One additional item in development is a smart band converting to an earpiece to expand the usability of function options.

There are also mergers happening between smart watches and smart clothing with most of them related to tracking health, fitness, and vital signs. Biomechanics is being embedded in clothing to create 3D representations, including tracking and feedback. Most of the fitness and health smart clothing collects biometric data, including respiration, blood pressure, and changes in body chemistry. Specifically, one smart clothing company is working on a device to detect early stage cancer. Smart clothing is being developed to protect babies by detecting their sleep, breathing, body position, and vital signs reporting through their parents' smart phone to give status notifications. Another company is working on a smart sock monitoring system to record vital signs for newborns.

A new smart earphone is in the works for listening to music using a head tilt to flip through songs, turn on/off lights and TV, and voice control of a smart phone. A new Bluetooth earbud or smart earring dot sensor is in the works to be placed behind the ear to allow one to listen, record, and answer their smart phone using hand or head gestures, in addition to normal voice control. Another project in progress is an earbud tracking dietary information. A new hearing aid is being developed to allow for video chats. Another earbud in development serves as a trainer with appropriately paced music. There is a Bluetooth headset for answering smart phones with improved hearing and audio output. One hearing device is enabled to record heart rate and core body temperature. Another start up hearing device is designed for performance coaching, training, recording breathing, and blood flow. One smart hearing device does all of the fitness metrics then reduces the music loudness level when answering a call.

The expectation is, consumers will eventually switch from tablets and mobile devices to smaller more discrete wearable technologies providing the same or improved expanded functions. Fashion forward smart jewelry is a good example of how the technology is becoming more hidden, yet within very close access. For example, there is a ring in development to provide messages, emails, text, apps, and call notifications discretely for important contacts without disturbing meetings with cell phones. Rings with sensors can be used in a variety of ways for different functions such as subway tokens, as a computer mouse, even gesture controls.

Miniaturization of computers, storage, electronic technologies, and improved communications all contribute to the abilities of smart devices. Wearable technologies take the smart devices to a whole new level of mobile convenience and usability. The sensors used in wearable devices can be electrical, mechanical, optical, chemical, or be any of a myriad of forms for capturing data from the body or local environment. This data can interact with human or machine through tactile, audible, visual, or haptic interfaces.

The primary objective of human computer interaction design is to increase the usability. We all want to have a positive, user-friendly, intuitive experience when using new wearable mobile technologies. We do not want to be confused, frustrated, lost, disconnected or unsupported, thus causing a very difficult user experience, making us want to throw the device in the trash. Not a good way to release new products, or to make happy customers, get additional sales, continue sales, nor a great way to increase the desire to buy upgrades or new products later on. Placing the needs and goals of users as the first priority is proving over and over again to be the best place to begin. Mobile wearable technologies need to be designed and tested to ensure enhanced augmented interactions.

We hope this book will assist you in getting a preview of the current state of wearable mobile technologies used in a variety of applications. We hope you will be inspired to jump in there and experiment for yourself, then contribute to this growing field with additional research, successful implementations, and profitable product development to ensure long term success. It is only by looking at the pros and cons, continued progress will be made for mobile and wearable technologies.

Janet Holland
Emporia State University, USA

Section 1

Mobile and Wearable Technology Introduction

Chapter 1
Purpose, Scope, and Technical Considerations of Wearable Technologies

Mehtab Mehdi
Northern Border University, Saudi Arabia

Abdulrahman Alharby
Northern Border University, Saudi Arabia

ABSTRACT

The purpose of this chapter is to concentrate on the technical considerations of wearable technologies. Approximately 50% of this chapter will be based on technical considerations while 30% is based on study areas, universities, degrees, and other information related to wearable technologies. For new readers of wearable technologies, we begin with a brief history and cover the scope of the field. The remaining 20% of the chapter covers useful information for interested readers of wearable technology.

INTRODUCTION

The terms "wearable technology", "wearable devices", and "wearables" all refer to electronic technologies or computers incorporated into items of clothing and accessories, which can comfortably be worn on the body. These wearable devices can perform many of the same computing tasks as mobile phones and laptop computers; however, in some cases, wearable technology can entirely outperform these hand-held devices. Wearable technology tends to be more sophisticated than hand-held technology on the market today because it can provide sensory and scanning features not typically seen in mobile and laptop devices, such as biofeedback and tracking of physiological functions.

Generally, wearable technology will have some form of communications capability and allows the wearer access to information in real time. Data-input capabilities are also a feature of such devices, as is local storage. Examples of wearable devices include watches, glasses, contact lenses, e-textiles, smart fabrics, headbands, beanies and caps, jewelry such as rings, bracelets, and hearing aid-like devices designed to look like earrings (Mann, 1997).

DOI: 10.4018/978-1-5225-0069-8.ch001

While wearable technology tends to refer to items put on and taken off with ease, there are more invasive versions of the concept as in the case of implanted devices such as microchips or even smart tattoos. Ultimately, whether a device is worn on or incorporated into the body, the purpose of wearable technology is to create constant, convenient, seamless, portable, and mostly hands-free access to electronics and computers.

The implications and uses of wearable technologies are far reaching and can influence the fields of health and medicine, fitness, aging, disabilities, education, transportation, enterprise, finance, gaming and music. The shared common goal of wearable technologies in each of these fields is to smoothly incorporate functional, portable electronics and computers into individuals' daily lives (Martin, 2007). Prior to their presence in the consumer market, wearable devices were primarily used in the field of military technology and had the biggest implications for healthcare and medicine (Le Tellier, 2009). In fact, just 10 years ago, medical engineers were talking about wearable devices to unobtrusively monitor the health and well-being of patients in the form of a "Wearable Motherboard™" or a "Smart Shirt," aimed at monitoring vital signs sending biofeedback information to a hub station in real time (Park & Jayaraman, 2003).

Even though wearable technology can potentially have the most impact in the fields of health and fitness, the technology also promises a great influence on gaming and entertainment. Augmented reality and wearable technology can combine to create a much more realistic and immersive environment in real time. The concept is not necessarily new, as augmented reality through the use of wearable devices has been discussed since the late 1990s; however, the prototypes are moving away from bulky technology such as large goggles and backpacks, to smaller, lightweight and more mobile systems.

If the more polished designs of mobile phones and digital cameras, currently on the market, are any indication for the future of wearable devices, then fashion, practicality, function and design will all be taken into account as these products advance. This consideration for both technology and aesthetics is already evident in devices such as Google Glass, which has a very sleek, lightweight, almost unobtrusive design.

As the potential uses of wearable technology in various fields continues to grow, the sociological and cultural impact of wearable technology in the future should not be minimized. Already, the current hand-held devices available to consumers, such as Smart Phones, iPods and tablets, have changed the technological and social landscapes on a global scale, such as, walking out in public and seeing an individual engaging with a hand-held device is commonplace. Such an image was nonexistent only 20 years ago. With that in mind, developers and analysts predict wearable technology will very quickly change the technological and cultural landscapes once again, and may even change the nature of mobile phones and other hand-held devices entirely (Mann, 1997; Davies, 2009).

BACKGROUND AND HISTORY OF WEARABLES

Today, wearable devices are exploding onto the market, with everything from smart glasses (Google Glass) to smart watches (Samsung Galaxy Gear) on the rise. As for smart watches, the technology isn't exactly new, however, as it began in the 1970s with the release of the first calculator watch. The calculator watch, specifically first released in 1975 under the Pulsar brand, became a widely popular tool for science geeks and math nerds everywhere. These early "smart" watches had their heydays through the mid-1980s and although their popularity went downhill, many companies still produce calculator watches to this day.

The calculator watch can be seen in pop culture, getting front cover prominence via the rock band "The Police" with the song titled "Wrapped Around Your Finger" (1983), worn by Sting. Calculator watches have also appeared in film, most notably in the Tom Hanks movie "Cast Away" (2000) and previously in "Back to the Future" (1985), with Michael J. Fox's character Marty McFly sporting one. In the hit game "Grand Theft Auto V" (2013), the protagonist Trevor Philips can be seen wearing a calculator watch as well (Smith, 2009). The dwindling popularity of the calculator watch was further crushed by the introduction of PDAs, smartphones and other techy products.

While the 1970s saw the production of the first modern-era wearable computers, the history of wearable technology may go back even farther. Due to the varied definitions of "wearable" and "computer," the first wearable computer may have been introduced as early as the 1600s, when the first abacus necklace was unveiled. Other early wearable computers include a sixteenth-century abacus ring, the first wristwatch worn by the Queen of Naples in 1810, and the first cheating devices worn in shoes at roulette tables in the 1960s (Lipman & Denise, 2014). In 1961, mathematicians Edward O. Thorp and Claude Shannon built computerized timing devices to help them cheat at the gambling game roulette — one concealed the device in a shoe, while the other in a pack of cigarettes. Various versions of this apparatus were built in the 1960s and 1970s (Eisenberg & Weickgenant, 2009). Thorp referred to himself as the inventor of the first "wearable computer." However, the specificities of the device were archaic and since the device relied on several other alternately placed components to work properly, it was deemed this was not a wearable computer. Others went on to perfect the device and built similar wearables through the 1970s, all for the purpose of cheating at the gaming tables.

Along with the early wearable calculator watches in the 1970s, came the introduction of a wearable system for the blind. Produced by C. C. Collins in 1977, this camera-to-tactile device converted images into a 1024-point, 10-inch square tactile grid on a person's vest.

In the 1980s, wearable computers started becoming more general-purpose and a better fit for the modern definition of "computer", by incorporating task-specific hardware into more general-purpose devices. Steve Mann built a backpack-mounted multimedia computer in 1981. Remaining active in the wearable computer field through the 80s led Mann to create and use the first wearable wireless webcam in 1994, becoming the first example of "lifelogging" (Clark, 2014).

Other wearable computer devices unleashed in 1994 include the first "wrist computer," developed by Edgar Matias and Mike Riucci of the University of Toronto, and the Forget-Me-Not device, developed by Mik Lamming and Mike Flynn at Xerox EuroPARC Forget-Me-Not recorded interactions with people and devices then stored it in a database for later query. DARPA's Smart Modules Program was started in 1994 to find a humionic approach to wearable and mobile computers. In 1996, DARPA hosted the "Wearables in 2005" workshop, bringing industrial, university and military visionaries together for the purpose of delivering computing to the individual.

As the world moved into the 21st century, wearable technology started to take off. In 2002, as part of Kevin Warwick's Project Cyborg, Kevin had his wife, Irena; wear a necklace electronically linked to her nervous system via an implanted electrode array. Her necklace color then changed dependent on signals from her nervous system.

Devices supporting augmented reality also got their start in the early-2000s. In the late-2000s, various Chinese companies began producing mobile phones on wristwatches, the descendants of which are, as of 2013, the i5 and i6, Global System Mobiles (GSM) phones with mini 1.8-inch displays. Approaching 2010, wearable devices started moving toward incorporating IEEE, IETF and other industry standards, including Bluetooth technology, leading to a variety of interfaces under the wireless personal area network (WPAN) and wireless body area network (WBAN) categories.

Wearable computing really took off in 2011 when Google developed the first prototype of what it now calls its Google Glass Project (smart glasses). The technology is based on military research of head-mounted displays starting back in 1995 (Tech Specs, 2013). In April 2013, Google Glass was released to a handful of Glass Explorers who were invited to try the technology. It was then officially released to the general public at a starting price of $1500 in May 2014. With the advent of Google Glass, numerous companies have made runs into the smart wearable market; including Apple (iWatch), Samsung (Galaxy Gear), Sony (Smart Watch), and others. Wearable technology has now evolved into numerous types of devices, including watches, glasses, headbands, wigs, rings, etc. Devices are being implemented for numerous applications, including personal and business computing, practical everyday tasks, fitness tracking, healthcare monitoring, etc. (Verma, 2013).

Of course, there have been concerns with such technology; the main issues being that the devices could be a huge threat to security and privacy, with some legislation already being passed to ban the use of smart wearables in some instances and locations (ABI Research, 2013; Pichai, 2014). Still, it will be interesting to see where and how wearable technology will travel with us as we move forward over time.

MAIN SCOPE OF WEARABLE TECHNOLOGY

Today, we can say, "This is the era of wearable technology. You can find them everywhere and see how wearable technology is rapidly dominating in every field. We present here some of the important areas where wearable technology is growing the fastest (Ror, 2014).

Scope in Medical Science

The year 2014 may well go down as the year of wearable technology. The impact of wearables is already being felt in education, communication, navigation, and entertainment; but perhaps the greatest potential lies in healthcare. Wearable technology started to revolutionize healthcare by assisting doctors in the operating room and providing real time access to electronic health records. The full potential of wearable technology in healthcare, though, goes well beyond directly assisting doctors (Pichai, 2014). Patients can now continuously monitor their own health. By 2014 many electronic showrooms in the USA, Australia, India and China introduced devices tracking everything from heart rate and blood pressure to a patient's O2 saturation. By 2018, the overall number of wearable devices shipped to consumers is expected to reach 130 million per year. With such acceptance on the part of the public, wearables are perhaps the perfect application for healthcare. Today, many of these wearable technology products include devices such as blood pressure, pulse monitors, body posture, walking activity, running and many others. Amongst the more novel products include baby monitors and smart shirts or biometric smart wear to record heart rate, footsteps, and calories burned per day. These devices are showing their increasing importance in the field of Medical Science (Afshar & Peterson, 2014).

Scope in Entertainment

Smart wearable devices such as miniature cameras can easily be worn on a garment while being controlled via a mobile app. This development is largely due to the availability of better touch screens and higher resolution cameras. Hence, this opens up another area of mobile app development wherein smart mobile apps will be created specifically to control various wearable devices. Such devices have embedded software to control the device via commands received from the mobile app. These apps will be helpful in developing a variety of mobile games.

Scope in Business

Right now much of the hype around wearable technology focuses on consumer devices – think Apple Watch, Google Glass, and devices from Pebble, Jawbone and Fitbit. But many industry watchers believe it will be businesses and not consumers driving the wearables trend. Among them is J.P. Gownder, an analyst for the IT market research firm Forrester Research. “The market for company-provided wearables will be larger than the consumer market in the next five years,” he predicts. “Wearables,” he says, “have the potential to change the way organizations and workers conduct business” (Twentyman, 2014).

Scope in Education

Technological costumes are not just clothes and offer a wide range of applications for their use in e-learning projects. Their potential is currently in full exploration. Numerous projects are being developed in collaboration between academia and corporate fields, as they work together to further study and understand the possibilities of these emerging innovations. The Oracle Applications User Experience (OAUX) team recently conducted a pilot for its employees. The company team wanted its employees to learn to design wearables (also develop an outreach program for customers and partners), then to share lessons learned in building such solutions. The supporting evidence demonstrates the growth of this trend, Deloitte revealed in a 2014 report, 10 million device will be sold annually worldwide including watches, bracelets and glasses; all including; wearable technology. Along this same line are the studies developed by the prestigious consulting firms of Canalys, ABI Research and IHS Global Insight. According to the J.P. Medved - Editor in Chief at Capterra, “I think the power of wearable devices for education is tied very closely with their mobility. Much like mobile devices such as smart phones, wearable technology can be with the learner at all times, even when they’re not physically present in a classroom or other formal training environments. This makes wearable technology ideal for performance support e learning content, delivered at the “point of need” out in the field. Where I think wearable technology has a leg-up over traditional mobile devices is allowing for a lot more immersion within the learning experiences and can, like Google’s Glass technology, be utilized in a hands-free fashion. One example would be to allow a service tech to study a blueprint of a machine while at the same time using both hands to repair it” (Cyr, Horn, Miao & Specter, 2014).

On the flip side of the coin, if we add wearables for engineering science, it will be helpful to increase the creativity, modify the logic, improve programming skills; so overall we will be able to create a modern technological environment among our students.

Purpose of Wearable Technology in Education

Most everyone has seen children’s shoes with lights turning on when the child walks around in the shoes. Many other examples of electronics being embedded into clothing exist in more experimental works such as, on stage and in the military. As embeddable technologies develop, it is likely more clothing will include this as an option. Increasingly, artists/designers are expected to work with sensors, controllers, LEDs, motors, and other electronics within their designs. The benefit of including wearables in the curriculum for the bachelor and master degree levels will be through increased creativity among students. Clothing is only one of the options for utilizing electronics in the art/design field. Engineers also benefit from working on and solving authentic real world problems. Whether working in a commercial setting or in a classroom, when working with an artist/designer, engineers have the opportunity to collaborate on the design of the final product and to implement a unique solution to specific problems. For most engineers, fabric is an unusual construction material; meaning, they will be required to apply prior knowledge about other materials to solve the problems involved in these particular projects.

For engineering students, a course or program of studies can provide the opportunity to apply prior knowledge of embedded electronics and programming in a new and creative way. It also provides an understanding of the creative approach for projects taken by the art/design students. Design students can gain proficiency in programming and electronic circuit construction while bringing their ideas to reality. The course could be taught with equal emphasis given to the creative and the technical aspects of wearable technology design. An important component of the course is the development of 'soft skills' required when working with those in different discipline areas. Students are asked to listen carefully to each other, to ask why particular things are important to the other students, to be respectful, provide support for the project from their knowledge base, take responsibility for completing their part of the project in a timely fashion, and work to ensure each project is a success.

The Targeted Learners could be upper class students in Engineering/Engineering Technology and Imaging Arts and Sciences. An equal division of technical and design students would be ideal.

There are many universities who are currently offering Bachelors, Masters and Doctorate degrees in wearable computing. We will focus on only some of the universities worldwide offering such programs.

Just as Silicon Valley business itself is inventing new forms of wearable technologies, a handful of business school professors and students in the US are already working out how these inventions can help them in the classroom. Below, you will find a list of examples.

1. The Haas School of Business at the University of California, Berkeley, the novelty and wide range of wearable technology itself is the key to using it as a learning tool, as is the chance to explore the possibilities of a new field, says Steve Blank, a lecturer on entrepreneurship.
2. On the other side of the US, at the New York University Stern School of Business, the administration is using wearables to give incoming students an early lesson in project management and teamwork.
3. At Harvard Business School, a number of professors have used wearable trackers to measure their steps. The aim was to test activity level in class, to see how it might correlate with how well students absorb course material.
4. Prfysgol Aberystwyth University of the United Kingdom offers courses on wearable technology.
5. Victoria University of the New Zealand offers a variety of courses on wearable technology.
6. King Abdullah University of Science and Technology (KAUST) Saudi Arabia is working to design courses based on wearable technologies for their students.
7. University of Petroleum and Energy Studies (UPES) India Offers a M. Tech for embedded systems with specialization in wearable technology.

As one can see, this short representational list of various universities offering specific courses/programs in wearable technology is growing worldwide. In addition, the information is changing day by day as viewed on the Internet related to this topic, because every day the numbers of users are increasing.

The Impact of Wearable Technology on Society

In this section, we describe the impact of wearable technology on society. Nowadays, very few people are using this technology. At this time, it is not only creating life saving devices but also becoming increasingly helpful to the intelligence agencies, media, entertainment, education, accounting, business, management, etc. So in this section we provide a brief overview regarding the influence of wearable technology in society.

Architecture of Wearable Technology

For wearable technologies, many researchers have proposed their own framework. In the section you will be able to learn about them. The first architecture proposed by Joremiah Nugrohu is shown below (Figure 1). This architecture shows how wearable devices function and are connected by the human control system. In this architecture he described how Art+Fashion+Computing=Wearable Technology.

Figure 1. An Interconnected Architecture of Wearable Devices (Nugrohu, 2013)

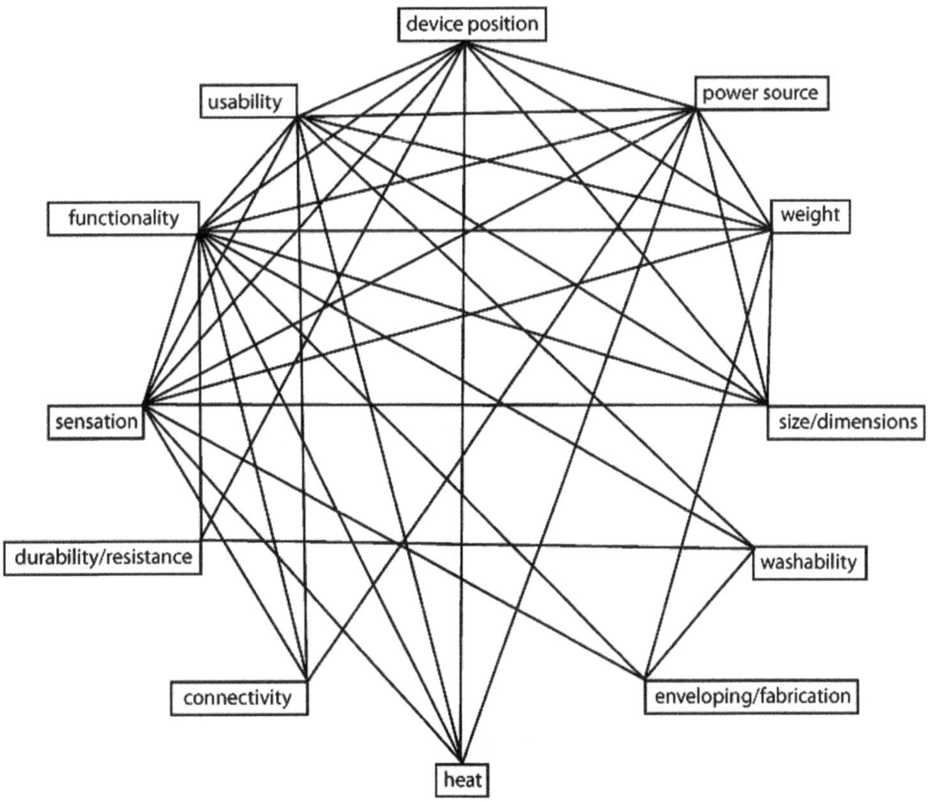

Another Simple Architecture of Wearable Technology

A second architecture example, below (Figure 2), shows three key parts. They are defined in the following list.

1. Data: Data is in binary format and needs storage prior to transferring the data from one place to another
2. Power: It has three main parts, i.e. how the power is generated, how it is transferred from one place to another and where the power is stored.
3. Social: Wearable devices are used at so many different places, and play a role on the impact of these devices on our culture, social connections, and relationships (Maidment, 2014).

When we compare the figures we understand all wearable devices have these three common aspects included. These are all interconnected from one to another as shown in (Figure 2).

Wi-Fi Architecture of Wearable Devices

Even if operating conditions can ensure local connectivity to a wireless hub, the link from the hub to the Internet to the cloud then back may still introduce unacceptable uncertainties. This is one of the structural challenges of the IoT, in general. Given this situation, when it is necessary to move computing tasks out of wearable devices, it may make sense to place them in a local wireless hub rather than in the cloud (Figure 3). Of course this precludes the use of just any commodity Wi-Fi hub. Integrating a computing node into a Wi-Fi hub significantly increases the flexibility for the system architect. Hubs are usually comparatively unconstrained in space and power, so one can put substantial computing and storage resources there. Short-range Wi-Fi links can provide reliable high-bandwidth, predictable latency connections, allowing the hub to participate in critical control or human-interface loops where a latency outlier would be a problem. Finally, a hub equipped with a multitasking CPU and appropriate accelerators can serve exported processing tasks from a number of remote wearable devices.

Figure 2. Simple Architecture of Wearable Devices (Mehdi, 2015)

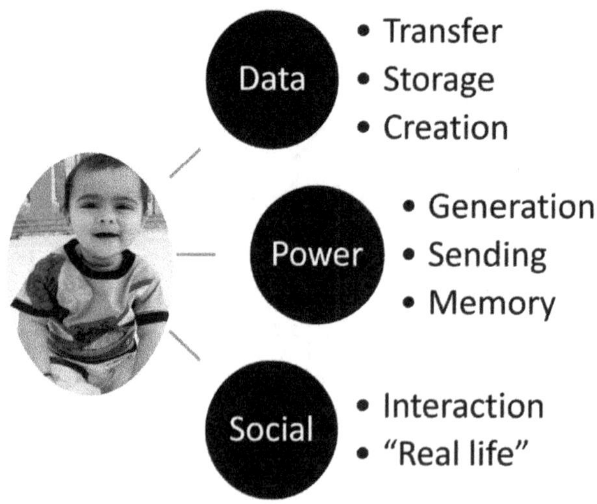

Figure 3. A Wi-Fi Architecture of Wearable Devices (Altera, 2014)

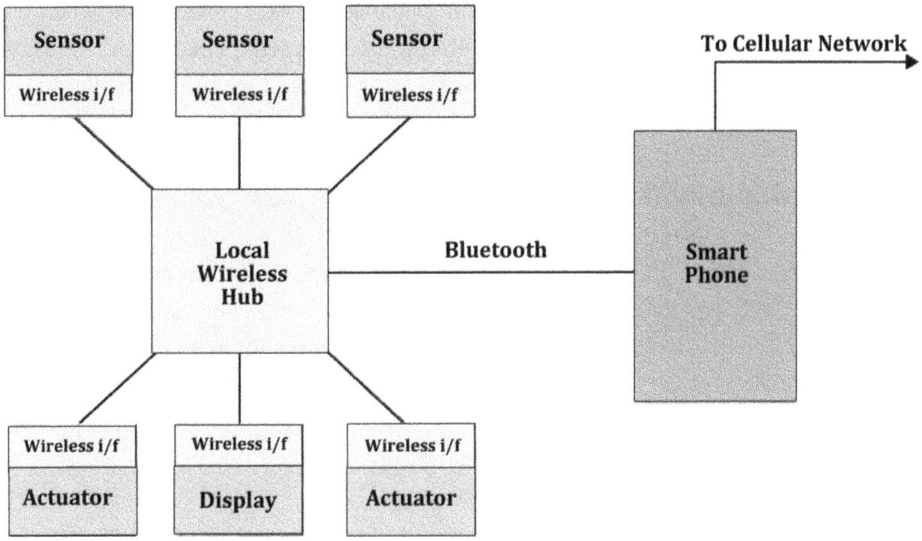

Technical Considerations

Implementation

Wearable technology is not based on the software programming, so anyone who wants to learn about wearable programming must first know all about those things related to the hardware. Wearables are not software, rather they are embedded devices! Get used to programming close to the metal, debugging with an oscilloscope, using programmer/emulator tool chains, and managing 512 bytes of RAM. Your code needs to be conscientious of battery life, memory concerns, graceful failure, hardware exploits, and mixed-reality UX. Now, more than ever, is the time to learn more about the technical implementation considerations due to the rapid growth.

Security

Like everyone else, I'm excited by the promise of instant health data being conveniently available at our fingertips, wrists, and ankles. But, I am a Metadata era blogger, and while we strongly believe you can't manage what you don't monitor, all of the monitoring comes with a lot of new data, and data privacy and protection is always top-of-mind.

In the sections below, we discuss privacy and data security issues to keep on your radar:

Inconsistency

As users of the health monitoring devices, we're often contributing health information to a centralized database maintained by the wearable maker. Most of us don't want outsiders looking at our data, but many fitness trackers' privacy policies are vague and ever-changing, with platitudes beginning with "We respect your privacy" and ending with "We may share your information with third parties…" Unless you live in a state treating this data as PHI, Protected Health Information, expect they can legally share your sensitive medical data without your permission, because HIPAA's extensive privacy regulations (see below) don't yet apply to this new industry. Ambiguously worded terms of service may give these companies just enough wiggle room to sell your un-PHI (Unprotected Health Information) once you've clicked on the 'accept license' button (Cyr, Horn, Miao & Specter, 2014).

Privacy and Data Security Issues

Following, you will find a list of three primary privacy and data security issues.

1. **Trust Issues:** We've entrusted these companies to gather our personal health information, but what measures will the company or 3rd party partners take to ensure our unPHI is safe and secure? Many privacy policies indicate they "protect your personal information from unauthorized access, use, or disclosure," but what does that really mean? Do they encrypt the information? Do they periodically review who has access to it? What about monitoring?
2. **Privacy Concerns:** The companies have a social networking aspect, and subscribers can choose to publicize and share their information with others. Unfortunately, it's not unusual for the default privacy setting to be set to public, allowing profiles to be found in search results. In 2011, one vendor was criticized when sexual activity was tracked– yes, you can learn a lot from accelerometers –showing up in Google search results. If you don't want your unPHI data searchable online, make sure you triple-check all of the default privacy settings and turn off anything you're not comfortable sharing publicly.
3. **Who Owns Your Data?** Who owns the personal data you generate with the wearable device — you or the business compiling your vitals? As the owner and creator of your very own health data, I was alarmed by one company's privacy policy regarding accessing your own data: It begged the question, "How can we get more control over our own data?" Despite the privacy issues, I wore a sports band for one week to see if the potential benefits outweighed the risks. I thought the device was accurate until the log reported I lost more calories during my 30-minute leisurely walk to work than in 1.5 hours of swimming, which normally takes everything out of me. Perhaps, as we wait for wearable technologies to mature and collect better data, it would also be a good time to figure out how to keep the data private and protected.

Design Considerations

In an increasingly connected world where almost every appliance imaginable is being connected, no market seems to be hotter than the wearables market. From a few simple devices available just a short number of years ago, the number and type of devices has been growing at a pace expected to reach a value of $70 billion by 2024, according to a January 2014 IDTechX report. Products range from relatively simple accelerometers packaged in increasingly ornate packages to devices of near medical-grade electronics designed to gather vital signs and other parametric measurements.

In this world of rapidly expanding devices, here are a few simple design considerations to keep in mind when tackling a new development. While this discussion may make it appear these elements are sequential, in reality they are all part of the whole and must be considered in a holistic fashion. Below is a list of six key design considerations.

1. **Understanding User Needs:** This discussion centers on design considerations as opposed to the very real obstacles associated with legal, regulatory, and business issues confronting the designer and are a necessary part of any project. As we focus on design, the first consideration is to understand the user and their specific needs, since this drives identification of what will be measured. This information guides the selection of sensors and sets the architecture for the remainder of the system, determining everything from data collection and processing circuitry to power consumption.
2. **Choosing the Best Sensor Location:** The second item to consider is derived from the first and it is the location of the sensors. While much of today's attention is focused on the tried and true wristwatch location, wearable devices can be worn all over the body with each location having its own unique set of considerations to optimize for both form and function. Certain parameters may be only obtained from specific locations on the body, constraining the design to fit within those limits.
3. **Considering Form and Function:** A third key attribute pairing is form and function. In today's fast-paced consumer-oriented environment, sometimes what a product does or how well it performs a function isn't as important as how good it looks. If your customer won't wear a product, they won't receive its benefits -- nor will you. In a medical device this may be less important as there is a medical necessity driving the use, but even devices like insulin pumps have begun to take on a form factor familiar to those with popular smartphones. As wearable devices continue to become more common, it will be important for the form to be as appealing as the technology.
4. **Displaying Data:** Once the data is captured, it needs to be utilized. User interface considerations are important here. Displays need to be tailored to different audiences and interactions. An active user may want instant feedback on performance, while a medical device may have no display at all for immediate use, choosing instead to relay the information to a secondary instrument for subsequent storage and analysis.
5. **Connectivity:** An integral part of the device design process is the connectivity. Just as the display decisions impact the usability of the wearable, the communications technology contributes to a seamless and easy experience. The choices are varied depending upon the application, ranging from the use of Bluetooth for short-range connections, to smart phones to Wi-Fi for facility-based devices, and cellular connectivity for roaming applications. Use case scenarios drive the decisions made at this stage with an eye on the ultimate end application, with major implications for device size, battery power, and utility.
6. **Translating Data Into Action:** A final consideration is the analysis to be performed to transform the gathered raw data into useful information and ultimately into action. Decisions made here center on where the processing of the data takes place, whether local or remote, and who the ultimate end user will be. Fundamentally it becomes an issue of turning collected data into value. While a wearer might be interested in all the details of their life as gathered by their wearable devices, their physician is not. Data analytics transforms information into value, a key to long-term adoption of wearable devices.

A number of additional situational based factors will likely need to be considered when designing a wearable, but hopefully this overview highlights some of the common key elements. However, in today's environment much of the data gathered, by the vast variety of wearable devices, exists as islands of non-interoperable information of limited utility. The real challenge is realizing the use of wearables, as part of the Internet of Things, lies in linking these islands with information correlated in ways to provide new insights into ourselves and the world around us.

Data Storage

The data stored in wearables is very similar to data storage in cloud computing. In today's tech-filled world, transferring data from one device to another is simple – sometimes as simple as just touching two devices together. Another concept for dead-simple data moving is a suite of products called in Touch from the business technology group VTT. The VTT Technical Research Centre organization is working to simplify storage framework issues in wearable technologies. Their products include a ring, a "functional fingernail," and a wristband. By wearing one of these objects and touching the screen of the source device, users can temporarily store the data on the wearable object. Transfer can then occur just as easily, by touching the object to the destination device – a friend's smartphone, a tablet, or even your car. VTT says the technology could also be used in a healthcare setting to help doctors keep track of patients' vital signs. Eventually, the wearable objects could act as passwords for online accounts or even to unlock your car or house doors effortlessly (Trend Micro, 2013).

SOLUTIONS TO SOCIAL, EMOTIONAL AND ETHICAL CONSIDERATIONS OF WEARABLE TECHNOLOGY

Social Implication of Wearable Technology

We will analyze the effects of wearable technology on every aspect of the social order. The smallest social component is the individual, with each having his/ her own unique internal features and external behaviors. Then the individual begins to interact with others to form communities. Each community then has internal features, with its own unique external behaviors and interactions with other communities. We typically do not refer to a specific device, but rather to the most extreme type of wearable device, which obstructs ones" vision, hearing or both and records and displays information. Google Glass is the closest, to date, of a marketable device aligning to our vision.

The Self – Internal

As a result of natural selection, humans have acquired many cognitive capabilities for dealing with and achieving survival. Our brains do not want to work at full capacity all of the time, so they often use heuristics in order to determine what is important and what is not. Our memories are not stand-alone objects in our brain, but are fully integrated with our cognitive functioning. Wearable technology will enable us to extend our cognitive capacities, but at a cost.

The Self – External

The depths of what resides inside us are shrouded in mystery. There is much about ourselves we are not even aware of and much more so when it comes to others! We behave in certain ways and cover our bodies with particular items we believe will help others to get a sense of who we are, and even to remind ourselves. Since wearable technology is worn on the body, it will not only affect our internal behavior but our external behavior and image, as well.

The Self Meets the Community

Mammals are unique; from the moment they are born we are dependent on others. While some of this dependency wears off, it still colors our experience of the world. Humans have evolved many methods for interacting with one another, and many complex yet often unspoken social norms govern the lives of individuals. While these norms vary across societies, there are still certain fundamental features in common. The goal of socialization is to foster bonds between individuals, and a collection of individuals has a greater chance of survival than a lone individual.

Wearable Technology in Ethics

Since wearable technology is likely to have an effect on so many aspects of our lives, there are many ethical considerations to take into account. Wearables are unique from other technologies because they directly affect our senses and cognitive capabilities, so it is important to consider the ethical implications of these differences. Also, since these technologies affect us as individuals, the way we interact with one another, and our communal interactions, we must analyze the resulting effects through an ethical framework.

Virtue Ethics

Virtue Ethics is an ethical system focusing on the character traits of the individual. There are various formulations of virtue ethics and what it means to be a virtuous individual is complex. Commonly recognized virtues are wisdom, courage, and temperance. Wearable technologies can pose a threat to all of these virtues and many more. Wisdom is not attained swiftly. Wisdom is not the result of acquiring vast amounts of data or being able to perform certain cognitive functions faster than others. It is the result of thinking deeply for a long time on various predicaments, and being able to hold and manage various apparently contradictory opinions in one's head. Courage is achieved when one is in a difficult situation and faces it head on. It necessitates being able to know one's situation and environment, to accept it, or to decide to fight against it.

Utilitarianism

Utilitarianism focuses on the consequences of one's actions rather than the virtues of the person. The actions should have an optimal balance of good outcomes over bad outcomes. So, many wearable devices are based on this phenomenon.

Practical Solutions

For various reasons it is difficult to conceive of solutions to problems wearable technology may pose. First, the effects of wearables are not always direct but are speculative, and so it is difficult to blame the wearables directly when so many other factors can play a role.

Required Skills to Use Wearable Technology

Often times there is not a specific set of knowledge or skills to use wearable technologies. However, training can be helpful in using it, especially if the device is not intuitive.

How Wearable Technology is Helpful in Education

In this section we cover some wearable technology tools implementation in the field of education. In one college at Abu Dhabi, wearable technologies were implemented with students and staff for the purpose of improving their overall education program. After many assessments they concluded the use of wearable technologies are beneficial in the process of teaching students.

We will examine the potential benefits of adding wearables to the curriculum, while considering the target audience needs in learning how to use the technology. In addition, we will look at business areas using wearable technologies.

FUTURE OF WEARABLE TECHNOLOGY IN EDUCATION

Wearable technology has steadily become the centerpiece for innovative startups looking to change the scope of healthcare communications with recent reports indicating the industry is primed for rapid growth in the coming years. Twenty percent of American adults own a wearable device of some kind, such as Fitbit bracelets, and the adoption rate of these technologies is expected to rise.

A PwC's Consumer Intelligence Series report focused on 1,000 consumers, wearable tech influencers, and business executives to gather a range of perceptions on the devices, including extensive monitoring of social media posts. For wearable tech to be successful, it has to interpret big data while delivering valuable insights to both users and physicians. In addition, the user experience needs to be simplified to ensure ease of use and help achieve specific health goals, such as weight loss or reducing blood pressure. Seventy-two percent of the study's participants indicated customer service was especially significant for wearable tech. Rewards were also a big part of the solution, as 50 percent of millennial respondents stated they would be more motivated to use a device if it had apps or features rewarding users for frequent use (Mann, 1997).

Despite the optimism, wearable devices still face an uphill battle against smartphone apps already collecting and reporting health data, such as Apple's Health app. Yet, top executives from around the industry believe the momentum for wearable tech will be hard to stop.

In education many wearable devices have been created to improve the teaching and learning experiences. Some examples of wearable technology use in the classroom include: Autographer, Keyglove, Muse, VR, Smart Watches, GoPro, and Google Glass. Autographer can be used to allow students to capture direct notes to ensure complete note taking. Keyglove uses wireless gloves for gaming, design, art, music, data entry, device control, and 3D object manipulations. Muse tracks students' brain activity on a smartphone or tablet to detect what activities might keep students focused on studying. From a career perspective, wearable technologies are a good choice for students.

Predictions Show Bright Future for Devices

Wearable technologies cover a range of categories, from glasses recording video to bracelets monitoring heart rates. While it is unlikely these devices will replace smartphones, the market is growing at a fast pace. Nike, Fitbit and other fitness brands control roughly 90 percent of the wearable device market, but businesses have technologies in the pipeline with even more features available. For example, Sony has SmartBand Talk, to monitor everything from phone calls and daily movements to logging physical activity and personal fitness goals.

Because the industry is relatively new, there is plenty of opportunity for someone to develop a device setting itself apart from the competitors. Fitness bands cost much less to produce than smartphones or tablets, making them more cost-effective as wearable technology. Health care could experience the greatest surge in devices as users look for affordable and in-depth biometric tracking and physical activity monitoring. Simon Segars, CEO of microprocessor manufacturer ARM Holdings, believes this inexpensive model behind wearables will open the door for new companies to get involved in the industry.

Workplace Might Lead Adoption of Wearable Technology

Adoption of smart-watches and fitness monitor technology might experience its biggest gains in the workplace. A global survey conducted by Harris Interactive of more than 9,100 adults, aged 18 or older showed, 75 percent of respondents saw potential benefits in efficiency, productivity and safety at the office when using wearables (Le Tieller, 2009). The research was coordinated on behalf of the Workforce Institute at Kronos, a human resources software and services company. The survey reached adults from a range of countries, including Australia, China, Mexico, India and the U.S. One of the most surprising findings was that U.S. adults displayed the least amount of optimism, as only 48 percent of respondents saw the benefits of wearables. By comparison, Mexico (96 percent), China (94 percent) and India (91 percent) expressed the greatest optimism for these devices (Smith, 2009).

While some U.S. adults might not be overly optimistic about the potential impact of wearable technology, the data gleaned from these devices could be integral to improving health services. Vitals and exercise results could be transmitted directly to hospitals, bolstering healthcare communications between doctors and patients.

Limitations of Wearable Technology

More Immediate Limitations

The use of many apps often requires some planning ahead of time. If you need information without delay, alternate directions in lieu of a detour for example, you'd better already have the app you require booted up and ready to go. Otherwise, you might be in for a bit of a wait. Likewise, wearable tech can, hypothetically, provide you with the information you require immediately, responding to your environment while keeping you updated with whatever you might want to know on the fly. No time wasting stumbling through your phone for apps or even getting it out of your pocket and unlocking it.

Allow You to Stay Better Engaged With Your Environment

For the most part, the usage of a smartphone or a tablet is a solitary activity, the screen forms a dividing wall between you and the outside world. Wearable tech largely eliminates the wall. Sure, glancing down at a watch face is similar to looking at a screen, but it's much less involved. When it comes to devices like Google Glass, the level of immersion possible is quite remarkable. Doing away with standard screens entirely, Glass and its ilk present an incredibly exciting and revolutionary way of engaging with personal technology. Indeed these pieces are generally just more convenient than smartphones or tablets.

More Discreet Devices

This is kind of a tricky topic, with the relative discreetness yet highly visible presence of devices like Google Glass could lend the product abilities to some less savory applications, too. However, for the most part, the discreet nature of wearable tech is a positive thing. For example, there is a bracelet that can alert you to the existence of any and all Facebook messages? Well, it can actually alert you to quite a bit more, including phone calls the user has received. Previously, to learn the same information the student would need to pull out their phone or computer, if so allowed, potentially disrupting the class (or, at the least, their own learning) while earning the scorn of their teacher. Now, a student can learn essentially the same information in a much quieter, far less obtrusive manner. This is one example of the discreet nature of wearable tech benefiting not only the user, but their environment as well. (Figure 4)

Potentially More Fashionable

One of the most interesting ways portable technology has progressed is in the area of aesthetics, an outgrowth of Apple's design aesthetics impact on the whole tech world. For example, many buyers seem to care just as much, if not more in some cases, about a device's sleek look as they do its performance or capabilities. Tech companies have learned they must set out to make their devices smaller, sleeker, lighter and more intuitive. As such, tech like pads, pods, and, of course, phones, since they have almost become fashion accessories to some. However, because they are replacing items like standard glasses and jewelry, wearable tech has an even more centralized focus on being fashionable. If these items don't look

Figure 4. Smart watches (Mehdi, 2015)

as good as possible, their appeal will be greatly limited. After all, who wants to wear an ugly bracelet or a pair of glasses not complimenting their face? A few won't care much for the look of the pieces, buying them solely for their technological capabilities, but for a vast number, aesthetics are very important, and the companies who make these products will need to recognize that (Deffeo, 2013).

CONCLUSION

In today's era, wearable technology is one of the burning issues for researchers, Google and other companies are working on many projects with this type of technology. There are many technology-based devices we can wear in accessories like shirts, shoes, rings medical devices etc. In short, the technology used to create these devices do not require much input from humans. They are small computers, enough to take the input on their own. This is the reason smart devices are being used and integrated into the development of wearable technologies.

These devices are used in many ways, as outlined earlier. Like every technology, specific framework integrates all aspects of System Analysis. It is hoped, by clarifying what goes into existing wearable technologies, anyone can easily understand designing new systems. Security issues, in regards to wearable technology, are to open up for discussion and for additional research and studies, while warning users about potential concerns when using such devices. In education many wearable devices have been created to improve the teaching and learning experiences. Some examples of wearable technology use in the classroom include: Autographer, Keyglove, Muse, VR, Smart Watches, GoPro, and Google Glass. Autographer can be used to allow students to capture direct notes to ensure complete note taking. Keyglove uses wireless gloves for gaming, design, art, music, data entry, device control, and 3D object manipulations. Muse tracks students' brain activity on a smartphone or tablet to detect what activities might keep students focused on studying. From a career perspective, wearable technologies are a good choice for students.

The abundance of conferences held on wearable technologies reflects the high interest in the topic. What developments in the future with respect to careers, jobs, business, society impact, children's education and health issues are still to be determined. The limitations become our challenges to work towards overcoming. We hope the technical information provided will assist in broadening your understanding and interest in wearable technologies.

REFERENCES

Afshar, V., & Peterson, D. (2014). Wearable technology: The coming revolution in healthcare. *Huffington Post.* Retrieved from http://www.huffingtonpost.com/vala-afshar/wearable-technology-the-c_b_5263547.html

Cyr, B., Horn, W., Miao, D., & Specter, M. (2014). *Security analysis of wearable fitness devices (Fitbit).* Cambridge, MA: Massachusetts Institute of Technology.

Davies, C. (2009, July 1). Pocket tweet app turns your shirt into a Twitter bubble. *Slashgear.com.* Retrieved from http://www.slashgear.com/pocket-tweet-app-turns-your-shirt-into-a-twitter-bubble-0148404/

Deffeo, C. (2013). Wearable tech: 5 obstacles to going mass market. *Entrepreneur.com*. Retrieved from http://www.entrepreneur.com/article/229545

Denise, R. (2014). The brains behind the hoodie that texts. *Forbes.com*. Retrieved from http://www.forbes.com/sites/deniserestauri/2014/08/14/the-brains-behind-the-hoodie-that-texts-its-not-who-you-think/

Eisenberg, A. (2009). Inside these lenses, a digital dimension. *The New York Times.com*. Retrieved from http://www.nytimes.com/2009/04/26/business/26novel.html?_r=0

Le Tellier, A. (2009). And you thought the Jawbone Headset was stylish. *LA Times.com*. Retrieved from http://latimesblogs.latimes.com/alltherage/2009/07/and-you-thought-the-jawbone-headset-was-stylish.html

Lipman, V. (2014). 71% of 16-to-24-year-olds want wearable tech. Why don't I even want to wear a watch? *Forbes.com*. Retrieved from http://www.forbes.com/sites/victorlipman/2014/09/22/71-of-16-24s-want-wearable-tech-why-dont-i-even-want-to-wear-a-watch/

Maidment, D. (2014). Advanced architectures and technologies for the development of wearable devices. *ARM*. Retrieved from http://www.arm.com/files/pdf/Advanced-Architectures-and-Technologies-for-the-Development-of-Wearable.pdf

Mann, S. (1997). Wearable computing: A first step toward personal imaging. MIT, Cambridge, MA. *IEEE Computer Journal*, *30*(2), 24–28.

Martin, H. (2007). Stash your gadgets in these tech-ready togs. *LA Times.com*. Retrieved from http://www.latimes.com/travel/la-trw-roadtest24jun24-story.html

Park, S., & Jayaraman, S. (2004). *E-health and quality of life: The role of wearable motherboard*. US National Library of Medicine, National Institute of Health: IOS Press.

Pichai, S. (2014). Sharing what's up our sleeve: Android coming to wearables. *Google Blog*. Retrieved from https://googleblog.blogspot.com/2014/03/sharing-whats-up-our-sleeve-android.html

More than one million smart watches will be shipped in 2013. (2013). *ABI Research*. Retrieved from https://www.abiresearch.com/press/more-than-one-million-smart-watches-will-be-shippe/

Ror, L. (2014). Moto 360: it's time. *Motorola.com*. Retrieved from http://motorola-blog.blogspot.com/2014/03/moto-360-its-time.html

Smith, D. (2009). The rise of the virtual trainer. *PDDnet.com*. Retrieved from http://www.pddnet.com/news/2009/07/rise-virtual-trainer

Big data, wearable devices create anxiety about storage practices. (2013). *Trend Micro*. Retrieved from http://blog.trendmicro.com/big-data-wearable-devices-create-anxiety-storage-practices/

Twentyman, J. (2014). *A business magazine report based on wearable technology*. Retrieved from http://raconteur.net/technology/welcome-to-the-wearable-workplace

Verma, D. (2013). *Google finally reveals glass specifications. MyGlass app now live*. Retrieved from http://techlomedia.in/2013/04/google-glass-specification-revealed-15817/

Weickgenant, J. (2009). Plenty of spinning, but more than just the D.J. *The New York Times*. Retrieved from http://intransit.blogs.nytimes.com/2009/07/15/plenty-of-spinning-but-more-than-just-the-dj/?_r=0

KEY TERMS AND DEFINITIONS

Cognitive Capacity: Cognitive capacity is the total amount of information the brain is capable of retaining at any particular moment. This amount is finite, so we can say our total capacity is only ever 100%.

GoPro: This is an American corporation developing, manufacturing, and marketing high-definition action cameras, often used in extreme-action videography.

PHI: Protected Health Information (PHI) is any information about health status, provision of health care, or payment for health care that can be linked to a specific individual.

PwC's: Price Water House Coopers is a network of firms in 157 countries with more than 195,400 people. It had total revenues of $34.0 billion in FY 2014, of which $15.1 billion was generated by its Assurance practice.

Roulette: Roulette is a casino game named after the French word meaning *little wheel*. In the game, players may choose to place bets on either a single number or a range of numbers, the colors red or black, or whether the number is odd or even.

Solitary Activity: Any activity completed by one person.

VTT: VTT Technical Research Centre of Finland is the largest multidisciplinary research organization in Northern Europe. It provides high-end technology solutions and innovation services. VTT is a non-profit research organization.

Chapter 2
Education Technology Disposable Information

Rosa Iaquinta
High Secondary School, Italy

Tiziana Iaquinta
University of Catanzaro Magna Graecia, Italy

ABSTRACT

The educational comparison between past and present, in information technology, makes obvious how much innovation influences didactics and the relationship students have with knowledge and competences. Tools accelerate relationships and information, often, though; obsessive use of technology determines an impoverishment of one's personal knowledge. This ties young people even more to Internet connections to fill these gaps. Wearable technologies might represent a concrete possibility of overcoming the knowledge gap, by offering, in any place, at any time and in any situation, some definite supports to reassure us about any of our occasional or well-established forms of interaction lack. How much of what we know and take care of, our personal heritage, contributes to determining who we are, our way of acting, thinking, planning, interacting and loving? The article aims to reflect on wearable relationship to education, by offering some data resulting from an Italian experience a secondary high school.

INTRODUCTION

If we turned the question: "Where can we find information?" The answer would be obvious: "on the Internet". The Istituto Nazionale di Statistica (ISTAT) or National Institute of Statistics reports on "citizens and new technologies, (ISTAT, 2011). They confirmed the most frequent activity practiced by Internet surfers, after electronic mail, is searching for information. Certainly, among those using these services we find teachers and students, for whom research includes finding resources, useful for teaching and studying, underestimating the reliability of those contents as well as their trustworthiness. The knowledge society generates new reflections on educational institutions, traditionally designated to learning purposes and to transfer knowledge that would be useful for a lifetime.

DOI: 10.4018/978-1-5225-0069-8.ch002

Technology can be an information distributor, which is a lot faster and less demanding than traditional educational paths. It challenges the educational system, which for ages was presented as the main if not the exclusive cradle of knowledge (Regni, 2004). The extensive presence of technology imposes a rethinking of the role of the school and a definition of erudition (not considered separate from a pedagogical project for learning) or the fundamental competences. The technological component, which in the last twenty-five years has occupied an ever more important space in the world of education, has produced a sort of dispersion of knowledge and ability in students, within a context, sometimes disoriented and overloaded.

Can new technologies be effective or are they just instruments, capable of giving in to restless students and providing teachers the illusion of carrying out innovative teaching? The pairing "more technology equals more learning" appears to be way too superficial, it would be best to pass to a phase characterized by more careful evaluations; technologies should be introduced whether it is clear the "added value" they can give. Starting from the 80s, with the development of the digital revolution, all technologically advanced countries launched programmes to introduce *Information and Communication Technologies* in schools. The program was supported by massive funding and equal expectations, but the process is quite complicated. On one side the schooling system remains at the margins of the decision-making processes, on the other technology is not capable of becoming a resource with a non-capital nature (Desinan, 1998). The objective of the contributions are to demonstrate how the use of technology in the educational environment does not necessarily represent a resource for developing competences; instead, it develops a pseudo-knowledge in young people, who consider the tools a solution to every problem.

BACKGROUND

The educational system of the past was always considered a listening place. Culture was passed on orally with the student expected to listen carefully, repeat and memorize what the teacher explained. This type of schooling had its cornerstone in the traditional lesson with the teacher teaching the whole class at one time. In the beginning, it was characterized by reading, explaining and commenting on the classical texts which were considered to be of great importance; or, more recently, narration, descriptions and explaining facts, environments, characters or events (Laurillard, 2012). During the history of education the lesson method gradually changed, becoming even more profound above all when psychology started to show its influence in the pedagogical environment. Lessons therefore became more conversational/colloquial, more dynamic and active, placing beside the teacher's oral presentation an iconic support. Also the manuals alongside the words inserted the images carrying out both the function of clarifying and of conceptual integration. The succession of changes has opened the way to operational activities on behalf of the student, in different forms: playing, direct observation, working with one's hands, and the manipulation of material objects. However, the changes did not influence the basic structure of the lesson, which remained the same. The method still rotated around the content to be taught instead of the student's needs. At the center of the teaching/learning process was the teacher, who structured and organized the presentation according to the logic behind the subject, and his/her personal preparation and mindset. This established one-way educational communications, during which the student had very few opportunities of intervening and becoming in some way an active part of the learning process. He or she continued learning by listening. An incisive criticism towards this educational setting was raised by J. Dewey who wrote: in the ordinary school classroom, with its rows of desks set in geometric order, "everything is done for listening (…), everything guarantees mental dependency of a mind on another.

The aptitude to listening means comparatively speaking, passiveness, absorption (…); the student must retain mentally as much as possible, in the least time possible (…). If everything is based on listening, there has to be uniformity in material and method. The ear and the book, which reflects the ear, constitute the didactic means for everyone. There is no way of adapting it to the various capabilities and requests (Dewey, 1964). It is only with Activism that the school of listening opens up in a decisive way also to "doing" and to active operating on the students' part. Teaching, as written by Ferrière, must be based on the child's personal activity and above all on his or her direct facts and experiences, for example personal observations, experiments, inquiries, visits, etc. (Ferrière, 1967).

The type of school, which is taking shape gradually in the environment of the activist movement, is one of a "laboratory", in which a strict association between intellectual and manual work is set; the students work mainly in a group; learning takes place by following the research method.

In the sixties and seventies of the last century the selective and meritocratic features of the school system had been criticized for its lack of attention towards the social world, for the hollowness of the content, for the absence, or uselessness/irrelevance of innovation, and educational experimenting (Bottani, 1986). Today we are living a more troublesome phenomenon, a transformation destroying the roots of the youth's world and the essential behaviour defining being a student. "Impossible classes" are more frequent and are resistant to every teacher's request or stimulus. In these cases it is not a matter of pedagogical techniques or of a major or minor availability on the teachers part, what is missing is the basic humus on which interaction is constructed, where codes and languages remain reciprocally unfamiliar (Vincenzi, 2005). One of the causes of such a metamorphosis is to be referred to the new "digital competence", due to influence of the new communication technologies, dominated by the pervasiveness of the audio visual component and predominance of fast interactivity corresponding to different thinking styles and behaviors that alter regular social practices. Most of the studies conducted on education start from technology and ask themselves, which could be their contribution to the school system. We have to begin from the school and ask ourselves if this should be enhanced or not by technology.

The current situation in terms of technologies is constantly changing. While in the recent past the computer and, more generally, technologies interested a small minority of people, now their spread is wide and not just among the youngest. Teachers, who were perhaps the most reluctant, now write in large numbers on the computer, have an e-mail account, surf the Internet, download digital photos, etc., Children and young people are also much more keen to use/have blogs, facebook and other social networks that allow them to communicate their points of view, while the most advanced area is the use of electronic games in their spare time. It's all right? Not so much. The issue is that the use of new media and technologies is predominantly consumerist. The danger is that it may be considered normal images and sounds bad definition, the impoverishment of writing, with a tendency to a jargon of phrases, creating an exclusive world impenetrable by adults. As a result we have a derangement instead of creating new meanings.

Postman (1995) vigorously admonishes the educational system for neglecting the goal of education and concentrating on the tools. The school system must have a clear vision of its values. Considering the society in which it interacts, the role of a school is not to adapt to technology but to control this process in a way to attain the social values desired. Any type of technology, once introduced, comes to modify the social place welcoming it. If we consider the specific school environment, in its sense of physical space of study and learning, in the moment it accepts new technologies, it suffers a series of multiple changes on which we need to reflect on. Before entering the school structure, we are obliged to ask ourselves which reactions in general are aroused by the introduction of technology in a social environment. Referring to the theories spreading in the sociological environment, we can identify two different attitudes. Some claim a new technology, just for the fact it is more advanced with respect to another, will obtain success and will be necessarily adopted and widespread. According to a similar concept, the higher technological level of an innovation is reason enough for its' introduction in society: this is defined as a determined approach to the phenomenon. Other theories instead demonstrate a more problematic attitude: they reveal in some cases, a technology, even if more advanced than another, may not be successful in a particular society. So, a more advanced technology is not the only factor determining its good results and its introduction in a social context. Within the first group, we can distinguish two conflicting expressions: one actually defined utopian determinism (optimistic), the other specified as dystopic determinism (pessimistic). Utopian determinism considers every new technological instrument as a step ahead in progress; dystopic determinism lives, instead, in the belief that the new technologies dehumanize man. They carry him or her far from his other essence, constituting a threat for all humanity. The extremes are represented, on one side, by the so-called "fanatics", on the opposite, by the supposed "irreducible" conservatories. The fanatics of new technology would buy any new technological instrument simply because it is new, convinced that it will contribute to being useful for them and their personal and global progress. The irreducible conservatives, instead, claim the profound uselessness and dangerousness of new technologies. In a social group, there are those who are enthusiastic to the limit of mania and those who instead shun away from the incentives. Seeing in them the worst threats, there are those who decide to accept the new support use, not giving up for this reason a constant approach of criticism, aimed at considering benefits and disadvantages derived from a new situation. This point of view allows us to avoid obsolete stiffness, passiveness and resignation, supporting an attitude of participation and activism that makes it possible to grasp the characteristics the new technologies must have in order to be used as well as foreseeing their complexity, evaluating their compatibility with one's job, testing the effects and pondering their positivity. The school institution, in particular, in front of its innovation, has to give itself the opportunity of trying grand fulfilling attempts, with the scope of having a glimpse of the technological potential without being tyrannized by false claims and from unreal anticipations.

The look on the world of education and technologies cannot disregard the cognitive results on basic skills. The international comparison data on students at the end of their obligatory schooling are known (OCSE-Pisa, 2014). They show Italy's low levels compared to the other industrialized countries. The possibility of confronting the results in the diverse cycles is one of the main goals of PISA. It is an international survey promoted by Organizzazione per la Cooperazione e lo Sviluppo Economico (OCSE) on an Organisation for cooperation and economic development) to accept, every three years, the competences of fifteen year old students. OCSE has the general goal to verify to what extent fifteen year old students have assimilated some basic skills to carry out a conscious and active role in society and to continue learning for a lifetime. It is characterized by a three-year period. The environment mainly focuses on surveys in each cycle, allowing us to follow the evolution of the system's results and its' impact on school policies over a period of time. The purpose is to promote focused interventions in instruction and education systems on behalf of the political decision makers for the fifteen year olds learners, since this is the age which comes before the end of compulsory schooling in almost all countries and environments of study, reading comprehension, math and sciences.

Having data available for a certain period in each country participating allows us to monitor how one's own educational system is proceeding, above all, and in comparative terms. Furthermore having this data at hand gives an idea of the effects of education policies initiated in a given moment and context, stimulating consideration of them for eventual adjustments or future in depth studies.

TEACHING AND TECHNOLOGY

The school environment is a particularly delicate place and needs an important clarification: in this environment, with the term technology we intend an area of culture and formation of prime importance, which has to find a precise place in the education of people. It is a significant fact the school system is opening up to technological innovation with the intention of incorporating it in its curricular paths, and not making it just an island to land on once in a while, as often happens. Even more important is anchoring use to the acquisition of competences in students, freeing technology from a mechanical and recreational use.

It is essential that educational innovation adapts, in its structure and in its system, to the social context which is strongly influenced by computer revolution and, in addition, it has to accept and use the new technologies assessing their efficiency and considering them as tools to improve teaching and learning. So far education technology has often been underestimated in acquiring technological knowledge, identified only as a set of different types of knowledge or as a set of competences in the function of a job. Actually, it is revealed as a guarantor of purposeful behaviour, imaginative and creative, active relational behaviour, of benefit to traditional and daily activity if the teacher predisposes class work to integrate technology with reflection on knowledge and skill acquisition.

Without any doubt, the arrival of a new instrument in the classroom means a state of technical and social uncertainty. The fears the machine may be too complicated, non-comprehensible, or not reliable are frequent social fears. The introduction of new technologies, in fact, may give life to atypical dynamics, causing concern and in particular putting into question the role of the teacher.

Teaching the "new alphabet" is a part of the school mission; it is up to the school to contemplate in the curriculum, specific formation paths bringing the conscious use of technology (Cartelli, 2006). Digital competences are generally included in a framework of basic competences including not only possessing procedural ability, but also more complex components, such as: the capability of communicating, problem solving, data analysis, use of symbol systems, capability of discussions, bettering one's own ability to learn and other similar competences now considered on the basics of entering the labour market.

Teachers are probably those who are mainly affected by the role held by the new technologies in school reform. On them falls the major responsibility of absorbing the changes triggered. The role of teachers in the development of innovative aspects in school is important: a teacher holds a fundamental role in forming his/her students' culture and education. They must reflect on the changes induced by new technologies, in the way students communicate, share, cooperate and learn. The instructor finds themselves reflecting on new knowledge needing to be acquired. In the swift dynamics of technology and social-economics, schools that have chosen to be included feel the necessity of constant and regular learning. There is a continuous questioning of the traditional didactic assumptions, of openness towards the new and unknown, becoming the basic element of the teacher's functions. It is a necessary condition to bridge the gap created between education and society. This implies an undeniable transformation, invading the usual isolation in which teachers find themselves in carrying out their job. Information and communication technologies (ICT) bring teachers to recognize themselves as a part of a practicing community, sharing approaches, methods, techniques and tools (Unwin, 1972). This favours in a sensitive way interactions with other professors, both of one's own subject and of diverse subjects. Resorting to technology may sustain significantly the creation of interdisciplinary work, encouraging the exchange of information, sharing material and common projects. So what is primarily requested is the constant effort to open up to new ways of learning and new tools for producing thought and cooperation constituted of information technology systems, and to assume consciously the changes new technologies induce in the subjects/disciplines, eventually modifying in part the way they are transmitted. The teacher finds himself/herself reflecting on their capability of using technology to implement and simplify the process of learning of their own subject for their students and organizing the class so that it can benefit from this in an effective way. Naturally, alongside all of this grows the fear of teachers in front of the fast evolution of technology and the right way to transmit it. Besides being aware of one's own limits, uncertainty about one's role may grow: the teacher is in front of a class made up of youth born in a technological context, already familiar with managing these tools or understanding the features. The major problems are not only practical; actually, they arise from a wrong attitude, on the basis of the teacher posing as the holder of an absolute knowledge, giving a flawless image of themselves. In front of technology we encounter deficiency, uncertainty, and incapability in answering all the questions or curiosities the instruments create, having to accept at times, not knowing or being proven wrong by a student, it is not rare that they know more. It is necessary that one allow themselves to be open to this state of things, embracing a serene outlook: where traditional knowledge is closed within the school, use of the web opens this context, it introduces an element of unexpected, and if this is managed well by the teacher it can be a great advantage and a great opportunity. Technologies bring us to a situation in which we no longer have total control of the lesson, we are induced to face the incentives reaching us continuously, in which we have to accept the change of an initial project, we have to be open to new ideas. This carries with it giving up the demonstration of knowing and of being capable of doing, bringing us to concentrate on the process and on the participation and comprehension of the students, instead of on the final product. Therefore, the teacher assumes a new role of guide and support, with the responsibility of giving the guidelines for the project, the didactic contents, and allowing students to elaborate freely. The space of uncertainty is vast between knowing what is needed in order to learn and knowing what is needed to teach: the planning environment has not created a bridge yet between the two banks. The technology available in the educational field has drastically multiplied and the planning of teaching and learning has become even more complex. The technologies of learning seem to have particular potential exactly because they are interactive, communicative, and controlled by the user, they are well inserted in the logic of social-constructivism and active learning. Empirical work on what is happening now in education has demonstrated in reality we are very far from the promises. At the end of the school year, the teacher measures the students' parameters in output in terms of benefits and the marks reveal how the use of the PC for every student was not very effective. This revealed how much the revolution promised by technology, in terms of competences is only an illusion (Mazza, 2015).

As far as the students are concerned, the introduction of the PC in schools offers them the feeling of being part of an innovative context, keeping the pace with the times. What we encounter is an increase in curiosity and enthusiasm, strongly dictated by the solicitation given by multimedia, since diverse competences are requested and activated, and they are not just the ones in the program. Often the use of technology reveals a balancing of the class to the same level, not necessarily upwards: students with difficulties in concentrating or in obtaining good results demonstrate, in the activities supported by the use of the calculator, an improvement. This is sensitively significant above all in delicate cases due to a psychological or physical disadvantage. Students with problems in movement find the use of a keyboard easier than the use of a pen, they become faster in finishing an assignment and this increases their efficiency. Students who have problems with their eyesight are helped by increasing the dimensions of the characters and by the use of the monitor. Adolescents with relational or social problems willingly accept a cooperative situation and being included in a group project. Many are those who jump from one application to another show difficulty in concentration, in doing profound reading, of mentally linking information if not with a "link". The disadvantage is the students must be capable of interpreting accurately the technical representations and integrate them with the meaning of the text, and this is a significant and complex assignment already on its own (AA.VV, 1970). While the multimodal theory celebrates the increase in multiple forms of representation, made possible by new technology, the cognitive psychologists demonstrate how this can make text comprehension and learning through appropriation more difficult and not easier. Attractive images and dynamic animation may seem to learners the most important aspects, but they may not be the most suitable for comprehension.

L. Reid in one of his essays, reported the results of an American study on the new tools of the time used for instructing in high school. He expressed comments about what, up until then characterized the relationship between the rich availability of means and their effective use also referring to his English situation. The author added "it would be correct to base didactic technology on behaviour sciences and concentrate it on the study of instruction methods and its systems, instead of on the equipment" (Reid, 1972). On one side the point of view is strictly instrumental in the sense it distinguishes technology from technological and it is not based on the use or the particular instruments; on the other hand attention is moved to the criteria of how the methodology is set, and the fulfillment of the didactic procedures. Therefore, technology is "the means born from the communication revolution), in other words the tools used for teaching purposes along with text books, boards and teachers" (Desinan, 1998). Technological is "the systematic way of planning, producing and evaluating the global process of human learning and communication, with combined use of human and non-human resources for creating more efficient instruction" (Perkins & McMurrin, 1971).

In the survey of the competences, knowledge and ability of students using technology in class every day, we did not observe an increase in school advancement within diverse subjects, during the midterm assessment and at the end of the year. Instead, there was a decrease in learning the information and mastering it. We observed the written and oral performance capabilities decline in vocabulary grammar and lexical errors. Furthermore, there is a filter between the teacher and the student created by the machine/technology. This weakened the between interpersonal relationships in the sense of an impersonal and neutral demand and with this the danger of going towards a form of dehumanization of the educational relationship and humans. Skinner expressed in an analytical way his criticism, which synthesizes the interpretation of teaching as "systematic organization of the strengthening contingencies that induce students to learn" (Skinner, 1976). He ensures that "just like the apparatus necessary for research has not eliminated the researcher, the machine for teaching will not eliminate the teacher. But both, if they want to do an effective job, must have the appropriate instruments" (Skinner, 1976). Technology is valid as support for an innovative commitment in school, but it cannot be itself the cause of innovation: "to clarify the concept with an analogy, the technical and organizational innovations of the industrial revolution took place at the same time as the new needs, certainly not for or only for one of their inescapable rationalities" (Robinshon, 1976). These are limits that on another level and in other terms Pellerey (1978) recalled, defining the role of technology as "a sort of interface, a noose, between educational sciences and educational action. In fact, it is found to be in a close interaction with the first ones and has as a specific assignment, that of giving consistency and effectiveness to the second… However, the relationship between educational technology and educational action is reciprocal… A careful observation is necessary and a valid analysis of the operation results in order to have enough data elements about the goodness of the plans and of the educational procedures."

SCHOLASTIC EXPERIENCE

ICT research in education has already demonstrated that technologies alone are not a determining factor for the success of an innovation-oriented option, considering also that they evolve very rapidly and that none of us can affect market strategies and the choices of hardware manufacturers as well as digital contents. Instead, what we can try to reasonably verify is the impact technology can have or not have on various aspects of teaching and learning.

In the scholastic year 2013-2014, an experiment was launched in a high school, grade 12, with the purpose of verifying the effect of the impact of technology on learning. The students, 12 males and 8 females, were presented with two didactic modules from the literature, one developed through the search of information, films and explanations on the web; the other performed through active didactics using both the teacher's lesson and reading the book being studied (Rivoltella, 2014). The organization of the work was constituted mainly of planning the didactic paths and starting from the competences students were supposed to develop, from the abilities requested for carrying out the assignment. The objectives were defined taking into consideration the measurability, the reachability, the time limit and ways of assessment, in relation to the didactics for competences and to the eight European key objectives. During the phase we determined the activity details and related organizing aspects: structure of the classroom; assigning the student's homework; definition of the steps used in the experimentation related to the single modules; timing the distribution of questionnaires; preparation assessment grids; and analysis of questionnaires. Programming of the activities takes into consideration the sequencing of the two modules; the identification of the resources (humane, material and structural); determining the methods and the duration of the activities done within each module. For each activity, a schedule was made.

From the experimentation in class, which lasted two months, the results registered contradict the technological enthusiasm. At the end of the experience, the students underwent a test (midterm and final), for each one of the two modules. The results emerging were completely different. The module performed following didactics with a traditional imprint gave students good results for 75% of the students; 15% obtained a sufficient result and only 5% a result, which was just under sufficiency. The tests completed by the students over the module using technology demonstrated 45% of the students who were certified a little bit over sufficiency; 40% did not reach sufficiency and the remaining 15% registered results gravely insufficient. The most significant difficulties found with students who used the technology arose from the incapability of discussing, showing knowledge, and being capable of moving to the historical period of reference.

Although it is possible to identify some points remaining the same, on the reflections of the technological innovation in education and formation they seem to converge.

Five Essential Steps:

1. Technological innovation does not have a value on its own but assumes meaning only if it becomes a vehicle and/or occasion of organizational innovation and methodological-didactics.
2. Technological innovation cannot be carried out sporadically, but it implies virtual interaction between all of the actors who interact in a homogenous context, continuity in time and consolidating good practices.
3. The most significant innovations are those which are capable of becoming a shared asset on behalf of all people involved (teachers, students, managers, directors) and they trigger processes oriented towards implementing the quality of the whole organization in which they are placed.

4. Any technological innovation implies a vision of the system and requires permanent support strategies to be enacted for the individuals involved and flexible policies of literacy teaching, lifelong learning and updating competences.
5. The most significant methodological innovations the technologies may facilitate or sustain are those taking back students to the center of the learning process and push him/her to be "actively involved".

The experience conducted demonstrated any didactic activity calling for the use of technologies should be guided and managed in a coherent way. It is not the technology used, in itself, increasing the learning processes. But rather, the way in which they are placed in context and used in function of precise didactic goals, posing the primary objective of verifying if and how much the use of mobile and versatile tools for reading and re-elaboration of knowledge helps the cognitive processes and increases students' capabilities of criticism.

The open questions are different, and teachers should reflect on them. Firstly, understanding whether educational technologies may be used effectively in a personal learning and studying environment.

Secondly, it is about verifying and measuring the learning settings results as more effective in an integrated context, or understanding whether educative technologies can integrate among themselves with respect to the other educative technologies used by students and teachers/educators, regardless of the contexts in which they operate (formal, as schools and universities, or informal as in some scenarios of vocational and continuing training).

An ulterior goal consists in trying to understand if and in what measure integrated use of the technologies experimented as learning settings are capable of determining a significant change in the methods of study, in the strategies of teaching and in the results obtained, above all in qualitative terms. In the end, it is about trying to verify if the set of interactions between devices and dynamic digital contents used helps the problematic approach, the thought of criticism and the predisposition to reading, and therefore can represent a true factor of methodological-didactic innovation.

The starting point of the experiment is a preliminary survey on the phenomenon perceptions. It consists of trying to understand the way technologies are perceived and experienced by the individuals involved (students, teachers/educators) and if and in what measure the perception encountered initially changes over time. As the study proceeds one can try to measure two essential implications: the epistemological impact and the methodological – didactic impact the technologies have eventually produced in the context and in particular on students and teachers/educators. To evaluate the epistemological impact of the devices one will have to try to understand if and in what measure the use of this type of educative technology modifies the relationship between person and knowledge, or does it change the way each person defines and organizes knowledge and the semantic configuration of the knowledge field.

This also means the chance to explore possible new configurations in the definition of traditional disciplines and the relationship between the disciplines. As well as to imagine an hypothetical new curricular structure within more formal learning process, or the identification of the most open and dynamic new reference "maps" within more informal paths or oriented to vocational and continuing training. In order to evaluate the methodological and organizational impact of the devices one will have to try and understand if and in what measure the way of teaching changes and if and in what measure the student's method of study and strategy of learning is modified. Finally, one must try to understand if eventually cognitive fallouts are perceivable, at least in terms of efficiency in studying, effectiveness in learning and quality of the results obtained by the students, as an effect of the change on the epistemological level and also on the methodological-didactic level. In order to do this type of evaluation, the quantitative data is not enough (in other words, the results corresponding to the students' performance with respect to the assessment tests). However, it will also be necessary to reflect on the evolution of the competences matured and on the identification of eventually new competences as a result of the diverse approach to the didactic problems made easier by the technologies used and by their epistemological and methodological impact.

DIGITAL NATIVES, TECHNOLOGIES AND EDUCATION

We live a real technological revolution that transformed our habits, thoughts as well as our way of acting and relating. Such a revolution is marked by a general process of convergence to digital, redefined by a shift from mass media to personal media, leading to several pros and cons. The latter seem to weigh particularly on young people and children who, because of the extensive use of television, computers, video games, cell phone, etc., are more exposed to phenomena of addiction and dependence. We must consider that the digital natives are always in contact with immersive and hypertext screens and therefore they are characterized by new learning styles and interactive dynamics. According to the French anthropologist Noemy Paymal, who has studied the characteristics of the digital natives of 33 countries all over the world, an early, direct and immersive contact with technologies would cause a "modification of the brain". In particular, the most affected is the right hemisphere with an increase in creativity, in the ability to implement fast connections and in the development of potential overlapping intelligence. There is no denying that the use of technology has accelerated a process of metamorphosis investing the whole system of symbols, leading to new forms of representation, perception and evaluation of reality by human being. This is true both in human relationships and between man and nature so much that, in 1987, Toffler stated the birth of a new civilization that, emerging in our lives, brings with it new family styles profoundly transforming the way people work, love and live. While new economies are born, new political conflicts arise and, lately, a profound change in the conscience of man.

We must consider that technology is not only the use of specific mechanical tools, but rather the operating moment in which takes place the choice of man and his critical judgment. Behind any technological tool there is always the man with the ability to use the information and organize them, especially with the ability to find new and original solutions to problems. According to McLuhan, technologies should not obscure the critical capacity and the will to understand what is happening. Anthropological and social scenarios, given by technological tools, outline the image of a subject that uses them for every aspect of his life, leading to a new way of relating, to experience, to enjoy free time, to work, to think. Such scenarios are setting up a more and more complex world because of the ability of the web to interface with other media channels and other languages. Bolter states that in addition to the computer, which is the characterizing instrument, a tool that does not just transform and improve the environment but adds to this a cultural perspective, there is the presence of multiple technologies integrated with each other. The multimedia and hypermedia systems, indeed, contain within a single medium the characteristics of multiple instruments, simultaneously integrating the hearing, vision, communication, perception, images, sounds and gestures. These multimedia products, according to Calvani, have a lattice structure, which allows connection between pages, links, etc. that, basing on an associative logic, allows a surfing and a consultation which opens in more directions. The computer is, in particular, the tool with such characteristics and that has completely revolutionized the lives of children and young people of our time, it allows the subject to express himself through three actions: interactivity, activity and multimedia. With the first activity the subject makes choices that have as an answer a feedback from the computer itself; with the second one he actively expresses his creativity; with the third one, finally, he gives vent to multi-task, using the computer as a synthesizer of other media, in the function of a book, telephone, television and implementing different symbolic systems and performing various activities. Frabboni, while managing with due caution new information technologies states that "many signs foretell that with the new millennium will be back the child and adolescent "golden lock". An event that will occur when the students will be able to write with multimedia alphabets (and not monomedia, responsible in part for their disappearance), their universe of things and values. When they will be manufacturers of languages, thoughts, imaginary. The impetuous and spectacular advent of new electronic and computerized languages will offer, for the first time, to both children and young people the opportunity to be citizens of their own historic season, main characters of this age of change. " It is clear that the chances of action and interaction allowed by the new generation of media, have made the world of multimedia increasingly attractive to the younger generation so that now even in schools, and thus in teaching, these "wish instruments " made their entrance, demanding a more massive presence. And after the computer entering the school now also computer (educational) games are arriving. The moment of computer games seems to be arrived, removed from what seemed to be the only intended use: gaming activities and entertainment. They are moving in some way to be ennobled by a different use, at least in the theories of researchers and in the still limited experiments results.

NEW SCENARIOS: WEARABLE TECHNOLOGIES

In the Italian language, the word technology has many meanings. It can be intended in a very general sense, the combination of what technical processes or methods have produced in different fields of human activity. Technology is the discipline dealing with scientific applications towards the solution of practical problems, involving a rational analysis of techniques and a systematic reflection on them. Technology is also a way of intervening on reality and on the environment to modify them, to resolve practical problems and those for the development of production. In education, the term technology has been posed and proposed in all of its meanings. Therefore technology in its general sense is the set of tools and procedures the teacher enacts in the exercise of his/her activity: from the book to the MP3, from the lesson to mastery learning. Technology is also the construction and application of particular didactic tools in which the contribution of the science of learning and systematic working of their applications constitute an essential part (for example, the systems of objective assessment or methods of teaching assisted by the computer). Didactic planning is also technological starting from the problems present in educative reality; it works on operation projects with the scope of producing specific changes desired. Furthermore, the term variously accompanied to education, didactic, teaching, and instruction, has produced a semantic area quite variegated within the diverse expressions, used in a promiscuous way and at times with emotionally and culturally ambiguous connotations, often makes the meanings blend one into the other, and leaves confusion. Following the evolution means going through a particular angle of the historical profile of the development of the pedagogical discourse – trends, problems, emergencies – in the alternating stories of its innovative drive and call towards tradition, in the continuous dialogue of its internal debate between scientific dimensions and humanistic needs. And in reference to this it cannot in any way be underestimated as, directing the development of the dialogue in a pedagogical and didactic key, they have contributed in an important way to the incessant development and the extreme transformations of communication and information. In regard to the use of acceptations and meanings in the field of education they have assumed different weights and values under the influence of particular sociocultural and scientific contexts, tending to make behavior and assessments change and along with these varying terms of discussion and the voices of defense (Maragliano, 1994). An interpretative analysis and criticism of the technological dimension of the technical role in problems and educational facts is interconnected closely to all of this. It has induced pedagogy to stay on two sides since the beginning, and they are not quite solid, but sufficiently clear enough to allow the identification of a scheme, according to the diverse moments and based on diverse reasons, lively supporters and more or less careful detractors.

The tendency of contemporary culture is based strongly on swift, permanent and widespread information, demands constantly integrating artificial intelligence in our daily environment and in many recent technology discoveries brought to experimentation. Registering, analysing, filing, sending and showing through wearable technologies brings a new dimension of high-tech exceeding the traditional basic functions of what is being worn. After consolidation in specialized fields, sanitary, sportswear and military applications, wearable technologies are starting to be interpreted openly. Today one of the major applications is for monitoring health and medical applications: reducing costs of assistance, even if the service quality is maintained. The future will expand the relationship between body and information, and it will do so in complex ways: things and processes will be overwhelmed and involved in diverse social contexts in life. In addition, we need to learn to face our cultures of connection. It is the Internet of Things advancing, and coupled with a dimension of wearable technology. To say it all, the "Internet will disappear" (Servaes, 2014). This was claimed by Eric Schmidt, president of the board of administration for Google, at the World Economic Forum: Internet will disappear because it "will be a part of our presence for the whole time" (Servais, 2014). The scenario Schmidt anticipates is this: access will be possible from everywhere and at all times, to the majority of people. We will no longer perceive the infrastructures behind it, we will be a part of it.

With the cognitive and cultural invisibility we will be confronted with during the next few years, it is necessary with urgency we educate towards technology instead of educating with technology, since one of the educational problems is of transforming information into knowledge by activating the competences of the students. Worries come from those who see education and school from a sociological point of view. We must once again, go back, to reflection (Patassini, 2015). A technology spreading mainly images is centered on images lasting a few seconds. The image is always concrete, it does not show concepts but things, it represents reality avoiding abstraction, so what is shown is neither true nor false, but only beautiful or ugly. Postman juxtaposes the image and word abstract, since it does not present things but ideas about the things. It needs the transition from concrete to abstract and from abstract to concrete. Words are linked to writing and therefore to a book. Without wanting to demonize technology, without which, much knowledge would be precluded to us, we must go back to education supporting reflection, analysis, and language. The school system must register the changes, taking place in the technological environment and compensate the one-sided exaggeration. In front of the invasion of images schools must reinforce the possession of words; it has to contribute by giving the historical dimension, studying the past and present.

DIRECTION OF FUTURE RESEARCH

Increasing the impact of TIC in the didactic context is one of the branches towards which research should be oriented. The new technologies tend to foster some general changes in the didactic context that in their most typical form can be detailed as follows: moving attention from the teacher to the student, shifting attention from learning to doing, implementing motivation in students, and emphasis towards cooperation among students.

Moving attention from the teacher to the student teachers can reduce their centrality, not appearing anymore as the only distributor of knowledge etc.; their role tends to organize, becoming now "director", coordinator, and facilitator.

Moving attention from learning to doing the new technologies foster learning by doing. Whether we are talking about a writing environment, a simulation, or a virtual reality setting, they all give emphasis to doing in a context/situation. In this, they are not dissimilar compared to other opportunities, which lead to workshops or laboratory activities. Nevertheless, they can make "active" also abstract perceptions traditionally acquired only through books.

Implementing motivation in students maybe the most evident effect, according to some scholars the only true factor keeping the innovation of technology standing (because of interactivity and multimodality). It is for emphasis towards cooperation among students, the presence of computers in school gives value to forms of cooperation and peer tutoring. Technology continuously poses problems, from ones related to functioning to more specific ones proposed by the software; therefore inviting students to lean on other students who are more experienced or experts, to receive useful suggestions. Let's think, for example, about the phenomenon of computer clubs for the young, vast chains of mutual assistance created spontaneously as well as the meaning of "Open Source".

Even if these ideas appear to be positive changes, none of them really are. Implementing motivation in students could go side by side with minimizing the activities, "doing" with a reduction of reflection, cooperation may implement gregariousness, giving up thinking autonomously and so on. What gives this change more or less positive value and implications depends on context factors, in particular from methodology and pre-existent didactic situated climate.

IMPACT OF TICS ON THE COGNITIVE PROCESSES

The relation between cognitive processes and use of technologies may be characterized on the basis of various and complicated ways. First of all it is naive to think the effects on the processes of the mind stabilize through alternating and brief use of the tool. Every activity has an impact on the mind and media, before giving a place to any structured "forma mentis," and demands long and repeated activities (in many cases practices in prolonged use for generations).

We can pinpoint many possibilities.

The first two represent disturbing factors, weighing negatively on learning

- The first is cognitive saturation. The machine (and the software interface) can absorb attention, disperse, produce an overload of information distracting from the problem, and subtract from reaching the goals through absorption on the machine itself (dispersion, saturation).

- The second is cognitive deactivation and depending on the machine it may disable important cognitive processes. Imagine the case where the software automatically provides a solution, skipping elaboration and processes in the middle. Instead, it is significant the students do the problems on their own. Trivially this is the case of calculators in relation to mnemonic operations.

Among the possibilities with positive implications, we pinpoint four concepts listed below:

1. Internalization, taking place when the mind "incorporates" the tool's functions. An example is the systematic use of software interfacing with hierarchal organization of data which may foster a mindset, more oriented to the hierarchal structuring of information (texts, archives etc.).
2. Synergy, from another point of view, may be considered acknowledging the effects following a synergic action of a mental-medium. An example is the possibility of doing more sophisticated calculations allowing the solution of complicated problems, otherwise unsolvable.
3. Consolidation is about situations in which cognitive structures or existing abilities find the way of being practiced, refined and broadened through technologies. For example, writing with a word processor may contribute to develop elaboration and editing skills.

Surfacing, from another angle, we can identify new attitudes and cognitive processes that spring to mind on the basis of interactions with the new technological environments. Let's think about cognitive behaviors connected to hyper textual surfing that lead to involuntary discoveries (serendipitously) through the Internet, etc.

Forms of negative impact of technology occur when these conditions are not fertile to integrate with the teaching methods.

1. Technologic Innovation
    a. Can be inconsistent if it does not go over a certain critical threshold.
    b. Even when quantitatively it goes over a certain critical threshold, it could remain insignificant if not accompanied by a culture and by a technological policy lasting over time.

Almost like the impact of TIC on cognitive processes, there are multiple possibilities.

Wild impacts can also create a decrease in reflection and a deactivating of internal cognitive functions.

On the other hand, other possibilities for research opens up to reflection on education in regards to consolidating cognitive skills already possessed by the learner, to the internalization of technology control, to the pursuit of knowledge results otherwise inaccessible, and the surfacing of new mindsets.

CONCLUSION

Schools, that aim to innovate in a broad and confident way, bravely face the radical change that is underway, that is the one that concerns the legitimacy of knowledge and its possible hierarchy. In the complex system we live in, characterized by the enormous flux of information exchanged through the Web, the true challenge the school faces is this one. Which knowledge is valid? Which is not? How can we discern what is true from what is false?

The capability of judging knowledge cannot be other than individual. It is rooted in the awareness that each one of us can employ their own sense of criticism in front of information, incentives, and numerous proposals arriving through the web. The discretion of judgement is fundamental acquisition schooling must bring beginning with the first years of primary school. In addition, if in this process, technology is truly the great instrument of access to knowledge, schools must be the place and center in which this access is experimented, learned and consciously used. Willing to make this innovation a major engine of cultural growth for our society (Calvani, 2007).

Studies on mobile phones, always within reach of students, along with other researches, are significant for the personal, cultural, and civil growth of everyone, no one excluded, constitutes a concrete goal set in an organic plan with wider horizons. However, implementation of technological possibilities, Internet access and stable or mobile instruments is not sufficient or significant on its own. Simply possessing or using technologies does not guarantee better results in learning. ICT play a complex role in the process of learning but they do not have linear and automatic incidence on students' performance. To date, in fact an unambiguous relation does not exist between use of technology and learning. What makes the difference, however, is not the quantitative aspects, the frequency with which one has access to the laboratory or the number of hours spent in front of a PC or a LIM, but the quality, linked to effectiveness and the innovativeness of the learning proposal. In order to have positive effects registered on learning, many hours of using technology are unnecessary but use must be given meaning, it must make sense, be appropriate and stimulating. Intensive use but at the same time inappropriate or reductive use would seem to produce non-virtuous effects, to the point of being considered negative.

In the educational setting, different subject area topics, content, course materials, instructor, directions, instructional strategies, motivation, students, course design, etc. are all possible variables influencing classroom implementation and research study outcomes. When implementing new and emerging technologies it is important to ensure the technology selected aligns to the learning needs and goals for effective results. The technology in itself is not inherently good or bad, it is all in how we use it.

The crucial step forward, not precluding the possibility of revolutionary developments, it is to connect the virtual world to the real one instead of replacing the latter with the former. We need to return to an active use where the mind resumes its primacy through the exploration of reality. Technologies are useless unless they are based on the technology of the mind.

REFERENCES

Bottani, N. (1986). *La ricreazione è finita: dibattito sulla qualità dell'istruzione*. Firenze, Italy: Il Mulino.

Calvani, A. (2007). *Tecnologia, scuola, processi cognitivi. Per una ecologia dell'apprendere*. Milano, Italy: Franco Angeli.

Cartelli, A. (2006). *Un esempio di progettazione di attività formativa mirata*. Udine, Italy: Idea Stampa.

Centro Europeo dell'Educazione., (1971). L'alternativa tecnologica. Il rapporto Perkins Mc Murrin. Frascati, Italy: Centro Europeo dell'Educazione

Desinan, C. (1998). *Discutere la scuola. Ipotesi, contenuti e prospettive a confronto*. Milano, Italy: Franco Angeli.

Dewey, J. (1964). *Scuola e Società*. Firenze, Italy: La Nuova Italia.

Education at a Glance 2014. (2014OCSE. OECD Indicators.

Ferrière, A. (1967). *Trasformiamo la scuola*. Firenze, Italy: La Nuova Italia.

L'istruzione programmata. (1970AA.VV. Roma, Italy: Bulzoni.

Laurillard, D. (2012). *Insegnamento come scienza della progettazione. Costruire modelli pedagogici per apprendere con la tecnologia*. Milano, Italy: Franco Angeli.

Maragliano, E. (1994). *Manuale di didattica multimediale*. Bari, Italy: Laterza.

Mazza, S., (2015). Insegnare ai tempi del cloud. eBook.

Patassini, A. (2015). Google Drive e la didattica. Asiago, Italy: Coll@borare.

Pellerey, M., (1978). Compiti e insidie della tecnologia dell'educazione. Orientamenti pedagogici, 4, 7-13.

Postman, N. (1995). *The End of Education: redefining the value of school*. New York: Knopf.

Rapporto annuale. (2011ISTAT. Roma, Italy: Centro Stampa e Riproduzione.

Regni, R. (2004). *Geopedagogia. L'educazione tra globalizzazione, tecnologia e consumo*. Milano, Italy: Armando Editore.

Reid, R. L. (1972). *Tecnologia dell'educazione. Lo sfondo e la necessità*. Brescia, Italy: La Scuola.

Rivoltella, P. C. (2014). Smart future. Didattica, media digitali ed inclusione. Milano, Italy: Franco Angeli.

Robinshon, S. B. (1976). *Curricula scolastici come fondamento di ogni riforma*. Roma, Italy: Armando.

Servais, J. (2014). *Technological Determinism and Social Change: Communication in a Tech-Mad World*. Lanham, Maryland: Lexington Books.

Skinner, B. F. (1976). *Il comportamento verbale*. Milano, Italy: Franco Angeli.

Unwin, D. (1972). *Mezzi e metodi. Guida alle tecnologie educative*. Brescia, Italy: La Scuola.

Vincenzi, A. B. (2005). *Conoscere la classe e lo studente: costruire e usare questionari pedagogici*. Trento, Italy: Erickson.

KEY TERMS AND DEFINITIONS

Competence: DeSeCo (Definition and Selection of Competencies) defines a competence as the ability to successfully meet complex demands in a particular context. Competent performance or effective action implies the mobilization of knowledge, cognitive and practical skills, as well as social and behavior components such as attitudes, emotions, and values and motivation of knowledge, skills and attitudes appropriate to the context.

Content Knowledge: a term widely used by educators, content knowledge refers to the body of information that teachers teach and that students are expected to learn in a given subject or content area, such as English language arts, mathematics, science, or social studies. Content knowledge generally refers to the facts, concepts, theories, and principles that are taught and learned, rather than to related skills - such as reading, writing, or researching -that students also learn in academic courses.

Education: in its broadest, general sense, the means through which the aims and habits of a group of people lives on from one generation to the next. Generally, it occurs through any experience that has a formative effect on the way one thinks, feel, or acts. In its narrow, technical sense, education is the formal process by which society deliberately transmits its accumulated knowledge, skills, customs and values from one generation to another, e.g., instruction in schools.

ICT: (Information and Communications Technology), Includes any communication device or application, encompassing: radio, television, mobile phones, computer and network hardware and software, satellite system and so on, as well as the various services and applications associated with them, such as videoconferencing and distance learning.

INVALSI: National Institute for the Evaluation of Educational System of Education and Training.

ISTAT: Istituto Nazionale di Statistica, or the Italian public research institution that deals with censuses and surveys the social and economic.

OCSE-PISA: Organization for Economic Cooperation and Development-Programme for International Student Assessment.

Teaching: in education, teaching is the concerted sharing of knowledge and experience, which is usually organized within a discipline and, more generally, the provision of stimulus to the psychological and intellectual growth of a person by another person or artefact.

Technologies: in the framework of the studies presented here, technologies are perceived as tools (such as mobile devices and web-based tools) to support learning activities with the aim of fostering links among the different learning locations.

Wearable Technologies: all those technologies portable and wearable, modeled around the body of the people, which is used as natural support for their operation. The detection and monitoring of internal and external signals to the body, even emotional in nature, to allow these technologies to become a valuable assistant to the user's needs, expanding also its sensory capabilities.

Chapter 3
Wearable Technology Spending:
A Strategic Approach to Decision-Making

Jason Ribeiro
University of Calgary, Canada

ABSTRACT

Very little research has examined the applications wearable technologies can have in education despite industry experts forecasting major growth in the global market for these innovative devices (Futuresource Consulting, 2015). Notwithstanding this profound challenge, in the next 3 to 5 years school districts will need to make multimillion-dollar investments into supporting these technologies as they begin to enter the learning environment. Given this gap in the research literature, this chapter explores the potential decision-making challenges school districts and their leaders will have to face when wearables become more commonplace in education. Using the Strategic Model for Technology Acquisition (Ribeiro, 2015), the author outlines both the innovative opportunities and potential problems wearables pose for both school leaders and stakeholders in the near future. Moreover, this chapter is meant to serve as an informative resource for education leaders and presents strategic and research-validated approaches for procuring and supporting wearable technologies.

INTRODUCTION

What we are experiencing today, I truly don't believe we've ever had anything quite like it. There aren't that many people who really understand the individual technologies; to manage to have them integrated is extremely complicated – Jerry Luftman (Lebeaux, 2013, para. 6).

Since this book showcases research on both *wearable* and *mobile* technologies, it is important to acknowledge the difference between the two categories of devices. The Educause Learning Initiative (2013) makes the distinction clear: "an Internet-connected smart watch or pair of running shoes with sensors in them are 'wearable technology,' whereas a smartphone or a tablet is simply 'mobile'" (para. 5). Despite the current fervor surrounding innovations in wearable technologies, a number of wearable electronic devices have been produced during the last three decades (albeit with limited applications).

DOI: 10.4018/978-1-5225-0069-8.ch003

Expensive, bulky, and most often associated with experimental and research-based tasks, these devices lacked an aesthetic appeal and ubiquitous purpose for consumers (Sultan, 2015). However since 2010, the rapid advancement of emerging wearable technologies and their subsequent consumerization has left educational stakeholders wondering if this will lead to yet another shift in the K–12 education landscape. After all, districts are currently being enveloped by new educational technologies (or ed-tech) significantly impacting today's classrooms (Bebell, O'Dwyer, Russell, & Hoffmann, 2010; Brooks, 2008; Hsu, Hung, & Ching, 2013; Jenkinson, 2009). These unique teaching and learning tools have provided school districts with the opportunity to modernize their classrooms and equip their students with critical 21st century skills (Brooks, 2008; Culp, Honey, & Mandinach, 2005; Kim et al., 2010). The Partnership for 21st Century Skills (P21; 2011) often refers to these proficiencies as the "4Cs" (p. 1) and outlines critical thinking, creativity, communication, and collaboration as essential for today's students. However, with districts struggling to gain a firm grasp over educational technology's impact on the classroom and with Smart Glasses, FitBits, and the Apple Watch looming, there is a critical need for effective technology management and leadership in education (Bellamy, 2007; Morrison, Ross, Corcoran, & Reid, 2014).

The responsibilities of educational leaders and practitioners are further problematized by the collective weakness of educational technology research and its inability to provide substantial empirical evidence to support technology integration as a means of improving student learning (Bebell et al., 2010; Jenkinson, 2009; Johnson & Maddux, 2008). Notwithstanding these profound challenges, publicly funded school districts have made multibillion-dollar investments in educational technology, with spending showing no signs of slowing (Futuresource Consulting, 2014). As investments from government and venture capital steadily increase, industry experts believe wearable technology is indeed on the horizon of the classroom. Technology research firm, Gartner, has even predicted wearable technology will be a leading electronic device category in 2015 (Pulley, 2014). According to research firm, Futuresource Consulting (2015), the global market for wearable technology was worth over US $8.9 billion in 2014 (with 56 million devices shipped) and is expected to continue growing. This figure does not even include spending on applications (*apps*) or information technology (IT) support needed to deploy, manage, and secure these purchased devices. With public and private decision-makers having long been interested in a sustained inquiry into technology acquisition (Heinecke & Blasi, 2004), this chapter is quite timely as senior leaders prepare for the burgeoning market's impact on their school districts. The key areas of investigation surrounding wearable technology acquisition involve the protocols and policies influencing spending by senior-level leaders, the academic impact of these devices on student learning, and their ability to generate data to drive decision-making.

The purpose of this chapter is to explore potential applications of wearable technologies in education, as well as the impending impact on school districts, and to draw upon relevant research surrounding technology acquisition and spending to determine if school leaders are ready to usher in a new era of educational practice. Using the Strategic Model for Technology Acquisition (Ribeiro, 2015), the author outlines both the innovative opportunities and potential problems wearable technologies pose for both school leaders and stakeholders in the near future. The chapter begins with a background of technology spending and leadership in education and a description of the strategic model, followed by three sections. The first section reviews wearable technology as a tool for improving student learning. The second section explores wearables and their role in producing data used to inform decisions (data-driven decision-making or DDDM). The third section examines technology procurement and spending in the wearables context. Finally, recommendations and considerations for future research surrounding wearable technologies are discussed.

BACKGROUND

But as we're learning in the U.S., it's important always to start with the education challenge, and then determine which technology, if any, meets the need and adds value to other solutions. It's tempting to make it about new, glamorous gadgets – but that can distract from simpler and more effective approaches – U.S. Secretary of Education, Arne Duncan (Department of Education, 2013).

In evaluating Duncan's address at the 2013 USAID Global Education Summit, it is clear the United States is still learning from past mistakes and missed opportunities in trying to marry education and technology. His emphasis on addressing the "education challenge" (Department of Education) is echoed in much of the new research surrounding technology implementation and pedagogy (Avidov-Ungar & Eshet-Alkakay, 2011; Fullan, 2013a, 2013b; Prensky, 2008). But while statements like this are promising in addressing key areas for improvement and focus, they seem to fall short for stakeholders. Many feel while their education system is addressing the need for improving educational delivery through investments in technology, they have failed to ensure these dollars consistently promote strong student achievement (Boser, 2013). The success of the wearable technology movement in education hinges upon school leaders, teachers, and technology providers being able to make distinct connections between student learning and the devices. Effective technology leadership and decision-making is not something to be desired anymore, it is essential.

Without research-validated approaches and professional guidance supporting technology acquisition, frivolous spending, wasted time, and technological chaos can result. In June 2013, superintendent of Los Angeles Unified School District (LAUSD; America's second largest school district), John Deasy, announced a US $30 million commitment to provide Apple iPads for the 650,000 students in the system (the total expenditure would cost roughly US $1 billion). A myriad of problems immediately plagued the one-to-one (1:1) laptop rollout including a rushed timeline, device hacks, poor Wi-Fi infrastructure, a lack of professional development for teachers, and intense scrutiny over the bidding procedures with both Apple and digital learning software provider, Pearson (Pickert, 2014). Simultaneously, the district and its chief information officer, Ron Chandler, began to revamp LAUSD's student information system (MiSiS) in the summer of 2014. This project was troubled as well, with inaccurate student transcripts, glitches, and lost records being reported throughout the board's 1,124 schools (Blume, 2014). By October 2014 the iPad contract had been cancelled, the MiSiS computer system had yet to be fixed, and both Deasy and Chandler had announced their resignations. The fate of the 100,000 iPads already purchased by LAUSD remains unclear.

Clearly the uninformed acquisition of technology can result in wasted funds and a loss of public trust. Will wearable technologies suffer the same fate as its predecessors under misguided leadership? It is quite likely a possibility. Therefore, given the substantial rise in technology expenditures by publicly funded school districts, a deeper inquiry into the acquisition and support of wearable technologies is vital to ensure these future expenditures are informed by the relevant research, grounded in best practices, and enhance student learning. Larry Cuban (2000), the preeminent scholar on educational reform and technology spending, writes, "Given the reasons for current expenditures in technology and the paltry evidence supporting these reasons, it is time to ask again if the dollars being spent are worth it" (p. 42). With limited resources and funding, school districts and their leaders need to ensure they are being fiscally responsible and implementing technology plans positively impacting student learning. Horror stories of districts that making the mistake of utilizing technology for technology's sake are becoming increasingly common (e.g., Los Angeles Unified School District, Fort Bend Independent School District, etc.; Michels, 2013). Too often the pedagogy has been ignored or undervalued, and the technology on its own is failing to accelerate and deepen learning (Fullan, 2013b; Militello & Friend, 2013). These are costly mistakes (both socially and financially), but more importantly they deny students and teachers opportunities those funds could have provided (e.g., professional development, program funding, etc.). Senior fellow at the Center for American Progress, Ulrich Boser (2013), has examined educational innovation extensively and addresses this very concern:

Are taxpayers getting their money's worth when it comes to technology in schools? We simply do not know the answer to this basic question right now. Study after study shows that technology in education can raise student outcomes under certain conditions. The question now is how we can bring these outcomes to scale and at what cost. Education leaders could be doing far more in this area, including close and careful studies of technology's return on investment. (p. 8)

Although a great deal of informal and industry-based research has been gathered by nonprofit organizations and professional associations (e.g., Educause, Consortium for School Networking, International Society for Technology in Education [ISTE], etc.), the formal scholarship investigating wearable technology use, leadership, and implementation rarely explores acquisition specifically. According to Bebell et al. (2010), this gap "has created a challenging situation for educational leaders and policy makers who must use flawed and limited research evidence to make policy and funding decisions" (p. 31). In fact, the topic receives more focused attention in higher education, where the market is more competitive and the expenditures in IT and mobile technologies are much larger (Botelho, 2014). Furthermore, the roles of senior leaders like superintendents or technology directors often get neglected in K–12 educational technology research in favor of school administrators (Militello & Friend, 2013), teachers (Prensky, 2008), and students (Palfrey & Gasser, 2013). Yet without senior-level leaders who are experts in wearable technology management and advocates for the use of educational technology in classrooms, the field of education will never be able to successfully deliver 21st century learning skills to students (Courville, 2011). The author recognizes the importance of such leaders in technology acquisition and specifically examines their vital role in the decision-making process.

The Strategic Model for Technology Acquisition

With educational stakeholders in mind (i.e. parents, teachers, students, etc.), the Strategic Model for Technology Acquisition (Ribeiro, 2015) was developed to break down the key elements of effective senior-level technology decision-making. Both academic and professional literature was consulted to identify the most important elements of the technology decision-making process. The model represents the best thinking about technology acquisition in school districts and identifies the following three areas of exploration:

1. Strategic Assessment (TIC)
2. Strategic Data Use (TAC)
3. Strategic Acquisition (TOE)

Figure 1. Strategic model for technology acquisition (Ribeiro, 2015)

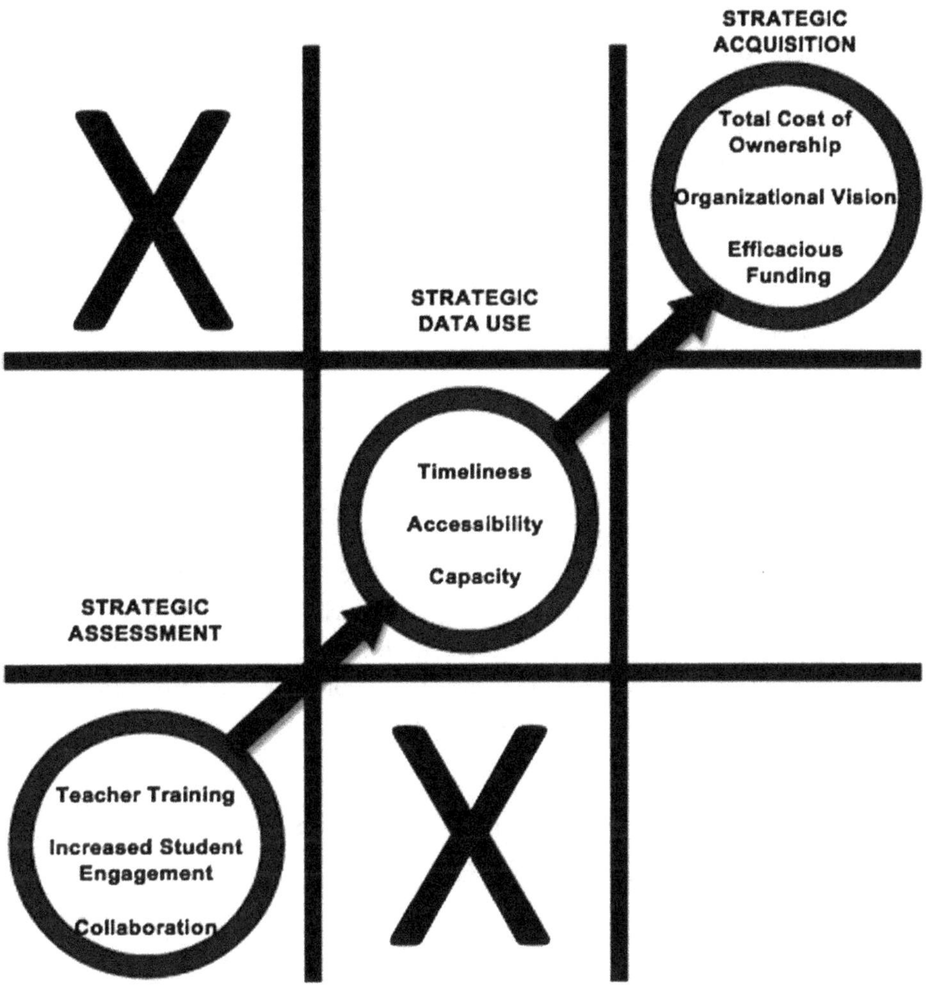

Furthermore, the literature surrounding the factors affecting student learning (Finkel, 2012; Marzano, Pickering & Pollock, 2001; McCombs & Whisler, 1997; Spector, 2013), data-driven decision-making (Ikemoto & Marsh, 2007; Mandinach, Honey & Light, 2006), and the essential conditions to support technology implementation (Bellamy, 2007; Finkel, 2012; Fullan, 2013b; ISTE, 2009) revealed three specific conditions in each area relating to technology decision-making in education. Figure 1 showcases the strategic model.

The model uses the pedagogical tool of *tic-tac-toe* to explain the importance of being strategic when making decisions related to technology in education contexts. Tic-tac-toe is a commonly played game dating back to the Roman Empire, in which two players, X and O, try to place three particular marks in a row (horizontal, vertical, diagonal) on a 3x3 grid before the opponent can (Zaslavsky, 1982). What is often overlooked is the distinct strategy to this game increasing one's ability to win based on the previously made move (rooted in statistical probability). However, more often than not the strategy is ignored and the game often ends in a draw with neither side prevailing. The literature shows us districts lack guidance when purchasing technology and often act impulsively when spending taxpayer dollars. The same can be said for all those tic-tac-toe games ending in a draw. Too often a player chooses the square with no regard for the strategy and simply settles for marking an open space that seems logical at the time. However, using relevant scholarship as one's *strategy* during technology decision-making can help districts avoid missed opportunities and costly mistakes. The Strategic Model for Technology Acquisition gives districts the opportunity to ensure each decision they make is aligned with the key theories and body of research. Each area of exploration and their relation to wearable technologies will be explained in greater depth in the next section.

MAIN FOCUS OF THE CHAPTER

Issues, Controversies, Problems

Notwithstanding the growing interest in wearable and mobile technologies in education, a deeper inquiry into how educational systems and leaders will support their purchase and implementation is greatly needed. The following sections use the Strategic Model for Technology Acquisition to explore wearable technologies' connection to student learning, data-driven decision-making, and acquisition. First, a breakdown of the relevant literature is presented, followed by a summary of the current trends, challenges, and opportunities for wearables in the area of exploration. It is important to note the model advocates senior leaders consider all nine areas before deciding to make a technology investment.

STRATEGIC ASSESSMENT: ACADEMIC IMPACT OF TECHNOLOGY ON STUDENT LEARNING

Teacher Training

For any technology implementation to have a high rate of success, the role of the teacher needs to be constantly considered (Avidov-Ungar & Eshet-Alkakay, 2011). In particular, professional learning and development for educators is universally identified in the literature as an essential condition for efficacious technology procurement and implementation (Blow & McConnell, 2012; Culp et al., 2005; Finkel, 2012; Greaves & Hayes, 2008; Hall, 2010; ISTE, 2009; Johnson & Maddux, 2008; Spector, 2013). This training can take on many forms and can be provided by individual teachers, administrators, corporate vendors, and even students.

Notwithstanding the shiny allure of wearable technologies, the objectives for teachers remains maximizing the technology to accelerate/deepen the learning of their students. In order for this to occur, districts need to make an extensive investment into professional development for their teachers. While early adopters may be highly adept at integrating these devices into their instruction, a great deal of training will be needed for instructors who lack that same level of comfort. Senior leaders should also anticipate resistance from both staff and parents alike. Historically speaking, the fear of students using the technology inappropriately, or increasing teacher responsibility has existed for centuries and often precedes major technological developments (Ribeiro, 2014). However, as we have learned from advances in educational technology over the last few years, turning a blind eye to the way students are interacting and learning in the digital world only hinders instructional practice. Foote (2015) writes, "I realize many teachers infuse technology naturally into classroom practices, but instructional habits in classrooms can be slow to change. So wearables force the question: Can we keep teaching as though students do not have instantaneous access?" (p. 13). In order for wearable technologies to have a lasting impact in education, educators must continue to develop their practice with their students and with the technology (if their instructional practice is to remain relevant and engaging). The literature reminds us if districts are unable to commit (financially) to training their staff on how best to use these devices in instruction, then they simply should not be purchased.

What then is the impact on stakeholders? If teacher training is limited (or non-existent) then instructors will be ill-prepared when issues concerning student usage of wearable devices begin to emerge (e.g. online bullying, academic dishonesty, privacy issues, etc.). However, the real issue lies with squandering the immense opportunity teaching with wearable and mobile technologies presents. The potential these devices have to improve student learning is examined next.

Increased Student Engagement

According to Finkel (2012), aside from standardized test scores, student engagement is the biggest measurement districts are currently using to assess the academic return on investment (ROI) on technology purchases. This claim is aligned with both the APA's (1997) learner-centered framework and Marzano et al.'s (2001) nine categories of instructional strategies affecting student achievement. Drawing from the developmental and social factors, the APA Work Group of the Board of Educational Affairs (1997) outlines *developmental influences on learning* as the consideration that "individuals learn best when material is appropriate to their developmental level and is presented in an enjoyable and interesting way" (p. 5). Marzano et al. similarly emphasize the importance of increasing student engagement among multiple strategies. For example, the researchers promote the use of *nonlinguistic representations* during instruction to engage students in different ways of thinking (e.g., elaborative thinking; Marzano et al., 2001). Prior to purchasing new technology, senior leaders should be assessing the technology currently being used in their schools and the resulting level of student engagement. Many times this can be accomplished by launching a smaller pilot study before making large-scale purchases. However, these pilots need to be monitored very closely and produce data and observations going beyond basic paper-and-pencil surveys.

Compared to some of the most recent innovations in technology in education (i.e. tablets, Massive Open Online Courses [MOOCS], etc.), wearable technologies have the potential to engage students in ways previously not possible. For example, wearable cameras like Google Glass (and other technologies) can enable students to not only create first-person videos but also engage in unique point-of-view experiences. Why open up a textbook to learn a foreign language when you can harness the power of augmented reality to travel to the country and engage with local citizens? Wearable technologies like Oculus Rift (a virtual reality device) can quite easily allow students and classrooms to engage in experiential learning by being transported to virtual 3D worlds. While it is still unclear how these experiences can be brought to scale and consistently incorporated into curriculum and instruction, wearables also present practical advantages to learning as well. According to the Educause Learning Initiative (2013), "these devices might offer powerful assistance to those with visual, auditory, or physical disabilities. Experts can monitor students with learning disabilities remotely and recommend interventions, and students struggling with language issues can access immediate translation" (para. 10). Students who struggled to become engaged with the educational content of their classroom can now receive highly personalized feedback to support their development. However, the onus now shifts to providers to continue to design wearable devices less about flash and more focused on the heightened learning and engagement they can bring to students (Foote, 2015).

With school districts having varied levels of success using current educational technologies to engage learners (i.e. SMARTboards being used like chalkboards), school leaders will be hard-pressed to justify an investment into wearables at this time. After all, stakeholders are still unsure as to whether these devices will create a more immersive learning experience or further isolate those students who lead largely online lives (i.e. constant texting, web surfing, gaming, etc.). Fredrick (2015) states, "The ubiquitous nature of wearable devices provides added possibilities for distraction and for isolation as individuals have instant connection to data. Just as we look for answers for mobile device use, the challenges will continue as wearable use grows" (p. 26). Ultimately, the successful implementation of wearables in the classroom hinges on districts being able to successfully pilot these devices to ensure the data emerging is used to refine decision-making (more on data use later in this chapter).

Collaboration

Implementing technology in classrooms can facilitate collaboration in a number of ways previously not possible. Providing students with equitable access to technology can allow *all* children to participate collaboratively in the learning process. Additionally, flipped classrooms, online tutoring, instructional software, and so on can provide relevant and engaging instruction in which students can collaborate with both teachers and their peers at any time (i.e., both at school and at home). Most important, an emphasis on facilitating collaboration to increase student achievement is present in the research. Once again both the APA's (1997) learner-centered framework (*social influences on learning*) and Marzano et al.'s (2001) nine categories of instructional strategies affecting student achievement (*cooperative learning*) mention learning through collaboration. The APA (1997) found through their research "learning can be enhanced when the learner has an opportunity to interact and collaborate with others on instructional tasks" (p. 5). While globalization and the knowledge economy have allowed people to connect with one another via the Internet and share information at any time, the ability to work collaboratively with others is still a *learned* practice.

When it comes to collaboration amongst students and teacher-student communications, wearable and mobile technologies are very powerful tools. Their applications to both traditional classroom environments and virtual/distance settings have been well documented. Foote (2015) states, "The positive potential of being able to connect with our students with positive messages or to build academic relationships with them is tremendous. What would it be like if we were on their wrists, helping them via FaceTime with a lesson or question?" (p. 13). The recent release of devices like the Apple Watch and Samsung Gear has certainly made this a real possibility. Similar to mobile devices (i.e. smartphones, tablets, etc.), wearable technologies have the ability to be with the learner at all times. The biggest challenge for educators moving forward will be the concept of time. Early adopters face waiting for students, parents, or senior administration to bring the technology into classrooms (Hargis, 2014). However, this should not be used as an excuse to stop innovating. A myriad of collaborative practices can be implemented with the existing technology/software as stakeholders wait for the wearable movement to take greater shape. Policymakers and teachers unions will also have their hands full trying to determine if the use of these devices outside of traditional school hours for instruction should result in increased compensation for educators. These are major issues and greater thought should be given to these potential problem areas as wearables begin to become more commonplace.

Consequently, in order to gain an academic return on a potential wearable technology investment, senior leaders need to ensure collaborative capabilities of the devices play a large part in the instruction and student learning in their district. The same can be said for the mobile features these technologies possess. If the education challenge does not require a mobile or collaborative solution, then the strategic model (and literature) advises against the investment.

STRATEGIC DATA USE: DATA-DRIVEN DECISION-MAKING

Timeliness

Simply put, analyzing standardized test results once a year is not a timely analysis of student performance data. In their analyses of data-driven decision making, Ikemoto and Marsh (2007) found "individuals in many districts across both studies commonly complained that state test data were not timely" (p. 120). Similarly, Greaves and Hayes (2008) discovered on more than one occasion the results from paper-based standardized tests arrived only after the next school year had already begun. This is unacceptable. Providing students with opportunities for online instruction and assessment can produce immediate results. With the Internet and technological tools providing students with real-time feedback in their personal lives (e.g., video games, text messaging, etc.), teachers and administrators need to be given the tools to deliver the same feedback in classrooms (Mandinach et al., 2006).

As technology's role in education continues to increase, there is greater pressure on districts to use data to drive their decision-making. After all, every time a student logs on the school's Wi-Fi or accesses a document on the district's learning management system (LMS), data is generated instantly. This information can then be used to guide senior leaders when they make technology decisions. For example, if a district's IT staff notices more students are using their own devices in the classroom than school-purchased technologies, then perhaps they will focus their next investment on creating a robust wireless infrastructure. This exercise would save districts unnecessary purchases on devices their students may not need. While this is data-driven decision-making at its most basic level, wearable devices can focus in on even more accurate and specific types of data surrounding student behaviors. The Educause Learning Initiative (2013) highlights the different ways wearable technologies can harness the power of data:

Wearables can be networked or might store data that can be transferred later to other devices. In many cases, the technology need not be activated; it simply functions as part of the item. Wearables can gather data—from the body of the wearer or from the environment—or provide information, or both. (para. 5)

The wealth of data wearables can generate may assist teachers supporting student learning in a more timely fashion. The common application is in physical education classes where teachers can use wristbands to track students' activity, fitness, and wellness instantly. Adjustments can then be made to the day's activities based on whether students met a particular fitness goal or mark. A similar exercise could be replicated in the classroom where an instructor modifies their lesson spontaneously based on the instant data generated by student feedback to questions posed to them through their device. The learning and instruction both become adaptive. The timeliness of the feedback will be attractive for district leaders who often struggle when making connections between data and student learning (Ribeiro, 2015).

Accessibility

Along with providing timely feedback, data tools need to be accessible to relevant stakeholders (i.e. teachers, students, principals, senior leaders). Mandinach et al. (2006) define accessibility as "how easy the tool is to access and use" (p. 10). Similarly, Ikemoto and Marsh (2007) found if the access to data is not easy, it will not be utilized. This is a significant barrier to employing the use of data-driven decision-making. The strategic model advocates districts purchase technologies creating *easy* data outputs (aside from some hardware lacking the capability) which can be used to evaluate student performance and inform future purchases.

With wearable technologies still being 3 to 5 years away from large-scale implementation (Futuresource Consulting, 2015), providers need to dedicate themselves to creating applications able to synthesize these vast amounts of data. In order for teachers to utilize student information the point of access needs to be simple. While a great deal of emphasis is often placed on access to student (and school organization) data, the use of wearable devices also raises several questions about privacy and security. For example, who owns the data these devices collect and who in particular will have access to it? Will schools continue to collect data off of students' wrists or glasses when they are outside of school? As more wearables begin to enter the market, this issue will become highly problematic (Foote, 2015). Fredrick (2015) implores stakeholders to consider both the pros and cons of data use stemming from wearable technologies. The author writes, "The ease of data use, however, needs to be balanced by a need for privacy and security of each individual. Just as current systems can be hacked, so too will wearables face challenges in security" (p. 26). District technology leaders need to be aware that investments in wearables (and the ability to provide data access to a number of stakeholders) will most likely require a substantial upgrade to schools' IT security plans. A major policy update regarding acceptable use practices will also be required. Given the health care field has adopted wearables at a higher rate than schools and universities (Sultan, 2015), perhaps education leaders should play close attention to the security issues beginning to emerge (e.g. privacy of patient data).

Capacity

One of the biggest challenges in employing the use of data-driven decision-making in school districts is a lack of technical capability. While a school board's IT department has access to a variety of student data (e.g., contact information, academic records, number of devices on network, etc.) and can analyze the data to inform decision-making, teachers and administrators may lack that ability. Mandinach et al. (2006) found "as a result, teachers' decision making strategies often lack systematicity, from student-to-student, class-to-class, and year-to-year, and may unintentionally be tinged with personal bias, and ignore key statistical concepts like distribution, variation, and reliability" (p. 2). The need for *quality* data collection was also found throughout the literature (Gomes, 2011; Ikemoto & Marsh, 2007; Mandinach & Honey, 2008; Mandinach et al., 2006). In order for this to occur, the strategic model advises districts either pursue technologies or partnerships with vendors to *build* the district's technical capacity to analyze/use data in their decision-making (e.g., procuring subject-specific software providing data analysis of student performance) or ensure personnel are being put in place to analyze the data emerging from existing technologies (i.e. *data coach*; Ikemoto & Marsh, 2007).

With curriculum departments and academic staff relying largely on anecdotal evidence to guide their recommendations, acquisition decisions are suffering. This sentiment is echoed by Digital Promise & Education Industry Association (2014). In their study of educational technology purchasing in the U.S. they stated, "The bad news? With a growing number of products and limited trusted information about them, many districts rely on informal sources instead of data and evidence to make decisions… Further, companies perceive little incentive to produce rigorous evidence" (p. 8). In order for wearable technologies to be used in support of student learning and not in tokenistic ways, senior leaders need to leverage the collection, analysis, and use of relevant data in their school systems. As stated previously, if an investment into wearable technologies is going to be made then the district will need to build its capacity in order to reap the benefits of the data the devices are collecting. This could entail lining up funding for an expanded data warehouse or hiring personnel who can quickly and efficiently help teachers draw meaning from the vast amounts of data.

STRATEGIC ACQUISITION: TECHNOLOGY PROCUREMENT AND SPENDING

Total cost of ownership (TCO)

The model draws on the work of Krueger (2013), Greaves and Hayes (2008) and highlights the need for senior leaders to be cognizant of total cost of ownership when they go to market. Krueger writes, "To get started, you must understand the cost of a technology initiative over the life of the project… Initial purchases, training, and implementation costs must be amortized or annualized, and ongoing costs must be added in" (p. 26). Greaves and Hayes also recognize "initial costs, maintenance costs, and TCO are all important factors for schools" (p. 97). With school districts moving to 1:1 computing models at a rapid rate (Jackson, 2004; Johnson & Maddux, 2008), it is imperative all costs are accounted for before hardware/software is purchased and continuously assessed throughout its rollout and implementation.

A large portion of this chapter has highlighted the importance of total cost of ownership. The author has reinforced the notion that investments in wearable technologies will most likely result in additional related costs for districts (e.g. enhancing Wi-Fi access, expanding data warehouses, providing professional development for teachers, etc.). Hargis (2014) emphasizes rigorous information and communication technology (ICT) infrastructure will be essential for wearables to be implemented successfully. Acquiring wearables at this time is further complicated by their early stage of development and technical shortcomings. Despite being evangelists for Google Glass, Robak and Ajmi (2015) take issue with the current iteration of the device. While highlighting their use of Glass over several months provided students with a rich and immersive learning experience, they wrote, "we have identified the device's serious hardware and software flaws that make it less than ideal in an educational environment in its current state of development" (p. 14). In order to avoid unnecessary costs (i.e. repairs), districts would be best served by avoiding large-scale implementations of wearables at this time and continuing their pilot studies until the technology has been refined.

Organizational Vision

The literature draws attention to an overall district vision as imperative for sound technology decision-making and leadership (Fullan, 2013b; Gomes, 2011; Hall, 2010; ISTE, 2009). If a technology does not fit within the broader vision and goals of the school board, then it should not be purchased. Furthermore, school leaders need to ensure policies they create align well with the technologies they are implementing. Finally, the overall organizational vision should be informed by the *entire* organization. ISTE (2009) focuses on "teachers and support staff, school and district administrators, teacher educators, students, parents, and the community" (p. 1). In a publicly funded institution, stakeholders should work collaboratively to ensure they are all working towards the same goals. Once this vision is established, leaders at various levels need to ensure the technology's implementation is monitored and any triumphs or challenges encountered are further explored (Hall, 2010).

Senior leaders and stakeholders need to constantly assess the vision or identity of their school district in order to ensure acquisitions are given the highest possibility of success upon implementation. While teachers play a large role in ensuring instruction is relevant and engaging for students, district leaders need to ensure their policies match with their overall goals for the organization. Foote (2015) notes, "With many schools still blocking social media on 'regular' devices such as phones and laptops, we are facing a situation in which schools are likely to be woefully disconnected from the technology lives of our students" (p. 12). Some districts have yet to meaningfully support Bring Your Own Device or BYOD initiatives in their schools. How can we expect that wearable technologies will be treated any differently? If districts wish to pursue wearables in their organizational vision it should be one of mobility and connectivity affording students the ability to learn anywhere and anytime. This may require investments in charging stations, learning management systems, and other hardware/software supporting student access to learning materials.

The perils of poor technology leadership often stem from districts acquiring mismatched technologies not complimenting one another or fitting within their existing infrastructure (i.e. purchasing tablets with low wireless Internet coverage). Google Glass poses a major problem in this area. Robak and Ajmi (2015) discovered Glass and its applications can only connect to wireless access points not requiring a password or single sign-on (SSO). For security purposes most districts have a password-protected wireless network could create issues for teachers and students who wish to use the full functionality of Google Glass (in its current form). Surely this is one of the issues Google is seeking to address after pulling the device off the market in January 2015.

Efficacious Funding

In order for school districts to avoid implementation failures, they need to determine whether the amount of funding they have (and will have in the future) is enough to successfully purchase, deploy, and assess a new technology (Boser, 2013; Cuban, 2001; Greaves & Hayes, 2008; Magolda, 2006; Sundeen & Sundeen, 2013). ISTE (2009) advises district leaders to acknowledge the importance of continuous and adequate funding in any technology acquisition as well.

One of the biggest concerns surrounding wearable technology (and educational technology for that matter) is the high cost of the devices. A FitBit currently costs around US $99 or more and Google Glass sold for roughly US $1,500 at the time it was taken off the market (Everson, 2015). The presence of sustainable funding will determine whether districts pursue acquiring wearable devices or wait until costs fall and students begin to bring them to school. It is unclear at this moment whether districts are even thinking about wearables considering the vast amount of funds tied up in current ed-tech ventures (e.g. tablets, Chromebooks, LMS, etc.). Unlike other organizations, school boards are not able to generate revenue and must conserve what they spend. Vendors of wearable technologies will need to make their products and services more affordable if they wish to work with the K–12 public sector. Another issue related to funding is equity of access. Foote (2015) writes,

And on the personal side, the proliferation of wearable possibilities may continue to add to the digital divide as students in wealthier communities have more access than others. Is this an area of technology equity that needs to be addressed when schools are still struggling with 1:1 adoptions? How can technology startups keep struggling schools in mind when rolling out new products that have potential for school use? Some schools will be facing these challenges sooner than others. (p. 13)

It is difficult to say whether a divide will form between families affording wearable technologies for their children and those who cannot. Districts may feel pressure from their communities to provide students with devices especially if the perception is one of transforming learning. However, it is simply too early to tell whether wearables will follow the trajectory of tablets and smartphones and be ubiquitous in the learning environment or remain a luxury for fitness enthusiasts and eager adopters (Hargis, 2014).

SOLUTIONS AND RECOMMENDATIONS

The previous section highlighted wearable technologies' connection to student learning, data-driven decision-making, and acquisition. The analysis emphasized the great potential and numerous challenges ahead for school districts. What follows are recommendations for district leaders and stakeholders to adequately prepare for the shifting nature of technology in the classroom. Once again, these solutions are rooted in the relevant literature. First, districts need to continue to make explicit connections between wearable technologies and student learning. Next, senior leaders should pay great attention to the current implementation of wearables in both the healthcare industry and higher education. Finally, the importance of planning a technology's purchase, implementation, and assessment is stressed.

Make Distinct Connections to Student Learning

Technology and business futurist, Dan Burrus once said:

People ask whether everyone's going to walk around with Google Glass, which is the wrong question . . . The better question is, 'Is there a time when people are standing, walking and moving when they could also benefit from using a computer?' (Everson, 2015, p. 20)

This is an excellent point and one that highlights the need for senior leaders to focus on the education challenge and how wearable technologies can help facilitate learning. Viewing wearables as a "silver bullet" for solving existing issues within a school district (e.g. low attendance rates) is highly misguided.

Technology leaders should learn from previous unsuccessful attempts to couple technology and education (e.g. LAUSD) and recognize the failure often stems from a lack of connection between the device and student learning.

Part of the problem stems from the acquisition process. Often times teachers (who are typically charged with harnessing the power of these technological devices) have a limited role in the decision-making behind a technology purchase (Maas & Lake, 2015). Recently, the Bill and Melinda Gates Foundation (2014) released a hierarchical study (surveying both students and teachers) called, *Teachers Know Best*, which addresses the "mismatch between the kinds of digital instruction tools teachers say they actually need and the kinds of products companies are creating and school districts are buying" (p. 4). The results emphasize the role of the teacher in K–12 technology acquisition process and how highlighting their voices can help strengthen investments and implementation. Some key findings indicate when teachers are given choice about the technology their students use, they are more likely to report the product was effective. In order for wearables to be tethered to both pedagogy and student learning, educators must take a leading role in the development, decision-making, and assessment of these digital tools.

Another issue lies with districts' inability to show the financial return on their technology investment (Maas & Lake, 2015). However, with validated ROI calculations (specific to education and wearable technology) not being prevalent in scholarship, Krueger (2013) advocates for a redefinition of return on investment focusing less on cost-related factors and more on student learning and community-building (i.e., developing 21st century skills, engaging parents, etc.). He writes,

But the mission of schools is not to increase revenues but rather to use public monies wisely in producing students who are thoughtful, productive citizens and members of the workforce. That is how schools "create value" for their communities. So, even though most aspects of schools' performance cannot be measured in financial terms, there are still measurable ways that you can demonstrate to stakeholders the value of the investment in technology in terms of advancing the educational mission of the district. (Krueger, 2013, p. 26)

Consequently, this chapter reinforces the notion of in the absence of traditional ROI calculations for technology spending, districts need to shift their focus to analyzing technology's impact on student learning.

Pay Close Attention to Healthcare and Higher Education Industries

As with many other technological developments, the use of wearable technology in K-12 schools will most likely follow the lead of healthcare and higher education, where they are consistently being experimented with and funding is easier to come by (Educause Learning Initiative, 2014). In fact, many medical schools are beginning to use Smart Glasses during instruction so students can benefit from viewing a first-person perspective during a surgical procedure (even if they are learning from a distance). When discussing wearables Sultan (2015) suggests,

Current practical and potential uses of this technology in healthcare (especially in patient monitoring and data reporting) and medical education indicate, as highlighted in this article, that a new era of interaction between people and technology is about to begin that could ultimately translate into our ability to provide healthcare systems that are more efficient and responsive than the current ones. However, for this to take place there are a number of technological, personal, social and cultural challenges that this technology needs to address and overcome (p. 526).

It was mentioned previously use of wearables poses a potential threat to the privacy and security of student information. Senior leaders would benefit greatly if they paid close attention to the management strategies used in the healthcare sector's collection and analysis of data. After all, patient records are often viewed as highly private. If hospitals begin to experience hacks and information leaks as a result of their investment in wearable technologies, education leaders should take note so the same mistakes are not repeated.

Plan, Plan, Plan

Districts need to take advantage of the fact that wearables are not quite ready for ubiquitous use and implementation at this point. ISTE (2009) suggests if a school district thinks they may want to pursue wearable or mobile technologies on a large scale in the future, then senior leaders and their staffs will need to begin creating an organizational vision immediately. A robust technology plan and infrastructure supporting a culture of access and mobility will likely take several years to plan and implement. However, if districts are able to begin this process early enough, the shift in the schools' operations would seem less drastic and it would allow stakeholders to get accustomed to the new protocols, policies, and expectations. The presence of a technology plan creates a standard to support acquisition and technology use in the district. While challenges may be experienced in getting stakeholders to adhere to the guidelines (i.e. catching teachers with personal Internet routers in their classrooms, which leaves the whole organization open to attack because it is not a secure wireless connection), having an overall technology strategy showing district leaders how the organization is developing in relation to what goals were outlined at the onset of the plan.

Undoubtedly the preparation for the wearables movement will be extensive. Hopefully, quality research will continue to emerge as time progresses and districts can constantly refine their thinking about wearables in the classroom. It is rare for a technology plan to remain unchanged over a 3 to 5 year period given advances in the technology cannot be predicted. In summation, districts need to highlight a technology's relationship to student learning, pay close attention to the healthcare and higher education industries, and place a large emphasis on implementation and strategic planning.

FUTURE RESEARCH DIRECTIONS

This chapter (and book as a whole) draws attention to a number of unknowns about the wearable technology movement in education. It is quite possible readers may walk away with more questions than answers. However, it is crucial scholars, practitioners, and technology providers feel comfortable collaborating and teasing apart educational issues with the goal of improving learning for students. While this chapter focuses on technology acquisition and spending, there is a wealth of opportunity for researchers to explore the numerous innovations currently being developed in the world of wearables. Early adopters need to continue to innovate, for it is only from their findings we may be able to establish best practices and find sustainable ways to introduce wearables into K–12 education. Of particular interest to the author is the shifting nature of social interaction between humans through wearable technology. Foote (2015) writes,

Will schools continue to block access to social media, or will it even be possible to block? If it is not possible, how will schools help students cope with potential bullying taking place via social media and landing directly on the students' wrist or jewelry? (p. 13).

Despite the excitement enveloping this current age of technological development, this is a real concern for student wellbeing. Are teachers going to be prepared when students' interactions with one another become even more personalized? Hargis (2014) seems to think so. In fact he believes wearables may have the potential to help in the area of student mental health. He posits,

On the near horizon technologies are currently proposed to allow learners to wear devices, which can monitor their physical health, perhaps someday their mental and intellectual health via the quantifiable self. The Internet of things will help connect the data and make available to teachers and other experts (p. 19).

Greater study should be done in this area (both on the development and evaluation side) to determine if student well-being is positioned to help or harm (or both) by the wearables movement.

With the prevalence of ICT in K–12 schooling rapidly increasing (through wearable devices, 1:1 computing, 3D printing, etc.), the amount of quality research exploring best practices in procurement must also increase. Future research should consider studying both centralized and decentralized procurement strategies and their respective advantages and disadvantages in acquiring emerging technologies. While very recent research has uncovered information about technology purchasing, governance, and assessment in school districts (Digital Promise & Education Industry Association, 2014; Morrison et al., 2014; Ribeiro, 2015), more comprehensive work needs to be done in this area. With larger school districts in particular struggling with technology acquisition, researchers need to intervene before costly mistakes begin to have a negative impact on student learning.

With the acquisition process constantly shifting in K–12 education, the Strategic Model for Technology Acquisition should be updated as more research begins to emerge. Today, very little formal scholarship exists on wearable and mobile technology procurement (specifically), and the model should be adjusted if innovations in technology alter the overall focus or mandate of publicly funded school districts and student learning.

Lastly, a review of the relevant literature reveals districts often struggle to conduct pilot studies with extractable data for the purpose of informing decision-making and assessing student learning through technology (most senior leaders sift through large amounts of anecdotal data; Digital Promise & Education Industry Association, 2014; Ribeiro, 2015). These statements should be researched further to discover how technologies are piloted in different districts and whether these problems are applicable to wearables (since they are able to collect a great deal of quantitative data easily) or only emerging educational technologies.

CONCLUSION

The purpose of this chapter was to investigate the potential applications wearable and mobile technologies could have in K–12 educational settings. Both the opportunities for improved instruction and data collection and analysis were highlighted. However, as we are seeing with educational technology acquisition in school districts, there are a number of challenges complicating the procurement process for school leaders and stakeholders. The author used the Strategic Model for Technology Acquisition (Ribeiro, 2015) to explore wearable technologies' connection to student learning, data-driven decision-making, and acquisition. Lastly, several recommendations and suggestions for future research directions were provided.

In closing, many of the innovations and advantages wearable devices can bring to the learning environment are only just beginning to be realized. As the technology becomes more commonplace and early adopters continue to establish and share their best practices with others, student learning in the classroom may become more engaging than it has ever been. However, the interaction between the devices, students, and teachers is only one part of the equation. Senior technology leaders, parents, and governments all need to embrace this vision and ensure it is less about "shiny" devices and more about exploring new channels of learning. While our excitement must be tempered for now as developers and providers work to create products designed specifically for educational stakeholders, it is no excuse to remain stagnant. We should continue to innovate, continue to experiment, and try to provide students with more engaging learning each and every day. However, when it comes to our schools parting with taxpayer dollars to explore larger implementations, we must remember to be strategic with our selections. The author readily admits adhering to the relevant scholarship and aiming for sustained technological growth in a school district may not grab them all the news headlines – however following the latest trends and relying on gut instinct when making acquisitions may end up grabbing them the headlines they never want to read.

REFERENCES

Avidov-Ungar, O., & Eshet-Alkakay, Y. (2011). Teachers in a world of change: Teachers' knowledge and attitudes towards the implementation of innovative technologies in schools. *Interdisciplinary Journal of E-Learning and Learning Objects*, *7*, 291–303.

Bebell, D., O'Dwyer, L. M., Russell, M., & Hoffmann, T. (2010). Concerns, considerations, and new ideas for data collection and research in educational technology studies. *Journal of Research on Technology in Education*, *43*(1), 29–52. doi:10.1080/15391523.2010.10782560

Bellamy, A. (2007). Exploring the influence of new technology planning and implementation on the perceptions of new technology effectiveness. *Journal of Technology Studies*, *33*(1), 32–40.

Teachers know best: What educators want from digital instructional tools. (2014). *Bill & Melinda Gates Foundation*. Retrieved from http://collegeready.gatesfoundation.org/sites/default/files/Teachers%20Know%20Best_0.pdf

Blow, D., & McConnell, S. (2012). Classroom technology rollouts. *Leadership*, *42*(2), 32.

Blume, H. (2014). Head of L.A. schools' technology division resigns. *LA Times*. Retrieved from http://www.latimes.com/local/lanow/la-me-ln-lausd-technology-resign-20141031-story.html

Boser, U. (2013). Are schools getting a big enough bang for their education technology buck? *Center for American Progress*. Retrieved from http://www.americanprogress.org/issues/education/report/2013/06/14/66485/are-schools-getting-a-big-enough-bang-for-their-education-technology-buck

Botelho, S. (2014). IT centralization: Institutional hesitation. *University Business, 17*(9).

Brooks, C. (2008). Alberta education technology policy analysis: Implications and future directions. *The International Journal of Learning*, *15*, 139–146.

Courville, K. (2011, November). Educational technology: Effective leadership and current initiatives. *Paper presented at the 2011 Louisiana Computer Using Educator's Conference*, New Orleans, LA.

Cuban, L. (2000). Is spending money on technology worth it? *Education Week*, *19*(24), 42.

Cuban, L. (2001). *Oversold and underused: Computers in the classroom*. Cambridge, MA: Harvard University Press.

Culp, K. M., Honey, M., & Mandinach, E. (2005). A retrospective on twenty years of education technology policy. *Journal of Educational Computing Research*, *32*(3), 279–307. doi:10.2190/7W71-QVT2-PAP2-UDX7

Improving ed-tech purchasing: Identifying the key obstacles and potential solutions for the discovery and acquisition of personalized learning tools. (2014). *Digital Promise & Education Industry Association*. Retrieved from http://www.digitalpromise.org/page/-/dpdocuments/marketplace/Improving_Ed-Tech_Purchasing.pdf?nocdn=1

Education is the only solution: Remarks of U.S. Secretary of Education Arne Duncan at the U.S. Agency for International Development (USAID) Global Education Summit. (2013, August 6). *US Department of Education*.

7 things you should know about wearable technology. (2013, November 8). *Educause Learning Initiative*. Retrieved from http://www.educause.edu/library/resources/7-things-you-should-know-about-wearable-technology

Everson, K. (2015). Learning is all in the wrist. *Chief Learning Officer*, *14*(4), 18–21.

Finkel, E. (2012). Getting the best ROI in technology: The latest, greatest tools won't buy you success without a plan. *District Administration*, *8*, 74.

Foote, C. (2015). Wear to learn!?!. *Internet@Schools, 22*(3), 12-13.

Fredrick, K. (2015). Dressing for the future: Wearable technology. *School Library Monthly*, *31*(4), 25–26.

Fullan, M. (2013a). Great to excellent: Launching the next stage of Ontario's education agenda. *Department of Education*. Retrieved from http://www.edu.gov.on.ca/eng/document/reports/FullanReport_EN_07.pdf

Fullan, M. (2013b). *Stratosphere: Integrating technology, pedagogy, and change knowledge*. Canada: Pearson.

Universe of education technology spend 2013. (2014, June 10). *Futuresource Consulting*. Retrieved from http://reports.futuresource-consulting.com/tabid/64/ItemId/177649/Default.aspx

Press alert: Global market for wearable technology worth $8.9bn in 2014. (2015, April 9). *Futuresource Consulting*. Retrieved from http://www.futuresource-consulting.com/2015-04-Wearable-Technology-7723.html

Gomes, W. (2011). Leadership in educational technology: Insights from the corporate world. *Journal of Leadership Studies*, *4*(4), 57–61. doi:10.1002/jls

Greaves, T., & Hayes, J. (2008). America's digital schools 2008: The six trends to watch. *The Greaves Group, The Hayes Connection*. Retrieved from http://www.lcps.org/cms/lib4/VA01000195/Centricity/Domain/52/Americas_Digital_Schools_2008.pdf

Learner-centered psychological principles: A framework for school reform and redesign. (1997). Group of the Board of Educational Affairs. Washington, DC: American Psychological Association.

Hall, G. E. (2010). Technology's Achilles heel: Achieving high-quality implementation. *Journal of Research on Technology in Education*, *42*(3), 231–253. doi:10.1080/15391523.2010.10782550

Hargis, J. (2014). Eager adopters in education: Strategic plan ideas for integrating instructional technology. *Turkish Online Journal of Distance Education*, *15*(4), 14–22.

Heinecke, W. F., & Blasi, L. (2004). *Methods of evaluating educational technology*. Greenwich, CT: Information Age.

Hsu, Y. C., Hung, J. L., & Ching, Y. H. (2013). Trends of educational technology research: More than a decade of international research in six SSCI-indexed refereed journals. *Educational Technology Research and Development*, *61*(4), 685–705. doi:10.1007/s11423-013-9290-9

Ikemoto, G. S., & Marsh, J. A. (2007). Chapter 5: Cutting through the "data-driven" mantra: Different conceptions of data-driven decision making. *Yearbook of the National Society for the Study of Education*, *106*(1), 105–131. doi:10.1111/j.1744-7984.2007.00099.x

Essential conditions. (2009). *International Society for Technology in Education*. Retrieved from http://www.iste.org/docs/pdfs/netsessentialconditions.pdf?sfvrsn=2

Jackson, L. (2004). One-to-one computing: Lessons learned and pitfalls to avoid. *Education World*. Retrieved from http://www.educationworld.com/a_tech/tech/tech197.shtml

Jenkinson, J. (2009). Measuring the effectiveness of educational technology: What are we attempting to measure? *Electronic Journal of e-Learning*, 7(3), 273–280.

Johnson, D. L., & Maddux, C. D. (2008). Introduction: Effectiveness of information technology in education. *Computers in the Schools*, *24*(3/4), 1–6.

Kim, P., Hagashi, T., Carillo, L., Gonzales, I., Makany, T., Lee, B., & Gàrate, A. (2010). Socioeconomic strata, mobile technology, and education: A comparative analysis. *Educational Technology Research and Development*, *59*(4), 465–486. doi:10.1007/s11423-010-9172-3

Krueger, K. (2013). Forget ROI, the future of technology investment is all about value. *T.H.E. Journal*, *40*(6), 25–28.

Lebeaux, R. (2013, October). Technology leadership quotes from the October 'CIO Decisions' e-zine. *TechTarget*. Retrieved from http://searchcio.techtarget.com/opinion/Technology-leadership-quotes-from-the-October-CIO-Decisions-e-zine

Maas, T., & Lake, R. (2015). A blueprint for effective and adaptable school district procurement. *CRPE.* Retrieved from http://www.crpe.org/sites/default/files/crpe-report-effective-adaptable-school-district-procurement_0.pdf

Magolda, P. (2006). Critical fusion – Technology and equity in secondary education. *AACE Journal, 14*(3), 287–311.

Mandinach, E. B., & Honey, M. (2008). *Data-driven school improvement: Linking data and learning. Technology, Education Connections (TEC) Series.* New York, NY: Teachers College Press.

Mandinach, E. B., Honey, M., & Light, D. (2006, April). A theoretical framework for data-driven decision making. *Paper presented at the annual meeting of the American Educational Research Association,* San Francisco, CA.

Marzano, R., Pickering, D., & Pollock, J. (2001). *Classroom instruction that works.* Alexandria, VA: ASCD.

McCombs, B., & Whisler, J. S. (1997). *The learner-centered classroom and school: Strategies for increasing student motivation and achievement.* San Francisco, CA: Jossey-Bass.

Michels, P. (2013, November 5). The high cost of iPad-mania in public schools. *The Texas Observer.* Retrieved from http://www.texasobserver.org/run-school-ipad-program/

Militello, M., & Friend, J. (2013). *Principal 2.0: Technology and educational leadership.* Charlotte, NC: Information Age.

Morrison, J. R., Ross, S. M., Corcoran, R. P., & Reid, A. J. (2014, September 22). *Fostering market efficiency in ed-tech procurement.* Retrieved from http://b.3cdn.net/dpromise/4e2869abb217c9062f_z6m-6v2v9u.pdf

Palfrey, J., & Gasser, U. (2013). *Born digital: Understanding the first generation of digital natives.* Philadelphia, PA: Basic Books.

Partnership for 21st Century Skills. (2011). *Common Core toolkit: A guide to aligning the Common Core Standards with the framework for 21st century skills.* Retrieved from http://www.p21.org/storage/documents/P21CommonCoreToolkit.pdf

Pickert, K. (2014, October 17). *How the iPad helped bring down the Los Angeles schools chief.* Retrieved from http://time.com/3514155/ipad-john-deasy-lausd-superintendent-resigns/

Prensky, M. (2008). The role of technology in teaching and the classroom. *Educational Technology,* 2008, 1–3.

Pulley, J. (2014). Exploring new frontiers with Google Glass. *Campus Technology Magazine, 27*(11), 31–33.

Ribeiro, J. (2014). Teacher technology adoption and the philosophy of fear. *Antistasis, 4*(2), 30–33.

Ribeiro, J. (2015). *Educational technology decision-making: A case study on technology acquisition for 746,000 Ontario students* [Unpublished master's thesis]. Brock University, Ontario, Canada. Retrieved from https://dr.library.brocku.ca/handle/10464/6318

Robak, M. J., & Ajmi, A. (2015). Google glass for the educator: A postmortem separating the reality from the hype and some thoughts for Google. *Computers in Libraries*, *3*, 12–15.

Spector, J. M. (2013). Emerging Educational Technologies and Research Directions. *Journal of Educational Technology & Society*, *16*(2), 21–30.

Sultan, N. (2015). Reflective thoughts on the potential and challenges of wearable technology for healthcare provision and medical education. *International Journal of Information Management*, *35*(5), 521–526. doi:10.1016/j.ijinfomgt.2015.04.010

Sundeen, T. H., & Sundeen, D. M. (2013). Instructional technology for rural schools: Access and acquisition. *Rural Special Education Quarterly*, *32*(2).

Zaslavsky, C. (1982). *Tic-tac-toe: And other three-in-a-row games from ancient Egypt to the modern computer*. New York, NY: Crowell.

KEY TERMS AND DEFINITIONS

21st Century Skills: A set of abilities believed to be critical for today's students in order for them to be globally competitive in both schooling and the workforce. Specific skills include collaboration, critical thinking, communication, creativity, etc.

Bring Your Own Device (BYOD): A technology model or policy encouraging students to bring their personally owned electronic devices to school so they can develop their 21st Century skills.

Consumerization: The process by which products (goods or services) are made suitable for mass consumption (with a focus on the end user).

Data-Driven Decision-Making (DDDM): A term used to describe the collection and analysis of vast amounts of data in order to make key decisions at an organizational level.

Information Technology (IT): A field of computer technology largely concerned with managing and processing large quantities of information (data).

Pedagogy: The practice or profession of teaching.

Pilot Study: A small preliminary study (with clearly defined areas of analyses and data outputs) conducted to evaluate the merits of a technology and whether or not a larger scale implementation should be pursued.

Return on Investment (ROI): The amount of value (academic or financial) extracted from a monetary investment.

Technology Acquisition: The purchasing of a technology for an organization and its stakeholders. Often times the purchase will abide by either a formal or informal decision-making process.

Technology Leadership: The management strategies and tools utilized during the purchase, implementation, and assessment of a technology.

Chapter 4
Health and Fitness Wearables

Mike S. Butler
Emporia State University, USA

Paul E. Luebbers
Emporia State University, USA

ABSTRACT

Health and fitness wearable technology has been around for about 45 years. The most common wearables initially included heart rate monitors, pedometers, and GPS systems. While most of these were initially available as stand-alone devices, many are now available in combination devices such as fitness trackers and smart watches. Health and fitness wearables are used in a wide variety of situations including sports, athletics, personal health, fitness, education, and medical settings. A brief history of each wearable is provided, basic functions are explained, selected research is presented, and potential future developments are discussed. Some information is provided regarding current models and features.

INTRODUCTION

This chapter provides an overview of wearable technology related to health and fitness. The focus is on the more common wearables including heart rate monitors, pedometers, accelerometers, and combination devices such as fitness trackers and smart garments. Some of the wearables discussed are devices and some are embedded devices. Research findings will be summarized and reviewed where possible. While some of the technologies have been heavily researched (such as pedometers), others – notably the newer technologies - have little if any research available.

BACKGROUND

Health and fitness wearables have been used in many settings for many different purposes and with many different populations. The more common applications include personal use, educational, sports, athletics, medical use, and research. See Sultan, 2015, for an excellent review on the development of wearables and their use in healthcare and medical education. Bächlin and Tröster (2012) provide a comprehensive

DOI: 10.4018/978-1-5225-0069-8.ch004

review on health and fitness wearables, describing their use in a wide variety of sports including cycling, diving, running, skiing, snowboard, ski-jumping, golf, rowing, football games, baseball, and table tennis. A review of the research literature on each wearable is included.

MAIN FOCUS OF THE CHAPTER

The objectives of this chapter are to provide an overview of the most common health and fitness wearable technologies including research findings as well as perspectives on the past, present, and future of these technologies and discussions of their applications. Issues, controversies, and problems, as identified in the research, are presented for each technology.

HEART RATE MONITORS

There was a so-called "fitness boom" during the 1970s and 1980s when aerobic activities such as jogging and dance greatly increased in popularity. This was in part due to an increased public awareness of the benefits of aerobic fitness in terms of decreasing the risks for some of the leading causes of death including cardiovascular disease. Bill Bowerman, co-founder of Nike, Inc. and head coach of track at the University of Oregon, along with a cardiologist, published a book called "Jogging" in 1966 selling over a million copies and was credited with helping to popularize jogging in the United States. Another influential book helped start what some called "America's fitness revolution" and further popularized running for fitness. The book was titled "The Complete Book of Running" by James "Jim" Fixx, published in 1977 and sold over one million copies. Many aerobic exercisers were taught the recommendations for the intensity, duration, and frequency of exercise as specified by the American College of Sports Medicine. The basic indicator of intensity is exercise heart rate but there was a challenge for exercisers in that usually, the only way to determine the exercise heart rate was to take it manually. This was not a problem if one stopped exercising, but stopping was disruptive unless one was at the end of a workout. Taking a pulse rate manually during a workout was often very difficult, particularly during running or swimming. One alternative commonly used was to have exercisers rate their "perceived exertion". There was a formalized system for this known as Borg's Rating of Perceived Exertion (Borg, 1970). Researchers frequently investigated the system in an effort to determine if it was valid and reliable. Given that the Borg system was mostly subjective it was not an ideal way to assess the intensity of exercise. This state of affairs set the stage for the introduction of the heart rate monitor. Heart rate monitors are often used to determine the intensity of aerobic exercise. Specifically, they allow users to track their heart rate during exercise to aid in maintaining the appropriate intensity or workload. Most heart rate monitors work by combining electrodes to sense the electrical signal from the heart. In some cases this is augmented with an audio device detecting the sound from the heart. The first wearable heart rate monitors used a chest strap sending telemetry on the heart rate to a watch worn on the wrist. The chest sensors usually needed to have good contact with the skin and enough moisture for the electrodes to pick up the signal. During high intensity activity there was usually adequate perspiration to ensure a good signal provided the fit was snug, but users often had to moisten the back of the sensor to get good signal when first putting it on. With some of the new sensors such as those in the smart watches and those involving ear clip-ons, the moisture issue may no longer be a concern. The American College of Sports Medicine is one of the recognized authorities on exercise recommendations and they have released specific recommendations for aerobic exercise in terms of the intensity, duration, and frequency of activity. The intensity recommendations are based on ranges of age-adjusted heart rate percentages (e.g., 40-60% of estimated maximum heart rate). Heart rate monitors make it easy for the wearer to know whether or not they are exercising at their desired intensity. In fact, many models allow the user to program the monitor to emit an audible alarm when the intensity of the exercise falls below or exceeds the recommendation.

Brief History of Heart Rate Monitors

The Polar Electro Company developed the first portable heart rate monitors in the late 1970s. They were originally developed to assist ski training, which is not surprising given the company is based in Kempele, Finland. The company was founded in 1977 and began selling their heart rate monitors to the public in 1983. The first model was the Polar PE 2000. Polar is still selling heart rate monitors over 30 years later and has developed an extensive product line. Polar maintains their own index of research conducted with their monitors and as of 1998, the company included 400 physiological and medical studies on humans and animals. They have 26 subsidiaries and 35,000 retail outlets. However, they now compete with many other vendors, such as Garmin, and Timex. By 2012, heart rate monitors were the most common wearable technology (Bächlin & Tröster, 2012). Heart rate monitors have evolved to accommodate diverse needs and applications. As examples of the diversity of applications, early developments included monitors for racehorses and monitors used by swimmers. Since then, many studies with a variety of animals have included heart rate monitors (Besch, Varosi, & Brigmon, 1992). Heart rate monitors have also been developed for group situations such as a physical education class or a sports team. Some packages offer the capability for a display of an entire group's heart rate on a large screen or wall. Other improvements include the ability to adjust the sample rate used to collect heart rate data and the ability to transfer or transmit the data to computers for storage and analysis. Early data transfer systems for heart rate monitors were not always reliable and would include mechanics such as placing the watch in a cradle connected to a PC with an RS232 serial connection. As computer technology evolved over time, so too did the data transfer systems, eventually transitioning to current forms including USB connections and wireless transmitters and receivers. When used in a group setting, such as collecting heart rates from students in a physical education class, interference between devices was sometimes a problem. If multiple users were in close proximity to each other and used the same model of device the watches would sometimes receive data from multiple chest straps causing errors. Many newer models have the capability to transmit on multiple or custom channels which essentially eliminate these types of issues.

Research on Heart Rate Monitors

A wide variety of research has been conducted with heart rate monitors. Much of the research has been done to determine the accuracy, validity, and reliability of the measurements taken with the heart rate monitors in various applications and with various populations. Diverse examples of the research are presented in this section. In an early example of a validity study, researchers compared the accuracy of nine commercial heart rate monitors with an electrocardiograph and found most provided suitable monitoring (Humen & Boughner, 1984). In a more recent example, the validity of seven commercially available heart rate monitors was examined and researchers concluded they were accurate compared to an electrocardiogram reference but the accuracy of the heart rate monitors tended to decrease at high heart rates and motion levels (Terbizan, Dolezal, & Albano, 2002). The majority of studies focusing on issues of validity and reliability have found them to be accurate and valid (Laukkanen & Virtanen, 1998). The research related to the application of heart rate monitors often has to do with their utility in terms of meeting aerobic exercise recommendations. In a typical example, researchers used heart rate monitors to continuously monitor elementary aged school children and found boys experienced more short sessions of intense activity than girls, but neither group experienced enough activity for the improvement of aerobic fitness (Armstrong & Bray, 1991). As previously mentioned, aerobic exercise recommendations are typically expressed relative to maximum heart rate. The maximum heart rate is often estimated from a simple formula of subtracting the age of the individual from 220. In an interesting research study on the use of heart rate monitors by endurance athletes, one of the conclusions found the estimated maximal heart rates were too variable to be of use and the true maximal heart rates needed to be obtained through testing (O'Toole, Douglas, & Hiller, 1998). Not all research on heart rate monitors has to do with establishing validity or using them for meeting exercise recommendations. Research has also been conducted on the utility of heart rate monitors for tracking sleep. In one such study, researchers found available heart rate monitors were useful in allowing sleep self-evaluation (Eyal, Dagan, & Baharav, 2012). Other sleep tracking wearables will be discussed in a separate section later in this chapter.

Some research has included devices merging the functionality of heart rate monitors with other health and fitness wearables such as motion sensors. An example of this is a study with a device called Actiheart© measuring heart rate and movement for the purpose of estimating energy expenditure. Researchers found the precise placement of the Actiheart device did not have a significant effect on movement count or energy expenditure estimates, and this alleviated a concern about placement of the motion sensor affecting the results (Brage et al., 2006).

Future of Heart Rate Monitors

Due to their utility and efficacy, heart rate monitors will likely continue to be a popular fitness wearable. This may be either as a stand-alone device, or as will be discussed later in this chapter, incorporated into multi-function wearables. It also seems likely the use of chest straps may decrease as other sensing technology becomes available.

PEDOMETERS

Pedometers are motion sensitive devices, specifically the motion of human gait, and are designed to record steps taken. It should be noted many of the devices in this category could be either mechanical or electronic. There are other devices performing the function of measuring steps, most notable accelerometers, but since they have other uses, they are discussed in a separate section. Trost and O'Neil (2014) point out that pedometers detect the acceleration of the body's center of gravity during the gait cycle and record it as a step when a critical threshold is exceeded. They also point out while pedometers can be used to estimate distance and energy expenditure, those estimates are dependent on the calibration of the device because each person's step distance can vary. To date, more research has been conducted on pedometers than any other wearable health and fitness technology although heart rate monitors have also been widely researched.

Brief History of Pedometers

The history of pedometers dates back to Leonardo Da Vinci who described a mechanical pedometer for military applications. Thomas Jefferson brought a pedometer from France to the United States, and in 1780, Abraham-Louis Perrelet introduced a pedometer in Switzerland measuring both steps and distance during walking. Yoshiro Hatano marketed a pedometer in Japan in 1965 and later published research suggesting that 10,000 steps a day was desirable for health and fitness (Hatano, 1997). While not universally endorsed, this recommendation has become a widespread standard. Berlin et al. (2006) point out that while a goal of 10,000 steps per day is often referenced as an acceptable level of exercise, further research is needed to see if that equates to the Surgeon General's recommendation of adults engaging in at least 150 minutes of moderate-intensity activity each week, and children and teenagers should engage in at least one hour of activity each day. Early pedometers were mechanical and required the user to wear the device in a particular location, which was typically the anterior hip, directly above the knee. This orientation generally provided the greatest sensitivity to the force and bodily motion produced with each step. This allowed components which were usually either a sliding ball or pendulum to most accurately register the steps. Inexpensive modern pedometers still utilize this design. Other modern versions use electronics instead of the mechanical sensors. High-end models include accelerometers with up to 3 axes for more accurate data collection. The best models are generally considered to be accurate to within 5% and produce fewer step recognition errors.

Research on Pedometers

Much research has been conducted on the validity and reliability of pedometers. In a review of 29 articles published since 1980, researchers concluded there is sufficient evidence to support both the convergent and construct validity of pedometers and this supports their use in research and practice (Tudor-Locke, Williams, Reis, & Pluto, 2002). In addition to research validating pedometers, pedometers have also been used to establish the validity of other methods of assessment. In one example, researchers in Japan used pedometers to establish the validity of a walking questionnaire and concluded the questionnaire provided reasonably reproducible and valid data (Tsubono et al., 2002). In another example researchers collected activity data from a large group of older adults in Amsterdam and compared a questionnaire and diary with data collected from pedometers (Stel et al., 2004). They found the pedometer data useful in validating the data from the questionnaire and concluded the questionnaire was easier to use than the diary or pedometer. As was previously mentioned, the recommendation of 10,000 steps per day has been questioned, and in one such study, researchers used an international sample of children between the ages of 6 and 12 from the USA, Australia, and Sweden. The researchers came up with criterion-referenced recommendations based on BMI measures and suggested girls should get 12,000 steps per day and boys should get 15,000 steps per day (Tudor-Locke et al., 2004). In an extension of the idea of how many steps per day are desirable for health benefits, researchers have also considered the idea of how many steps per day are too few. In a review of these studies, researchers generally use less than 5,000 steps per day as a cut-point that identifies those individuals as sedentary. In addition, researchers have created a profile of individuals likely to get less than 5,000 steps per day. The profile includes having a relatively low household income, being female, older, of African-American descent, a current smoker, and living with a chronic disease or disability (Tudor-Locke, Craig, Thyfault, & Spence, 2013). They also point out at this time there is little evidence for any specific value that would indicate a sedentary lifestyle for children or adolescents. Pedometers are widely considered to be motivational and studies have shown their use to result in lowered BMI and blood pressure. For example, in a study with midlife women between the ages of 45 and 64, researchers found pedometers function as important motivational tools for increasing daily physical activity (Gardner & Campagna, 2011). The use of pedometers in school settings has been widely studied. In a review of 30 articles regarding the use of pedometers by children in school, researchers found children accumulated between 35-45% of their daily activity at school with higher percentages from school during winter months (Brusseau & Hannon, 2013). They also found evidence that children in the United States accumulated fewer steps than children in other countries.

Future of Pedometers

The future of pedometers seems aligned with the future of heart rate monitors and other health and fitness wearables. They too are increasingly being integrated with devices equipped with multiple sensors used to monitor multiple aspects of the body during exercise.

ACCELEROMETERS

Accelerometers sense the acceleration of the body and "provide objective and direct measurements of the frequency and intensity of movements during physical activity" (Bouten, Sauren, Verduin, & Janssen, 1997). They also point out an advantage of accelerometers is that placement and orientation doesn't significantly detract from their accuracy. This is in contrast to the placement and orientation of mechanical pedometers. A good comparison of pedometers and accelerometers is presented by (Berlin, Storti, & Brach, 2006). Trost and O'Neil, (2014) point out some basic information about accelerometers. Accelerometers generally have a sample rate of 10 to 30 times per second and the signal is usually filtered, rectified and integrated to provide a generic measure of physical activity. They are thought to be less of a burden to participants than heart rate monitors because they don't require electrodes or chest straps. On the other hand they have a limitation because they are not able to detect increased energy expenditure caused by walking upstairs or an incline. Furthermore they are not always accurate when measuring movement in certain activities such as lifting weights and cycling. Accelerometers in general can also be categorized by how many orthogonal axes of motion it can detect, i.e., uniaxial, biaxial, or triaxial. The most common types of accelerometers are peizoresistive, piezoelectric, and differential capacitive. Both peizoresistance and piezoelectric accelerometers function based on what happens when pressure or stress is placed on the sensors. In the case of peizoresistance, the electrical resistance of the semiconductor changes are a function of the pressure or stress placed on it. In the case of piezoelectric accelerometers, the pressure or stress placed on the sensor causes a voltage or charge to be developed that has a linear relationship to the stress. Differential capacitive accelerometers are used most widely in sports and physical activity settings because of their ability to classify posture and movement, estimate energy expenditure, and analyze gait and balance control (Dellaserra, Gao, & Ransdell, 2014). While "accelerometer" may be a common term in the literature, it is likely not in the vernacular of the general public, despite their widespread use. Today, accelerometers are simply components of popular devices, rather than the device itself. There is a growing interest in fitness wearables worn on the wrist, such as fitness bands and smart watches, incorporating accelerometers. Health and fitness enthusiast are not purchasing accelerometers *per se*, but are in fact, using them at an increasing rate.

Research on Accelerometers

Accelerometers have been widely studied and have become one of the most common methods to assess physical activity in free-living participants (Stewart G Trost & O'Neil, 2014). Research has been done with a variety of populations including children, young adults, older adults, people with obesity, and people with various physical disabilities (Berlin et al., 2006). For example, one study had 69 patients with Parkinson disease wear an activity monitor on their wrists. Activity was continuously recorded for 5 to 6 days. The results showed males had lower activity measures than females and the researchers concluded this method of assessment offered an objective quantitative assessment (Van Hilten et al., 1993). In another example, researchers suggest accelerometry is a useful measure of function in patients with spine disorders (Tomkins-Lane & Haig, 2012). Accelerometers have also been recommended as the best option for studying the daily activity of persons with chronic pulmonary disease (Steele, Belza, & Cain, 2003). Steele et al. recommend accelerometers over pedometers for those persons because pedometers are unable to accurately measure slow walking speeds (under 2 mpg) and because accelerometers don't require subjects to record a daily value of pedometer data. In addition, they recommend accelerometers because they offer "the capability to document both the duration and relative intensity of movement" (p. 52) and this provides a unique quantitative index of the intensity of activity. Besides determining the usefulness of accelerometers with special populations, much of the research has focused on accuracy (as is the case with many other health and fitness wearables). In a typical example, a study comparing the accuracy of mechanical pedometers versus piezoelectric pedometers found them to be 96% accurate at three miles per hour but less accurate at slower walking speeds (Melanson et al., 2004). While on the one hand many motion sensors have been shown to have good accuracy, in some cases the accuracy has been shown to be questionable for specific applications. A 2008 review of research on using pedometers and accelerometers to monitor human mobility looked at 207 research abstracts and found "evidence-based clinical applications of these methods in older populations are in need of further development" (de Bruin, Hartmann, Uebelhart, Murer, & Zijlstra, 2008, p. 892). Their conclusions included the following:

The use of wearable systems for monitoring motion activities in people aged over 65 years has not yet reached its full potential. Future studies should evaluate older persons' satisfaction with wearable motion-sensing technology. The challenge for the various developed wearable motion-sensing technologies is evaluation in real life settings (p. 892).

In contrast to pedometers, accelerometers have been shown to be useful, although not perfect, for more intense versions of locomotion, such as running and sprinting. Hovsepian, Meardon, & Kernozek (2014) looked at two commercially available accelerometers and their accuracy at measuring running speed and found the accuracy was somewhat device dependent and the accuracy was influenced by running curves. However, inertial sensors have been found to be suitable to measure running gait outside of the traditional laboratory (Lee, Mellifont, & Burkett, 2010). Additionally, wearable inertial measurement units have been evaluated for the purpose of measuring temporal parameters of sprinting as an alternative to more common video analysis (Bergamini et al., 2012). While accelerometers lend themselves well to linear motion, they have been used to study human mobility in a wide variety of circumstances and situations. Examples include common activities such as ambulatory movements associated with age (Cavanaugh, Coleman, Gaines, Laing, & Morey, 2007), more specialized activities such as swimming (Magalhaes, Vannozzi, Gatta, & Fantozzi, 2014), and determination of bowling events in cricket (Mcnamara, Gabbett,

Chapman, Naughton, & Farhart, 2015). Some research has focused on one particular device. In the case of the Fitbit®, one of the more popular fitness trackers, the results of the research have been mixed. For example, one study concluded "the Fitbit Ultra may be a low-cost alternative to measure the stepping activity in level, predictable environments of people with stroke and TBI" (Fulk et al., 2014). However, in another study involving elderly who were rollator dependent, the Fitbit Ultra was found to have estimation errors greater than 60 percent due to slow walking speeds and gait disorders and it was concluded that the Fitbit was better suited for use by healthy adult or elderly populations (Lauritzen, Muñoz, Luis Sevillano, & Civit, 2013). Another study examined the validity of the Fitbit One model during treadmill walking concluded it was valid and reliable for measuring step counts in healthy young adults but also found the distance output of the monitor was inaccurate (Takacs et al., 2013).

Future of Accelerometers

As with heart rate monitors and pedometers, the future of accelerometers in health and fitness wearable devices seems to be integration into multiple purpose devices like fitness trackers and smart watches. This is already happening and many current fitness trackers, smart watches, and smart phones include not only a 3-axis accelerometer, but also a GPS sensor.

FITNESS TRACKERS

In their 2014 review paper, Dellaserra et al. (2014) refer to devices combining a GPS sensor, accelerometer, and heart rate monitor as integrated technology or IT devices. In this chapter, any of these integrated devices will be referred to as fitness trackers but it should be noted some individuals call these devices activity trackers. According to the popular press the fitness tracker market has been forecast to top five billion dollars by 2019.

While many can be worn on the wrist, some fitness trackers can be worn in multiple locations on the body. Dellaserra et al. (2014) speculate the future of fitness trackers may be in developing wearable clothing with the sensors built-in. They offer as an example that Japanese scientists have developed a jogging support system with GPS, heart rate monitoring, and accelerometers built in to clothing. Further information on integrating fitness tracking technologies into clothing is offered in a later section of this chapter called smart garments. Some of the sensors and technologies in fitness trackers such as heart rate monitors, pedometers, and accelerometers have been discussed previously in this chapter. However, there are other types, some of which are described below.

Additional Health and Fitness Wearable Technologies

Wearable Sleep Sensors

The sleep sensors discussed here are wearable and may either be specifically dedicated to sleep sensing, or part of a fitness tracker offering that functionality. Most sleep sensors of this description work mainly by detecting motion (typically accomplished with the use of an accelerometer). Some of these sensors are advertised as able to detect heart rate activity, respiration, body movements, and even body positions. Sleep science, sometimes referred to as polysomnography (PSG), has a long history and there are medical labs dedicated to study sleep, often with a focus on understanding and treating sleep disorders. These labs often measure many physiological parameters and may include the measurement of EEG, EKG, respiration, oxygen level, blood pressure, and body movement. It has been suggested it is necessary to monitor many of these parameters to properly understand sleep. Some of the wearable technology sleep sensors are typically worn on the wrist and measure body movement. This is referred to as actigraphy. In a 2003 review, Ancoli-Israel et al. provided some background on actigraphy. Although actigraphy had been studied for over 20 years and was found to provide useful information for assessing some sleep disorders, the American Sleep Disorders Association concluded due to a lack of clinical research addressing systematic methodological issues, actigraphy was not indicated for routine diagnosis or assessment of the severity or management of sleep disorders. Because of major advancements in actigraphy, those conclusions were reexamined. Their new conclusion was actigraphy is still not as accurate as PSG, since it allowed for continuous recordings for long periods of time, it was more reliable than both sleep logs relying on patients' recall ability and observations over short time periods. They recommend actigraphy for use with insomnia, studying the effects of treatments designed to improve sleep, the diagnosis of circadian rhythm disorders, and studying sleep patterns in patients who don't tolerate PSG well, including infants and some elderly. Other studies also provide evidence to support the feasibility of using a wrist worn motion sensor to detect nightly sleep periods and body postures. In one such study, researchers concluded the data collected from such a system could be useful to monitor sleep apnea disorders (Van Laerhoven, Borazio, Kilian, & Schiele, 2008). Another study supporting the use of actigraphy compared it with PSG and found actigraphy measurements of sleep more closely approximated those of PSG than did sleep diaries in a sample of depressed insomniacs (Mccall & Mccall, 2012). They also found the only significant difference between actigraphy and PSG was in latency to persistent sleep. As an alternative to actigraphy and motion sensors, it has been suggested a single-lead electrocardiography signal, not unlike a heart rate monitor could be useful. A study of this approach supported the idea this type of sensor could also be useful to monitor sleep apnea disorders and would be suitable for home use (Yilmaz, Asyali, Arikan, Yetkin, & Ozgen, 2010).

Wearable GPS Devices

The Global Positioning System (GPS) allows the device to locate its position. It works by triangulating time codes with multiple satellites. While this was originally designed in the 1970s for military purposes, it has become available for other uses. GPS systems have been incorporated into a variety of devices including some health and fitness wearables. The main purpose for including GPS functionality into a health and fitness wearable is to allow for accurate position tracking. If time data is recorded along with location tracking data, then a number of variables such as distance, pace, and average speed can be determined. Those sorts of data are typically of great interest to runners, cyclists, and others. According to a 2014 review of the use of integrated technology in team sports, the availability of GPS wearables has allowed for the collection of "real-time data on human locomotion to examine sport performance in a more convenient, efficient, and precise manner" (Dellaserra et al., 2014, p. 556). In one example of this, researchers fitted GPS sensors into the shoulder harness of players in the Australian Football League and collected data including total distance, time in various speed zones, maximum speed, accelerations, and longest continuous effort (Wisbey, Montgomery, Pyne, & Rattray, 2010). They found differences between midfield, forward, and defender positions and also an increase in physical demands between the 2005 and 2008 seasons. Devices with GPS capabilities are being used to develop mobile exercise games to take people outdoors. These may be an alternative to participation in traditional sports to help young people get their recommended physical activity (Boulos & Yang, 2013).

Popular Fitness Trackers

At the time of this writing, there are many fitness trackers on the market. Some of the most popular include the Jawbone®, Misfit©, Fitbit, Nike®, and Garmin© models. In 2013, Fitbit, Jawbone, and Nike accounted for 97 percent of fitness tracker retail sales (Dolan, 2014). The price range for most of these generally runs between about $75 and $150. One of the highest rated models, the Fitbit Charge HR, features continuous heart rate monitoring, information on caloric expenditure, an OLED display, and caller ID from a smartphone. As one of the higher-end models, it is at the high-end of the price range. Another highly rated fitness tracker is the Fitbit Surge. This fitness tracker is designed for runners and includes GPS. It can track steps, distance travelled, calories burned, floors climbed, elevation, and active minutes. The Surge also is able to track sleep thanks to having a built-in three axis accelerometer, gyroscope, and digital compass. With all of its sensors and capabilities, the Surge is more expensive than many other fitness trackers and sells for around $250. Although research is not yet available for some of the newer models, some research has been done with earlier models. For example, a recent study examined the accuracy of the Fitbit Classic activity tracker in assessing energy expenditure for a variety of activities including walking and running on a treadmill, and various normal daily activities including cycling, laundry, raking, carrying groceries, and climbing stairs. Findings indicated the Fitbit worn on the hip significantly underestimated caloric expenditure for many daily activities (Rowland et al., 2014). In one of the more recent studies on fitness trackers, researchers compared several models on their accuracy in estimating steps taken and energy expenditure during treadmill walking, treadmill running, elliptical exercise, and an agility drill (Stackpool, 2013). They included the Jawbone UP, Nike Fuelband, Fitbit Ultra, NL-2000i, Adidas MiCoach, and Body Media FIT Core models. The measurements from the devices were compared to step counts taken by direct observation and energy expenditures measured by a portable metabolic system. Their results suggested most of the step counts were reasonably accurate and within 10 percent of the criterion measure. However, they found the errors in measuring energy expenditure were unacceptably large. One of the highest rated fitness trackers for monitoring sleep is the Misfit Shine, which at present sells for about $99. It provides details on nightly sleep including the amount of deep sleep, light sleep, and REM sleep. Besides sleep tracking, the Shine can track steps, calories, and distance. It also includes a feature for tracking caloric intake if the user enters data into a food diary. One of the interesting features of the Misfit Shine is it can be worn in a variety of locations including the wrist, shoes, waist, pocket, or neck without compromising its function. Unlike many of its competitors, the Shine runs on a battery rated for up to 6 months of operation. It uses a 3-axis accelerometer and does not include a heart rate monitor. The Shine has a corresponding smart phone app, which is available for both the Android and iOS operating systems. For group situations, Polar offers a fitness tracker called the Polar Team Pro. It combines a heart rate monitor, GPS sensor, and accelerometer and data wirelessly transmitted to an iPad base station allowing a coach to monitor heart rate, caloric expenditure, speed, distance, and other variables of interest. The wireless transfer uses Bluetooth technology and has a range of 200 meters. The kit includes sensors and straps for 10 players and some accessories at the time of this writing costing $8,199 from Polar.

Smart Watches

Smart watches are computerized wristwatches and effectively, wearable computers. They typically have some of the features of smart phones and increasingly, are including sensors and technologies related to health and fitness. It could be argued smart watches are a combinations of a wristwatch, fitness tracker, and smart phone. However, it should also be noted at this time, many consider smart watches to be more of an extension of a smart phone than a standalone device. In other words, because of their current dependency on being tethered to a smart phone, many consider smart watches in their present form, to be a phone accessory. Although smart watches are relatively new compared to other health and fitness wearables, they have included a wide variety of sensors and features including a camera, accelerometer, heart rate monitor, thermometer, altimeter, barometer, compass, chronograph, calculator, cell phone capability, and touch screen, to name a few. Connectivity options may include Bluetooth, Wi-Fi, and GPS. Many currently available smart watches offer features similar to those in fitness trackers including heart rate monitoring, step counting, caloric expenditure, and sleep tracking. In 2009, Samsung released the S9110 Watch Phone. As the name suggests, it was basically a wristwatch interfacing with a cell phone. One year later Sony released a watch offering Bluetooth connectivity and the capability to interface with an Android Smartphone. Those early offerings paved the way for the current generation of smart watches. In 2013, a year many said was the "year of the smart watch", major companies reported to be developing smart watches included Acer, Apple, BlackBerry, Google, LG, Microsoft, Samsung, Sony, and Toshiba. One of the first smart watches to gain popularity was the Samsung Galaxy Gear. It was designed to work with a Samsung smart phone or tablet and originally used an Android operating system. In terms of health and fitness applications, it included an accelerometer and gyroscope that allowed for a pedometer app as well as some third party apps including MyFitness Pal and Runtastic. Another of the first smart watches to become popular was Pebble. It debuted in 2013 after a successful crowd-funded campaign. The Pebble features an e-ink screen and Bluetooth connectivity and currently costs $89. It offers notification features like a smart phone including texting and e-mailing, has up to 7 days of battery life, is water resistance up to 50 meters, and has fitness tracking capabilities. The fitness tracking is possible because the Pebble has a 3-axis accelerometer and compass built-in. An updated model called the Pebble Steel costs $149 and offers some additional features including voice control. Apple offered its first smart watch in June 2015. It was highly anticipated with estimates for Apple to sell 20 million in the first year of sales. The Apple smart watch contains an accelerometer, gyrometer, and heart rate monitor, but does not have GPS. Because it does not include GPS, the Apple watch relies on the paired iPhone for GPS information and distance tracking. In terms of health and fitness utility, the Apple smart watch tracks caloric expenditure and heart rate. Heart rate is measured through a sensor on the back of the watch facing the skin and uses LED-based optical technology. Heart rate is measured not only during workouts, but also every few minutes while not working out. Software updates are supposed to help the Apple watch calculate distance and pace during outdoor walking and running and improve caloric expenditure estimation for indoor cycling and rowing.

As of this writing, prices for the Apple watch range from $549.00 up to $1099.00. Apple also offers a special gold version priced at $17,000. Other brands range from $33.17 to $1400 for a Tag Heuer model. Typical prices range from under $100 to $300.

SMART GARMENTS

Smart clothing has been defined as those garments which sense biological and/or environmental signals and process the information, and provide feedback from which the wearer can react to or adapt behavior ((Tao, 2001); (Baurley, 2003); (Suh, Carroll, & Cassill, 2010). The sensing circuitry is often woven into the clothing fibers, either as fabric threads coated with conducting material, or as separate thin metallic or plastic conductive threads. Fiber optic thread is a more recent development appearing to have fewer restrictions compared to the traditional sensing threads ("Fibre-optic thread improves smart garments" 2015). However, like all technologies, smart garments have evolved into their current incarnations from more simplistic beginnings.

Brief History of Smart Garments

In the 1980s, smart garments emerged from the concept of a wearable computer. Due to the natural interface clothing has with its wearer, integration with garments was a logical choice. Yet, due to constraints of size and limitations of power supply, the few developments emerging during this time could generally be regarded as more "portable" rather than "wearable." Realizations of these devices/garments were limited, as most wearable technology remained in the conceptual or research and developmental stages. There were no commercial smart garments produced during this time According to Ariyatum et al. (2005), the first several years of the 1990s began much in the same way the previous decade ended, with the smart garments remaining in the research and development phase of technology innovation. However, towards the end of the decade, awareness developed in the fashion and textile industry itself, which compelled development forward. Collaborations between the electronics and fashion industries ushered in the first commercially available smart garments. The most notable may be the Phillips and Levi Strauss Industrial Clothing Design Plus Jacket (ICD + Jacket). This outdoor water-resistant jacket had both a cell phone and a web-accessible mp3 player, which could be controlled via garment integrated buttons. Due to limitations in the U.S. cellular infrastructure, the jacket was only available in Europe. The new century brought further miniaturization of electronics allowing them to be more mobile and increasing the ease of their integration into clothing. Also, it is during this time when there was a conscious move away from garments simply helping a wearer manage technological devices to garments essentially becoming devices themselves (Ariyatum et al., 2005). Some early examples of the types of smart garments common today are the Adidas Self-Adapting shoes and the Sensatex SmartShirt™. The Adidas shoes were able to sense surface conditions and automatically adjust its support/cushion for the user. The SmartShirt™ was capable of monitoring body temperature, heart rhythm, respiration rate, and movement. While it was intended more for the medical and military communities, it was a precursor of the smart athletic garments available today. Innovation continued with further advances in hardware miniaturization, increased sophistication of the environmental and biological sensors, and improved longevity and types of power supply. While there are several companies focusing exclusively on smart athletic and fitness garments (Athos, OmSignal, and Heddoko are among the industry leaders), most have products still in beta or in limited use. Below is a brief summary of the premier product of the Hexoskin Company (Carré Technologies Inc., Montreal, QC, Canada), which at present, appears to have made the most inroads into this field. Their smart shirt is commercially available to coaches, athletes, researchers, and fitness enthusiasts. It provides a representation of the smart garment industry as it currently exists. Also, given the advances observed in wearable technology over time, it affords a glimpse of what is likely to become commonplace in the near future. Hexoskin's biometric smart shirt is capable of sensing and recording a wide array physiological and activity data. This is done though the integration of three cardiac and two respiratory sensors, as well as accelerometers, into the spandex and micropolyamide shirt material. The data from these monitors are sent to a small, lightweight computer device (40 grams) placed in a small pocket on the side of the shirt, in a low midline location on the right-hand side. The computer stores information for later download and analysis, and transmits data via Bluetooth to a smartphone or tablet (iOS and Android apps available), for real-time data monitoring.

Monitoring and Data Collected by Hexoskin Smart Garment

- **Cardiac:** Heart Rate (BPM), Heart Rate Recovery, Heart Rate Variability (from which estimates of stress and fatigue are derived), as well as components of an electrocardiogram (ECG), which includes QRS event detection, and RR intervals.
- **Respiratory:** Breathing Rate [Respirations per Minute (RPM), Tidal Volume (L), Minute Ventilation (L/min), and Inspiration and Expiration Events.
- **Physical Activity:** Intensity, Peak Acceleration, Steps, Cadence, Sleep Duration and Position Changes. Estimates of caloric expenditure (Kcal) can be derived, as well as an estimate of cardiovascular fitness (VO_{2max}) via submaximal fitness tests incorporated into the Hexoskin mobile apps.

Features of the Hexoskin Smart Garment

- 14 hours of battery life, rechargeable
- Bluetooth connectivity with iPhone, iPad, and Android
- Free cloud server access and data storage
- Machine-washable, breathable, quick-dry, breathable, 4-way stretch wrap knit, athletic performance material

Research on Smart Garments

Academic published research on the modern smart garments is limited at this time. However, despite its relative newness, the Hexoskin smart shirt has undergone investigation. Researchers at the University of Waterloo presented findings at the Canadian Society for Exercise Physiology Annual General Meeting, concluding "the results support the utilization of the Hexoskin to quantify activities of daily living during postural changes and walking tasks" (Villar, Beltrame, Ku, & Hughson, 2013). More recently, Tobon Vallejo, Falk, and Maier, (2014) published a peer-reviewed article in the Institute of Electrical and Electronics *Engineering's Transaction on Biomedical Engineering* journal, which examined the ECG capabilities of the Hexoskin. Their experimental results demonstrated that the ECG quality index utilized by the Hexoskin outperformed conventional benchmark measures, particularly those that involved scenarios in which data was recoded in real-world environments. According to Hexoskin at the time of this writing, research is being conducted on and using the Hexoskin by hundreds of researchers in at least twelve countries, including the National Aeronautics Space Administration (NASA) and the Canadian Space Agency (CSA ASC). Other agencies of note are McGill University (Canada), the Massachusetts Institute of Technology (USA), Carnegie Mellon University (USA), the University of Bern (Switzerland), and the Brandenburg University of Technology (Germany).

Current Use of Smart Garments

While the Hexoskin is commercially available, its cost is likely a barrier to access for the average consumer, as a functional garment and data collection unit beginning at $400.00. Dedicated fitness enthusiasts and professional athletes are prospective users at this point. Elite sports teams such as Brooklyn Nets basketball team and the Paris Saint-Germain handball team have adopted the Hexoskin to aid in monitoring training and fatigue, as has the Cirque du Soleil acrobatic performance company. Olympic athletes competing in skiing, snowboarding, cycling, and the biathlon are known users of the Hexoskin shirt, as are professional cyclists, skiers, and athletic coaches, who use them with their athletes.

Future of Smart Garments

In 2014, Gartner Incorporated, an information technology research and advisory company, reported, of all of the wearable devices currently available in the fitness industry (e.g., wristband, smart watch, etc.), the "smart garment" is projected to have the greatest potential for growth (Table 1).

The rationale offered for their projection was relative to other forms of wearable technologies since smart garments are still in their infancy. They are essentially still in the testing phase and relegated to small niches such as athletes and coaches, usually associated with well-funded professional teams ("Gartner says," 2014). Their emergence into the larger conscious of the general public has yet to occur. As prices decline and availability becomes widespread, smart garment use will likely become more pervasive.

SOLUTIONS AND RECOMMENDATIONS

Health and fitness wearables are recommended because of their increasing utility in facilitating the pursuit of a healthy lifestyle. Fitness trackers and smart watches offer a multitude of functionality related to health and fitness. Most have been shown through research to offer valid and reliable measures of various health and fitness parameters. Trost (2001) pointed out major public health authorities including the Centers for Disease Control and Prevention, the National Institutes of Health, and the World Health Organization have all stressed children need more physical activity to achieve and maintain good health. He also points out researchers and practitioners need tools to provide valid and reliable measures of

Table 1. Worldwide wearable electronic fitness devices shipments forecast (2013-2016)

Device Category	2013	2014	2015	2016
	Millions of Units			
Smart garment	0.01	0.01	10.1	26
Smart wristband	30	20	17	19
Other fitness monitor	18	20	12	15
Chest strap	11	12.1	8	7.3
Total market	73.01	70.2	68.1	91.3

physical activity in youth. He reports second-generation activity monitors (heart rate monitors and accelerometers) have significantly enhanced our ability to objectively measure the frequency, duration, and intensity of children's physical activity (p. 36). Although Trost's comments were focused on children, many of the same concerns are applicable to a variety of populations. Advancements in health and fitness wearable technologies are increasing the availability of objective assessments for many populations. A different but perhaps complimentary consideration for recommending the use of health and fitness wearables is offered by Tremblay, Esliger, Tremblay, and Colley, (2007). The researchers discuss the potential importance of what they describe as incidental movement as part of total energy expenditure. They define incidental movement as movement that is non-exercise or non-purposeful physical activity and lifestyle-embedded activities such as chores and incidental walking. They speculate incidental movement is likely to occur throughout the day and can be a large part of the daily total energy expenditure. They also speculate incidental movement is vulnerable to reductions due to "increasingly ubiquitous mechanization and automation" (p. 208). Consequently, they recommend physical activity monitors to capture movements over the entire day. As wearables continue to evolve, certain trends are evident. Whereas in the past many wearables provided one function, increasingly they offer multiple sensors and technologies, while at the same time, are more simplified in user design. For example many new heart rate monitors do not require a chest strap. Another advancement has been in the ability of the wearables to wirelessly share data with smart phones, personal computers, and web-based fitness diaries. In order to obtain the greatest benefits of this new technology, users and practitioners need to have sufficient knowledge of relevant exercise recommendations, and be sure they are appropriate for the population they are being used with. For example, if physical educators are going to use heart rate monitors with their school children they need to know not to use the adult recommendations. Users and practitioners also need to understand any limitations the wearables might have for working with special populations. For example, research has shown some wearables (activity monitors) are not accurate for persons with certain gait disorders. The development and evolution of health and fitness wearables has been increasingly rapid, particularly as multiple technologies have been merged first into fitness trackers and now into smart watches and smart garments. This rapid development, along with the costs of these wearables, has made it difficult for educators to incorporate these technologies into student education. The research record shows at least to some extent, the feasibility of health and fitness wearables has been considered and explored for medical applications. However, at this time the responsibility for selection and proper use of these technologies seems to fall largely on the individual consumer. One possible exception to this might be in the area of athletics, where coaches or trainers might take the lead and use appropriate wearables to facilitate fitness or athletic training. Hopefully in the future, medical professionals will be able to improve both effectiveness and efficiency by better leveraging the ability of health and fitness wearables to monitor and collect a wide variety of health and fitness data. One issue likely to impact the ability of educators and others to make good use of the latest advancements in this technology is the degree and the way they work become standardized. If each new generation of health and fitness wearables change and companies elect to keep their technology proprietary, this will make it much more difficult for groups like educators and medical professionals to stay current and make good use of the technology. Hopefully the designers of these new devices will at least provide the ability to share data in a way that is standardized.

FUTURE RESEARCH DIRECTIONS

Future research directions have been mentioned for each of the types of health and fitness wearables discussed in this chapter. As the trend of embedding multiple sensors and technologies into new devices and clothing continues, research should continue to ensure the validity and reliability of the measurements and applications of the data.

CONCLUSION

Health and fitness wearable technologies represent some of the earliest and most popular wearable technologies. Beginning in the early 1980s with Polar's heart rate monitors, health and fitness wearables has grown and evolved into a wide variety of forms and applications. Popular health and fitness wearables include heart rate monitors, pedometers, accelerometers, sleep trackers, GSP sensors, and devices including multiple sensors such as fitness trackers, smart watches, and smart garments. Fitness wristbands and smart watches are becoming increasingly popular. The mobile trend continues to grow as wearable garments, such as smart athletic apparel emerges as the most recent entry into the field. In their 2015 annual worldwide survey on fitness trends, the American College of Sports Medicine found that wearable technology was the new number one trend and predicted smart glasses and smart fabrics would have sales reaching $2.6 billion by 2017 (Thompson, 2015).

Health and fitness wearable technologies have impacted both education and health-minded consumers. In education, we are seeing teachers challenged with finding ways to acquire new wearables and how to best use them within the curriculum. Health and fitness wearables occupy a prominent position in the marketplace and are likely to continue to drive more innovation.

REFERENCES

Ariyatum, B., Holland, R., Harrison, D., & Kazi, T. (2005). The Future Design Direction of Smart Clothing Development. *Journal of the Textile Institute*, *96*(4), 199–212. doi:10.1533/joti.2004.0071

Armstrong, N., & Bray, S. (1991). Physical activity patterns defined by continuous heart rate monitoring. *Archives of Disease in Childhood*, *66*(2), 245–247. doi:10.1136/adc.66.2.245 PMID:2001112

Bächlin, M., & Tröster, G. (2012). Swimming performance and technique evaluation with wearable acceleration sensors. *Pervasive and Mobile Computing*, *8*(1), 68–81. doi:10.1016/j.pmcj.2011.05.003

Baurley, S. (2003). Interactive and Experiential Design in Smart Textile Products and Applications. *Proceeding of International Textile Design and Engineering Conference (INTEDEC): Fibrous Assemblies at the Design and Engineering Interface,* Edinburgh (pp. 22–24).

Bergamini, E., Picerno, P., Pillet, H., Natta, F., Thoreux, P., & Camomilla, V. (2012). Estimation of temporal parameters during sprint running using a trunk-mounted inertial measurement unit. *Journal of Biomechanics*, *45*(6), 1123–1126. doi:10.1016/j.jbiomech.2011.12.020 PMID:22325976

Berlin, J. E., Storti, K. L., & Brach, J. S. (2006). Using activity monitors to measure physical activity in free-living conditions. *Physical Therapy*, *86*(8), 1137–1145. PMID:16879047

Besch, E. L., Varosi, S. M., & Brigmon, R. L. (1992). Technical note: A noninvasive procedure for measuring goat heart rates. *Journal of Animal Science*, *70*(11), 3371–3375. PMID:1459897

Borg, G. (1970). Perceived exertion as an indicator of somatic stress. *Scandinavian Journal of Rehabilitation Medicine*, *2*(2), 92–98. Retrieved from http://www.ncbi.nlm.nih.gov/pubmed/5523831 PMID:5523831

Boulos, M. N. K., & Yang, S. P. (2013). Exergames for health and fitness: The roles of GPS and geosocial apps. *International Journal of Health Geographics*, *12*(1), 18. doi:10.1186/1476-072X-12-18 PMID:23561306

Bouten, C. V. C., Sauren, A. A. H. J., Verduin, M., & Janssen, J. D. (1997). Effects of placement and orientation of body-fixed accelerometers on the assessment of energy expenditure during walking. *Medical & Biological Engineering & Computing*, *35*(1), 50–56. doi:10.1007/BF02510392 PMID:9136191

Brage, S., Brage, N., Ekelund, U., Luan, J., Franks, P. W., Froberg, K., & Wareham, N. J. (2006). Effect of combined movement and heart rate monitor placement on physical activity estimates during treadmill locomotion and free-living. *European Journal of Applied Physiology*, *96*(5), 517–524. doi:10.1007/s00421-005-0112-6 PMID:16344938

Brusseau, T., & Hannon, J. C. (2013). Pedometer-Determined Physical Activity of Youth while Attending School: A Review. *Sport Science Review*, *XXII*(5-6), 329–342. doi:10.2478/ssr-2013-0016

Cavanaugh, J. T., Coleman, K. L., Gaines, J. M., Laing, L., & Morey, M. C. (2007). Using step activity monitoring to characterize ambulatory activity in community-dwelling older adults. *Journal of the American Geriatrics Society*, *55*(1), 120–124. doi:10.1111/j.1532-5415.2006.00997.x PMID:17233695

de Bruin, E. D., Hartmann, A., Uebelhart, D., Murer, K., & Zijlstra, W. (2008). Wearable systems for monitoring mobility-related activities in older people: A systematic review. *Clinical Rehabilitation*, *22*(10-11), 878–895. doi:10.1177/0269215508090675 PMID:18955420

De Magalhaes, F. A., Vannozzi, G., Gatta, G., & Fantozzi, S. (2014). Wearable inertial sensors in swimming motion analysis: A systematic review. *Journal of Sports Sciences*, *33*(7), 732–745. doi:10.1080/02640414.2014.962574 PMID:25356682

Dellaserra, C. L., Gao, Y., & Ransdell, L. (2014). Use of integrated technology in team sports: a review of opportunities, challenges, and future directions for athletes. *Journal of Strength and Conditioning Research / National Strength & Conditioning Association, 28*(2), 556–73. Doi:10.1519/JSC.0b013e3182a952fb

Dolan, B. (2014). Fitbit, Jawbone, Nike had 97 percent of fitness tracker retail sales in 2013. Retrieved from http://mobihealthnews.com/28825/fitbit-jawbone-nike-had-97-percent-of-fitness-tracker-retail-sales-in-2013/

Eyal, S., Dagan, Y., & Baharav, A. (2012). Sleep in the Cloud: On How to Use Available Heart Rate Monitors to Track Sleep and Improve Quality of Life. In *Computing in Cardiology (CinC)* (pp. 329–332). IEEE. Retrieved from http://ieeexplore.ieee.org/stamp/stamp.jsp?tp=&arnumber=6420397

Fibre-optic thread improves smart garments. (2015). *Wearable Technology Insights*. Retrieved from http://www.wearabletechnologyinsights.com/articles/7623/fibre-optic-thread-improves-smart-garments

Fulk, G. D., Combs, S., Danks, K. A., Nirider, C. D., Raja, B., & Reisman, D. S. (2014). Accuracy of 2 activity monitors in detecting steps in people with stroke and traumatic brain injury. *Physical Therapy*, *94*(2), 222–229. doi:10.2522/ptj.20120525 PMID:24052577

Gardner, P. J., & Campagna, P. D. (2011). Pedometers as measurement tools and motivational devices: New insights for researchers and practitioners. *Health Promotion Practice*, *12*(1), 55–62. doi:10.1177/1524839909334623 PMID:19515864

Gartner Says in 2015, 50 Percent of People Considering Buying a Smart Wristband Will Choose a Smartwatch Instead. (2014). *Gartner*. Retrieved from http://www.gartner.com/newsroom/id/2913318

Hatano, Y. (1997). Prevalence and Use of Pedometer. *Res J Walking*, *1*, 45–54.

Hovsepian, D., Meardon, S., & Kernozek, T. W. (2014). Consistency and Agreement of Two Devices for Running Speed. *Athletic Training & Sports Health Care*, *6*(2), 67–72. doi:10.3928/19425864-20140306-02

Humen, D. P., & Boughner, D. R. (1984). Evaluation of commercially available heart rate monitors. *Canadian Medical Association Journal*, *131*(6), 585–588. PMID:6478343

Laukkanen, R. M. T., & Virtanen, P. K. (1998). Heart rate monitors: State of the art. *Journal of Sports Sciences*, *16*(Suppl. 1), 3–7. Doi:10.1080/026404198366920

Lauritzen, J., Muñoz, A., Luis Sevillano, J., & Civit, A. (2013). The usefulness of activity trackers in elderly with reduced mobility: A case study. In Studies in Health Technology and Informatics (Vol. 192, pp. 759–762). Doi:10.3233/978-1-61499-289-9-759

Lee, J. B., Mellifont, R. B., & Burkett, B. J. (2010). The use of a single inertial sensor to identify stride, step, and stance durations of running gait. *Journal of Science and Medicine in Sport*, *13*(2), 270–273. doi:10.1016/j.jsams.2009.01.005 PMID:19574098

Mccall, C., & Mccall, W. V. (2012). Comparison of actigraphy with polysomnography and sleep logs in depressed insomniacs. *Journal of Sleep Research*, *21*(1), 122–127. doi:10.1111/j.1365-2869.2011.00917.x PMID:21447050

Mcnamara, D. J., Gabbett, T. J., Chapman, P., Naughton, G., & Farhart, P. (2015). The Validity of Microsensors to Automatically Detect Bowling Events and Counts in Cricket Fast Bowlers. *International Journal of Sports Physiology and Performance*, *10*, 71–75. Retrieved from10.1123/ijspp.2014-0062

Melanson, E. L., Knoll, J. R., Bell, M. L., Donahoo, W. T., Hill, J. O., Nysse, L. J., & Levine, J. A. et al. (2004). Commercially available pedometers: Considerations for accurate step counting. *Preventive Medicine*, *39*(2), 361–368. doi:10.1016/j.ypmed.2004.01.032 PMID:15226047

O'Toole, M. L., Douglas, P. S., & Hiller, W. D. B. (1998). Use of heart rate monitors by endurance athletes: Lessons from triathletes. *The Journal of Sports Medicine and Physical Fitness*. PMID:9830823

Rowland, J. L., & Sasaki, J. E., Hickey, a, Mavilia, M., Tedesco, J., John, D., … Freedson, P. S. (2014). Validation of the Fitbit Wireless Activity Tracker(R) for Prediction of Energy Expenditure. *Journal of Medical Internet Research*, 149–155. doi:10.1123/jpah.2012-0495

Stackpool, C. M. (2013). *The Accuracy of Various Activity Trackers in Estimating Steps Taken and Energy Expenditure* [Master's Thesis].

Steele, B., Belza, B., & Cain, K. (2003). Bodies in motion: monitoring daily activity and exercise with motion sensors in people with chronic pulmonary disease. *Journal of Rehabilitation Research and Development*. Retrieved from http://www.rehab.research.va.gov/jour/03/40/5sup2/pdf/steele.pdf

Stel, V. S., Smit, J. H., Pluijm, S. M. F., Visser, M., Deeg, D. J. H., & Lips, P. (2004). Comparison of the LASA Physical Activity Questionnaire with a 7-day diary and pedometer. *Journal of Clinical Epidemiology*, *57*(3), 252–258. doi:10.1016/j.jclinepi.2003.07.008 PMID:15066685

Suh, M., Carroll, K., & Cassill, N. (2010). Critical Review on Smart Clothing Product Development. *Journal of Textile and Apparel*, *6*(4), 1–18. Retrieved from http://ojs.cnr.ncsu.edu/index.php/JTATM/article/view/702

Sultan, N. (2015). Reflective thoughts on the potential and challenges of wearable technology for healthcare provision and medical education. *International Journal of Information Management*, *35*(5), 521–526. doi:10.1016/j.ijinfomgt.2015.04.010

Takacs, J., Pollock, C. L., Guenther, J. R., Bahar, M., Napier, C., & Hunt, M. (2013). Validation of the Fitbit One activity monitor device during treadmill walking. *Journal of Science and Medicine in Sport*, *17*(5), 496–500. doi:10.1016/j.jsams.2013.10.241 PMID:24268570

Tao, X. E. (2001). *Smart Fibres, Fabrics and Clothing*. Smart Fibres, Fabrics and Clothing; doi:10.1533/9781855737600

Terbizan, D. J., Dolezal, B. A., & Albano, C. (2002). Validity of Seven Commercially Available Heart Rate Monitors. *Measurement in Physical Education and Exercise Science*, *6*(4), 243–247. doi:10.1207/S15327841MPEE0604_3

Thompson, W. R. (2015). ACSM's Health & Fitness Journal. *ACSM's Health & Fitness Journal, 19*(6), 9–18. Retrieved from http://journals.lww.com/acsm-healthfitness/Fulltext/2015/11000/WORLDWIDE_SURVEY_OF_FITNESS_TRENDS_FOR_2016__10th.5.aspx

Tobon Vallejo, D., Falk, T., & Maier, M. (2014). MS-QI:A Modulation Spectrum-Based ECG Quality Index for Telehealth Applications. In IEEE transactions on bio-medical engineering.

Tomkins-Lane, C. C., & Haig, A. J. (2012). A review of activity monitors as a new technology for objectifying function in lumbar spinal stenosis. *Journal of Back and Musculoskeletal Rehabilitation*, *25*(3), 177–185. doi:10.3233/BMR-2012-0325 PMID:22935856

Tremblay, M. S., Esliger, D. W., Tremblay, A., & Colley, R. (2007). Incidental movement, lifestyle-embedded activity and sleep: New frontiers in physical activity assessment. *Canadian Journal of Public Health. Revue Canadienne de Santé Publique*, *98*(Suppl. 2.). doi:10.1139/H07-130

Trost, S. G. (2001). Objective Measurement of Physical Activity in Youth: Current Issues, Future Directions. *Exercise and Sport Sciences Reviews*, *29*(1), 32–36. doi:10.1097/00003677-200101000-00007 PMID:11210445

Trost, S. G., & O'Neil, M. (2014). Clinical use of objective measures of physical activity. *British Journal of Sports Medicine*, *48*(3), 178–181. doi:10.1136/bjsports-2013-093173 PMID:24311601

Tsubono, Y., Tsuji, I., Fujita, K., Nakaya, N., Hozawa, A., & Ohkubo, T. … Hisamichi, S. (2002). Validation of walking questionnaire for population-based prospective studies in Japan: comparison with pedometer. Journal of Epidemiology / Japan Epidemiological Association, 12(4), 305–309. Doi:10.2188/jea.12.305

Tudor-Locke, C., Craig, C. L., Thyfault, J. P., & Spence, J. C. (2013). A step-defined sedentary lifestyle index : < 5000 steps / day (Review). *Applied Physiology, Nutrition, and Metabolism,* 38(2), 100–114. PubMed

Tudor-Locke, C., Pangrazi, R. P., Corbin, C. B., Rutherford, W. J., Vincent, S. D., Raustorp, A., & Cuddihy, T. F. et al. (2004). BMI-referenced standards for recommended pedometer-determined steps/day in children. *Preventive Medicine*, *38*(6), 857–864. doi:10.1016/j.ypmed.2003.12.018 PMID:15193909

Tudor-Locke, C., Williams, J. E., Reis, J. P., & Pluto, D. (2002). Utility of pedometers for assessing physical activity: Convergent validity. *Sports Medicine (Auckland, N.Z.)*, *32*(12), 795–808. doi:10.2165/00007256-200232120-00004 PMID:12238942

Van Hilten, J. J., Hoogland, G., van der Velde, E. A., van Dijk, J. G., Kerkhof, G. A., & Roos, R. A. C. (1993). Quantitative Assessment of Parkinsonian Patients by Continuous Wrist Activity Monitoring. *Clinical Neuropharmacology,* 16(1), 36–45.

Van Laerhoven, K., Borazio, M., Kilian, D., & Schiele, B. (2008). Sustained logging and discrimination of sleep postures with low-level, wrist-worn sensors. *Proceedings of the International Symposium on Wearable Computers ISWC* (pp. 69–76). Doi:http://doi.org/ doi:10.1109/ISWC.2008.4911588

Villar, R., Beltrame, T., Ku, R., & Hughson, R. L. (2013). Validation of the Hexoskin wearable body metrics vest to predict activities of daily living. Toronto, Ontario, Canada: Canadian Society for Exercise Physiology Annual General Meeting. Retrieved from https://cdn.shopify.com/s/files/1/0284/7802/files/CSEP_Hexoskin_Poster_-_University_of_Waterloo.pdf?11148

Wisbey, B., Montgomery, P. G., Pyne, D. B., & Rattray, B. (2010). Quantifying movement demands of AFL football using GPS tracking. *Journal of Science and Medicine in Sport*, *13*(5), 531–536. doi:10.1016/j.jsams.2009.09.002 PMID:19897414

Yilmaz, B., Asyali, M. H., Arikan, E., Yetkin, S., & Ozgen, F. (2010). Sleep stage and obstructive apneaic epoch classification using single-lead ECG. *Biomedical Engineering Online*, *9*(1), 39. doi:10.1186/1475-925X-9-39 PMID:20723232

KEY TERMS AND DEFINITIONS

Accelerometer: A device that measures g-force acceleration.

Activity Tracker: See Fitness Tracker.

Fitness Tracker: A device that allows for monitoring various fitness metrics such as distance walked, caloric consumption or expenditure, heart rate, sleep, etc.

GPS: A device that provides time and location information anywhere on Earth.

Heart Rate Monitor: A device that allows for monitoring heart rate in real time or for recording heart rate.

Pedometer: A device that counts steps taken by a person.

Sleep Sensor: A device that provides information on quantity and type of sleep.

Smart Garment: A garment with embedded technology.

Section 2

Mobile and Wearable Technology Device Extensions

Chapter 5
Digital Badges:
Tracking Knowledge Acquisition within an Innovation Framework

Talitha Hudgins
Utah Valley University, USA

Janet Holland
Emporia State University, USA

ABSTRACT

Digital badges provide credentials to demonstrate successful student acquisition of knowledge, skills, and experience. The digital badges can especially excel at demonstrating new and emerging technology experiences through the use of wearable devices tailored to prepare students for the modern work place. Digital badges become virtual tracking systems making it possible to show students' specific learning, skills, and ultimately their employability. Not only do they record data on the individual student, they can track the institution and issuer each student has interacted with. With each successive use of such badges, a virtual narrative can be recorded reflecting how the badge was earned, what criteria was used, what standards were met, the rigor, and the different levels of learning.

INTRODUCTION

This chapter discusses the potential uses of digital badges as the framework for assessment, along with ideas and guidelines for the integration of emerging mobile and wearable technologies. The use of digital badges is gaining momentum as a new representation for credentials and has the potential to become an accepted system in credentialing students for knowledge gained, skills learned, and achievements accomplished. A few higher education institutions are examining the use of badges for awarding non-academic credit. For example, Colorado State University (CSU) created a skills based digital badge program in gardening called: The Certified Gardener program. This program is based on the Colorado Master Gardener curriculum, which allows students to earn various badges as they complete the program.

DOI: 10.4018/978-1-5225-0069-8.ch005

The push for creating and implementing a different credentialing system is a result of higher education institutions being called into question regarding students' employability after graduation. Many institutions are required to demonstrate student-learning outcomes of knowledge and skills acquired throughout their college experience while employers are looking for specific skills and achievements that fit their needs. To acquire the necessary knowledge and skills many students are taking courses from online education providers; however, many programs lack the ability to provide the necessary credentials to verify what these students have learned, know, and can do. Students who are highly competent or proficient in skills not taught or assessed in schools lack a standardized way to demonstrate their abilities to others (Grant, 2014). Employees struggling to change careers or update skills face insurmountable obstacles when returning to school as adult learners, and without appropriate credentials to communicate their knowledge and skills find themselves unemployed or underemployed working in low-paying jobs.

Although some online courses may not offer degrees or diplomas, they do offer digital badges certifying specific skills and achievements students have learned. Through the use of digital badges, students and potential employers have the opportunity to access information anytime and anywhere 24/7. Digital badges are transparent, information rich, and bundled into one area. Employers in one click are able to access information about the organization, individual, or entity issuing the badge. They can also view student knowledge, achievements, and skills, how the badge was earned, what criteria was used, what standards were met, the rigor, and the different levels of learning. With this knowledge, employers are able to hire the right person for the job and not spend additional time training.

There are a number of organizations, educational providers, and communities supporting and using digital badges. For instance NASA, Purdue University, Penn State, the United States Air Force, and the City of Chicago, are all using digital badges to award and recognize a variety of skills and achievements. These organizations and other supporters of digital badges suggest that the use of badges certifying specific skills and achievements will one day either take the place of diplomas and/or augment them. They have found digital badges are more powerful because they identify accomplishments in knowledge, skills, quality, and interests illustrating a more complete picture and detailed set of achievements than simple paper certificates or degrees (Catalano & Doucet, 2013).

The promise digital badges hold for higher education is how they demonstrate each student's learning accomplishments. Badges are designed to make the invisible visible and validate learning in both formal and informal settings, providing the potential to help transform where and how learning is valued (Mozilla Foundation, Peer 2 Peer University, & MacArthur Foundations, 2012). The badging system is part of the 21st century digital age, for example; in the Computer Science Student Network (CS2N) badge system, evidence may include links to source code for a programming badge, online exam scores, or assessments generated by an artificial intelligence tutoring system (Bruckman, 2004). In other systems, badges may have an expiration date to signal skills that are subject to renewal each year. The badging system is aligned with the 21st century digital age because of the relevant information provided about student skills and knowledge which adds a layer of transparency to credentials. Digital badges capture student skills and competencies in detail across many different contexts; badges are collected and associated with students' online identity then displayed to key stakeholders to demonstrate the students' skills and achievements. Digital badges will make the accomplishments and experiences of students in online and offline spaces, visible to anyone and everyone, including potential employers, educators, and communities. As the learning landscape changes from formal and linear to informal and on-demand, and as employers seek faster and more reliable ways to connect with employees while adding the most value to their organizations by providing a secure, digital credentialing system it will become more of a necessity, instead of a novelty.

The Digital Badge System

Digital badges are similar to traditional badges in that they are earned and awarded for completing specific activities and achievements. The difference is, digital badges are virtual and collected, managed, and hosted across many different systems and platforms on the web. Digital badges are a validated indicator containing an online record of accomplishments, skills, quality or interest that can be earned in any learning environment. Any institution or organization can issue badges at any level of achievement, whether for a two hour workshop, a three day course, or a four year degree. Badges can represent soft skills like collaboration, or hard skills like math, and they can function inside traditional courses, alongside them or independently, whether offline or online (Grant, 2014). A badge may represent one assignment in a course, or it can represent successive levels of competency involving hundreds of hours of work. Badges carry a tremendous potential to transform student learning and are information rich, transparent, and portable, as noted earlier, across many different systems and platforms on the web.

In a well-designed digital badge, the learning, the criteria, the assessment, and the evidence will be visible in one place, associated with the student who earns the badge and the organization that issued it (Grant, 2014). The digital badge can contain videos, essays, assignments, or exam scores submitted by the student, as well as the instructor's testimony, peer comments, and perhaps a reflection that provides important context about what was learned and how. Badges can also be interconnected, which refers to an open data exchange or infrastructure allowing badges to be shared across multiple platforms or systems. For example, a digital badge earned for training on coding could be displayed on the student's Facebook page, or Wordpress blog, or it could be displayed in a digital backpack such as Mozilla, or a repository with other badges. Ultimately, the student decides how their badge will be shared, stored, viewed, and used. Students can earn badges after an experience such as completing a workshop, going through a museum, learning about the library system, completing a course off the web.

There are two critical conceptual approaches to digital badges: open source and metadata. First, the technology making it possible to share badges anywhere on the web is *open source*, ensuring no single organization owns the code. The open source platform includes tools to manage users, create and issue badges, add badge metadata, and track badges (Catalano & Doucet, 2013). Most badging systems use the Mozilla Open Badge Infrastructure (OBI) because the OBI acts as a common language so that different systems can communicate with each other, making it possible to push badges out to the web. The developers of the open source Firefox browser built the Open Badge Infrastructure (OBI) to support the primary role to direct the open source standards and technical specifications, allowing institutions to make their badges portable across the web.

Mozilla has been the most visible organization developing and advocating for the use of digital badges. The MacArthur Foundation partnered with Mozilla as a major funder for further development of digital badges. Additionally, the MacArthur Foundation held a competition and awarded funding to a wide variety of organizations for developing, designing, and issuing digital badges. The organizations who won the funding were: NASA, Disney-Pixar, the Girl Scouts, BadgeWorks for Vets (gives returning military personnel badges to represent military skills and training for use in applying for civilian jobs), and the National Manufacturing Badge System. Mozilla along with the MacArthur Foundation has been a strong advocate in moving forward the technology and standards behind digital badges.

Mozilla calls its digital badges "open badges" because the technology to create them is open technology, representing an open source model, meaning that improvements made by one partner can benefit everyone, from bug fixes to new features (Catalano & Doucet, 2013). After the first official version 1.0 release of the Open Badge Infrastructure in March 2013, it was found that an open digital badge offered a mechanism for representing skills, knowledge and accomplishments extending beyond a hobby or self-motivation. Furthermore, using the OBI, digital badges can be combined, or stacked, and easily shared with employers, social networks, and institutions. Students earning open badges can store them in a digital backpack or digital vault (a virtual lockbox) where they keep all their badges in one place to share, as needed.

The second critical conceptual approach to open digital badges is metadata. Every open digital badge contains metadata or data about data. Metadata can be viewed as a library card catalog containing cards with the metadata being about each book's author, the publication date, and the location of the item in the library. Another example of metadata would be personnel identification cards, having metadata about each employee, identification number, access code, authorization to certain areas, working hours, as well as any other information considered important by that organization pertaining to the employee. An open digital badge can contain metadata about each student, the badge description, the issuing organization, the date it was issued, criteria to earn the badge, the web address linking to evidence, and any other information that can help make it compatible with the OBI. The metadata includes a set of standards making it possible for other systems to process and recognize the badge, allowing it to travel outside the platform in which it was earned and issued (Grant, 2014).

The digital badge system is made up of three elements: 1.) Issuer, 2.) Earner, and 3.) Displayer. Issuers are the instructors or organization creating and awarding badges to Earners. Earners are students completing the achievement or skill required to obtain the badge. Displayers are web-based platforms facilitating public viewing of badges such as the Mozilla Backpack as an authorized data repository where students collect, manage, and display their badges.

Issuers range from academia, industry, and agencies, for distribution to individual websites. The issuers badge systems carry values within their own right; however, the true optimum value comes from these badge systems working together, with the learning at the center. Open digital badges should be collected in a way that ties them to the Earners' identity. Digital badges can be used to enhance digital identity, empower global perspectives, foster improved instructional design and curriculum transparency, facilitate better instructional management, define knowledge and/or skills, professional development, and establish new business models. Digital badges can be used to show course completion, evaluate student learning outcomes, and establish micro-credentials, while representing honors, showing event participation and community engagement. Digital badges can establish membership participation and/or attendance for badge holders. They are also used for assessment of learning or non-assessment of learning. For example, badges may be issued to students attending college events, completing volunteer services, traveling internationally, participating in academic clubs and student government as an indication of attendance and achievement. Alternatively, students may be required to achieve some measure of comprehension and retention (quiz or an assignment submission), or they may have different levels of a project to complete to earn a badge at each level.

Peer 2 Peer University (P2PU) partnered with Mozilla OBI in developing the School of Webcraft. The School of Webcraft is a free online community for social learning about open web development. They started off with fourteen value badges just to name a few; such as a hard-skill badge where students learn JavaScript, peer badges such as Good Collaborator, and participation badges such as Active

Responder. The goal was to use badges to capture hard and soft skills important for web developers. Various forms of assessment were used. For instance the hard skills were assessed through authentic challenges submerging students in technology. As projects were completed, students' work was peer assessed for basic-level skills or for expert level skills (Mozilla Foundation and Peer 2 Peer University, 2012). Anyone can join the School of Webcraft and start learning web development or use it to update needed skills. This model delivers engaging learning opportunities for students.

The OBI provides the underlying open technology and standardization to support badge issuers, badge displayers, via cloud-based repository for badge collection and management for students. The OBI includes specifications defining what information must be included with a badge when issuers push badges in and displayers pull badges out. The specifications ensure each badge provides the information needed since it serves as the gateway for evidence and value of what was completed (Mozilla Foundation and Peer 2 Peer University, 2012). Once the badge is earned, students add it to their Mozilla Badge Backpacks. The badge backpack is a personal badge repository where students can combine, manage, set privacy controls, and share their badges. Examples where badges can be shared are: blogs, social media sites, resumes, and webpages.

CHALLENGES

Providing value to digital badges, as in the value of a paper certificate or diploma, is an area needing to be explored by higher education. There are two types of endorsements when using digital badges: informal and formal. The informal endorsement is faculty members confirming a student is successful, capable, and competent in a given content area or skill, with evidence actually demonstrated. The instructor's act of endorsing the badge adds credibility to the process.

The formal endorsement will be through an official body or organization. The official endorser of the digital badge system signs it using a digital signature. The official's signature is a binding endorsement. Once badges are endorsed, then the badge will be displayed in a managed space. A support structure must also be created in order to house the managed space system. The support structure hosting digital badges is significant. Most institutions need a readily accessible electronic storage system to support the creation and deployment of badges, including all the storage of academic evidence, culminating with organization and distribution of badges. Instructors then need that place to store and display the criteria and to show the reliability of the badge with the ultimate storage and display requirement from students needing a place to store earned badges. Once students graduate, badges earned will follow them everywhere. It will benefit higher education to innovate creating transcripts supporting and showcasing digital badges.

The digital badge system can be a little more complex than other types of curriculum and assessment development. Badges can use learning pathways, resembling skill trees. Skill trees are custom configurations of learning pathways with branching so students make choices about how to progress toward competency (Gibson, Ostashewski, Flintoff, Grant, & Knight, 2013). For example, a student might choose to work more broadly across skill sets or focus more deeply to specialize in one area. Badge systems are suited to branching, constantly providing more pathway choices for students. This creates additional challenges because it requires thinking about multiple pathways across content areas while demanding more interconnected planning for courses (Gibson, et al, 2013).

The benefits from using digital badges are not only in providing multiple learning pathways across content areas but also providing a more complete picture of learning accomplishments or certifications for different skill sets and literacies outside of academia (Gibson, et al, 2013). Badges assist in validation of skills learned and at what level attesting to competency and participation. Digital badges foster better instruction by motivating participation and learning outcomes, providing instant feedback and rewards throughout a course and the learning experience, encouraging engagement and retention, and providing various pathways to learning while allowing students to spend more time developing specialized skills. Badges can facilitate better instructional management by supporting individualized learner support, capturing specific skills and qualities along with issue dates for each area successfully completed, and assisting in accreditation by capturing the learning path, linking student learning to skills demanded by employers, and meeting the documentation needs of accreditation agencies. Purdue University has used the badging system for the past several years and found it a better way for universities to acknowledge the full breadth of student learning (Tally, 2012). The system Purdue developed is called the Passport and Passport Profile (Figure 1). Passport and Passport Profile allows instructors and advisors to provide students with digital badges to indicate their mastery of skills. Purdue's Passport platform integrates with the Mozilla Open Badge system. This system provides information on the badges issuer along with how it was earned and when it was earned.

ASSESSMENT

The use and records of assessments is one of the most critical components of digital badges. Assessments are necessary for any badge system in which there is a claim about learning and a link to evidence. Badges

Figure 1. Passport (Purdue University, 2012)

are designed to recognize learning and represent assessment of that learning (Abramovich, Schunn, & Higashi, 2013). For that reason, assessment plays a large role in the way the badge system is designed and its potential impact on student learning and the institution.

Developing a digital badge system in the 21st century is an opportunity to think deeply about assessment practices in higher education (Grant, 2013). Assessment is often equated with schools but is also an integral part of all human learning, coming about as a result of social groups seeking ways to mentor and police participants (Gee, 2011). Assessments in traditional education tend to favor replicable and objective quantifiable measurements (Schmidt, Geith, Haklev, & Thierstein, 2009). This approach may work for some skills but not all, especially 21st century skills requiring different methods. For example, badges used and developed by Makewaves are designed to reflect real-world transferable skills, including social media creation, reporting, interviewing, editing, time management, collaboration, responsibility, motivation and mentoring of others. Forms of assessment in badge systems are designed to recognize collaborative work, deeper learning, and increased engagement. For that reason, assessment will play a large role in the way a badge system is designed and the impact it has on students.

In designing a badge system, instructors and institutions are provided more freedom to decide what and how to assess, along with what to credential and how. Having this freedom, it is important to think about the claims being made and the learning outcomes instructors and institutions want before choosing how best to assess (Grant, 2014). The digital badge system is truly transformative when providing an opportunity to design systems relevant and valuable to learners, educators, and higher education. This is so important today, since acquiring a college degree no longer guarantees a good job while having the potential of placing a major economic burden on many American families. We need to do everything we can to help our graduates to be successful by providing the clear and concrete evidence needed to demonstrate the alignment of the knowledge and skills to the current workforce. The digital badge accreditation component can challenge educators to think about the kinds of educational experiences we want students to have that will be important to them in the future whether it's in an academic setting or obtaining employment. As students receive badges for learning how to set up servers, work with customers, for creating tutorials or digital content, while exploring a full range of uses as a wearable technology; these benchmarks of learning are what those students' future employers or colleges are looking for.

How to engage and inspire students is bringing the assessment and/or the feedback much closer to their learning experiences. Assessment and feedback is useful when engaging students to get them to go further in their learning process while encouraging them to continue learning, enabling them to become lifelong learners.

FRAMEWORK

The final piece of the digital badge system is the framework used to support the issuing, collection and displaying of badges across contexts. Students learn across many contexts, across the Web, and across time, and when using digital badges it is important to give students control and to help them capture their learning. Providing a structure allowing this learning to be captured, wherever it happens, allowing each student to carry and display their badges wherever they go or feel is valuable requires an open and decentralized support system. This open badge framework should support independent-source badge issue, badge collection, badge display, and badge endorsement.

- Independent-Source Badge Issue
- Badge Collection
- Badge Display
- Badge Endorsement

There are several programs supporting this open and decentralized badge framework supporting badges from issuers across the Internet, manage badge endorsements and authentication allowing students' to collect their badges, and to associate them with a single open identity while allowing access to the badges across websites and experiences. The different programs are: Mozilla Open Badge Infrastructure, Class Badges, Google+Badge, HASTAC, Credly, and Purdue Passport.

WEARABLE TECHNOLOGY AND THE DIGITAL BADGE SYSTEM

The rapid development of new technologies also holds promise for providing a way to track innovative and emerging knowledge and skills needed in the modern work environment. With the tremendous growth of wearable technologies the badges can capture types of knowledge and skills not currently recorded by older traditional paths.

There has been a shift from personal computers of the 80's to the Web in the 90's that now is followed by a tremendous growth in e-learning, and mobile and wearable technologies. The reduction in size of the computer technology required combined with Bluetooth smart phone technology allows consumers to be portable with unrestricted continuous access. Mobile devices have some limitations due to the small screen sizes, limited processors, memory, storage, battery time, and slow connections. On the flip side, some of the benefits of mobile devices include the multiple inputs of handwriting, voice, and touch screens. Wearables take computing to a whole new level by offering hands free and sometimes gesture-based and augmented reality computing. Wearable technology is a category of technology devices that can be worn by consumers making it portable and operational simultaneous with hands-free and embedded sensors providing unobtrusive access 24/7.

With mobile and wearable technologies, we are seeing a shift towards lifelong learning requiring individuals to continually expand their knowledge, skills, and experiences with new and emerging technologies. Modern workplaces are increasing their use of these technologies, so it is important to develop the skill sets needed to use that same technology as a part of the continuous learning process. Teaching, training, and technology tools provide a context for the educational foundation needed now and in the future to be flexible and adaptable to the changing trends.

Wearable and mobile technologies framed within the field of education provide learning opportunities for individuals, joint collaborations, mobility, personalization, augmented reality, both synchronous and asynchronous delivery of content, research, problem solving, content generation, sharing, and even provide assistance for individuals with disabilities. All of these options can be paired with meeting the learning objectives, needs, goals, and standards while increasing new technology skills. Keeping everything else in mind, technology can be studied as a stand-alone topic or integrated with additional curriculum topics.

Technology rich situated active learning can provide users with an authentic real world context for solving problems (Dede, Brown-L'Bahy, Ketelhut, & Whitehouse, 2004). Learning within well-designed digital contexts can lead to future authentic applications of skills for personal or professional use (Murray, 1997). With the careful selection of technology, well aligned to the desired outcomes, and used effectively, technology can be an important tool for improving student proficiency.

From a study conducted by Educause Center for Analysis and Research, (2014) titled "ECAR Study of Undergraduate Students and Information Technology, 2014" found 9 of 10 students rate themselves as having favorable technology inclinations, yet their technology experiences and expectations suggest they lack the motivation, opportunity, or aptitude to use the full potential of technology for academic purposes. Even with the widespread use of technology, a surprisingly high number of students say they could be more effective if they were better skilled at using different types of technologies. The results challenge the idea of students inherently knowing how to use technology. These examples demonstrate the need to find meaningful ways to teach new technologies to promote student learning.

Digital Badge Technology Portfolio Skill Examples

The wearable technology badge credentials can be demonstrated through the development of e-portfolios. The portfolio can record a person's knowledge, skills, accomplishments, and experiences. It can show how it is relevant to the demands of the current workforce and showcase those accomplishments. Table 1 lists the top 10 skills employers currently desire. This is a good starting place for considering the types of skills to include in a digital portfolio combined and connected to technology experiences.

The demonstration of evidence can be shown through a variety of media examples in the learners' portfolios. For example, a portfolio could include the following artifacts such as awards, certificates, photos, and videos, written descriptions/examples, evaluations, creative projects, presentations, recommendations, and other sources. It is a way to capture and record evidence of learning accomplishments and experiences, telling an individual's unique story of who, what, when, where, why, and how it happened.

There are two primary types of skills to be demonstrated. The first are their technical skills. The second type is transferable skills. Table 2 below shows examples of both types of skills to be captured.

Transferable communication skills can come from a wide range of experiences such as: teaching, training, presentations, online learning, virtual conferencing, webinars, social media, and others. Transferable skill sources can be from either work or life experiences or both.

Table 1. The 10 skills employers most want in 2015 graduates (Forbes, 2015)

Number	Skill Abilities
1.	Work in a team structure
2.	Make decisions and solve problems
3.	Communicate verbally with people inside and outside an organization
4.	Plan, organize and prioritize work
5.	Obtain and process information
6.	Analyze quantitative data
7.	Technical knowledge related to the job
8.	Proficiency with computer software programs
9.	Create and/or edit written reports
10.	Sell and influence others

Having learners showcase their technology knowledge and experiences by bringing their work to life through a multimedia portfolio capturing what the learner knows provides evidence of the learning. In addition, it lets prospective employers know what level of knowledge the student has with emerging technologies and how poised they are for current and future job skills needed. Students can then stand out from the pack with in-depth presentations of their expertise proving they are professionals worthy of hiring and promoting.

Technology-Based Digital Badge System

So, how do we get students to use new software, wearable technology or other emerging technologies, learn new features, gain experience, and increase knowledge and skills? One way is to motivate student learning and participation with the digital badges. Using digital badges encourages students to engage in what is being learned. By awarding badges, educators can motivate the ways in which they want students to learn and participate. Digital badges may be awarded for completing an activity and/or project, how well students have completed the activity and/or project, or just for completing a task. For example, in school students take a wearable technology course based on instructor objectives, aligning each task with the learning outcomes, scaffold each step towards how to use the wearable technology, then finalize with a project based learning experience. Students would be assessed through a range of basic level skills to expert level skills then be awarded the wearable technology badge that has the potential to unlock different types of privileges in colleges and in employment.

One example of students earning a wearable technology badge would be learning to use the Microsoft HoloLens. The Microsoft HoloLens is the first holographic computer running Windows 10. It is completely untethered, which means no wires, phones, or connections to a PC. The Microsoft HoloLens allows a person to place holograms in their physical environment providing new ways to see the world. Students would earn a badge for learning how to wear and use the Microsoft HoloLens, and then earn a different badge for developing holographic apps to run on the Microsoft HoloLens platform.

For students to earn the HoloLens badge they would be given the task of completing the following: students would demonstrate 3D modeling and creation on the computer, explain Windows 10 and how it relates to holographic computing and Microsoft HoloLens, use HoloNotes in Skype, use HoloStudio to create a hologram, then perform the ways a person would communicate, create, collaborate, and explore

Table 2. Technical and transferable skill examples (Holland, 2015)

Technical Skills Mastered	Transferable Skills Mastered
1. Smart Glasses	1. Communication
2. Smart Watches	2. Teamwork
3. Smart Clothing	3. Creativity
4. Wearable Assistive Technologies	4. Adaptability
5. Mobile Technologies	5. Problem Solving
6. Augmented Reality	6. Obtain, Process, Edit, Analyze, Create
7. Virtual Reality	7. Leadership
8. Implanted Technologies	8. Plan, Organize, Prioritize

their environment using the HoloLens. Students would complete the different levels while interacting with experts from Microsoft. The instructor aligns each criteria with learning outcomes, scaffolds each task, and for the end product has the students submit a web page demonstrating each task learned and completed. Students would develop a web page demonstrating the use of HoloLens and use software such as Captivate, Camtasia or Articulate Storyline for demonstration purposes to be uploaded to their site. Students would then be assessed through a range of basic level skills to expert level skills before being awarded the HoloLens badge.

In awarding digital badges whether for wearable technologies, course completion, micro credentialing, event participation, or any skill it is important to remember and use the following steps:

1. The instructor responsible for creating or assessing the criteria assesses the assignment against the defined criteria before making the decision to award or not award the badge. The instructor may establish an assessment revision, where the student revises their learning until competency or comprehension is demonstrated.
2. When using assessed badges, students receive the badge once the comprehension or competency is demonstrated. For non-assessed badges, the badge is awarded for attendance, participation, and other non-assessed conditions.
3. The badge is linked to the instructor issuing it, along with the criteria and evidence required for completion. Linking the badge to the instructor, criteria and evidence of completion establishes the integrity of the badge. Anyone clicking on the badge may view all components: instructor, course information, program, criteria used, evidence, outcomes, wearable technology skills learned, and experiential learning tasks and project completion.
4. The student will be able to push out and publicly display their badge on the Mozilla Badge Backpack or other platforms - established webpage, e-portfolio, or resume - to facilitate career development, showcasing their skills learned in new applications and technology innovations.

For digital badges to achieve acceptance at mainstream institutions and in industry, there will need to be a badge infrastructure put into place showcasing instructor credibility and reliability of outcomes and content, badge ownership, university policies being updated, and a move from traditional assessment and credentialing to more innovative assessment. Digital badges have the potential to significantly change the way students are recognized in education while changing the way institutions and organizations evaluate their programs. Even though there are challenges to incorporating a digital badge system, the benefits to students and changes in assessment for the university outweigh the challenges. Digital badges provide a unique and currently positive innovation-taking place in education, technology, and the work force.

BADGES CURRENTLY IN USE OR DEVELOPMENT

The badge system is currently in use by: Colorado State University, Prude University (P2P), nanoHUB-U (https://nanohub.org/groups/u), MIT, NASA, University of California-Davis, Penn State, the National Oceanic and Atmospheric Administration, the U.S. departments of Veterans Affairs and Education, the Corporation for Public Broadcasting, Disney-Pixar.

Purdue University has led the way in developing, implementing and pushing forward the badge system. Purdue developed the Passport and Passport Profile. Passport is a learning and e-portfolio system using digital badges to demonstrate students' competencies and achievements. Students can show off the badges they have earned through the public Passport profile in Mozilla Backpack. Passport learning management systems enable students to earn badges by completing activities and/or projects presented as a series of challenges. The passport guides students through each task by providing a framework to submit documents, share links, complete quizzes, and/or gather approvals from instructors and/or experts. Instructors follow each student's progress and connect the badges with course objectives. As badges are collected they can be shared online using Mozilla Backpack, a mobile portfolio system, or an e-portfolio.

The Manufacturing Institute is creating badges that will become part of the National Manufacturing System. The badges are being used to connect new employees to potential employers looking for new talent. Penn State leads the way in developing an internal badging system that will interface with Mozilla's OBI.

RECOMMENDATIONS

Most of the open badges initiatives in designing, issuing, and displaying are at the pilot stages and the research is limited. Much of what exists in the research literature about the open badge system focuses on motivation and game mechanics, particularly among higher education. As the open badge system gains momentum, issuers will need to think through both their implications and applications. In addition, to solidify the use of digital badges as a widely recognized and trusted credentialing system employers will need to value the open digital badge as much as they do traditional credentials. Earners will then need to earn open digital badges representing their skills, knowledge or accomplishments. Especially those important for them professionally and that are clearly issued from trusted and recognized sources. Finally, career and social networking websites will need to make it simple for earners to display their digital badges.

Once barriers are overcome, open digital badging could have tangible and unique benefits. For example, employers should be able to more easily identify specific skills they desire in a large candidate pool. Issuers should realize a multitude of benefits such as generating revenue by offering assessments and recognition of subject matter concentrations, while being able to control and monitor the status of digital badges by building in expiration or recertification dates. Issuers should also continue investigating the use and value of digital badges, laying the groundwork for a new way of earning, managing, and sharing achievements and credentials.

As student loan debt continues to soar, college drop-out-rates rise, with open educational content, personalized learning systems, and massively open online courses disrupting higher education, the open digital badge may offer solutions to some of these issues. Evidence found in a digital badge will tell a more illustrative story about the learning and achievements associated with the award. Employers or graduate school admissions officer should have a much more information-rich way to evaluate student achievement. It is recommended that higher education systems consider the potential use and value of using open digital badges, while keeping in mind the badge system designs are competency-based and credibility-oriented which will add value to the traditional credentials they already use.

FUTURE RESEARCH DIRECTIONS

What's next in implementing the digital badge system in higher education? The value of the badge system is with various issuers pushing badges into higher education, students actively managing their collections, while displayers or consumers using badges on the back end connecting to students with real projects and results.

Further research needs to be conducted regarding policies and procedures, issuing badges, accepting badges from outside universities and what this means in higher education. What support structures do institutions need when hosting a badge system? At minimum, an institution needs a system that supports issuer creation and deployment of badges, earners storage of evidence, and a means for organization and deployment of badges (Friesen & Wihak, 2013). Issuers also need a place to store and display the criteria under which a badge may be earned. Earners need a place to store their badges and display them.

Further research needs to be conducted in order to understand the challenges to assessment and credentialing of the badge system. As open digital badging grows, it will force institutions to examine their assessment systems and credentialing structures (Olneck, 2012). Today, a university degree is seen as the only source of authority and competence; however, the open badge system could easily prove to be a disruption of this tradition. Questions to consider: will institutions accept students who acquire multiple badges from various external sources showing evidence of what they accomplished and/or learned? When employers begin to accept badges as evidence of competencies and learning instead of the traditional diploma what will higher educations' reactions be to the devaluing of a long established tradition?

Finally, authenticating digital badges is an area that needs to be addressed. How will organizations maintain digital badge security and validity? One area that will need to be thoroughly researched and developed is the use of authentic signatures. Digital signatures provide a way for demonstrating the authenticity of a digital message or document, while giving a recipient reasonable proof the object was created by a known sender and the message was not altered in transit (Catalano & Doucet, 2013). Without digitally signed badges, it is possible fake badges will emerge, invalidating the integrity of the entire badge system.

CONCLUSION

This chapter outlined the fundamentals of the digital badge system, including wearable technology, assessment, and a framework supporting a wider issuance, collection and sharing of badges. This framework supports the development of a high quality badge system supporting and encouraging learning, while promoting and recognizing important skills and fostering real outcomes.

With the tremendous growth and use of emerging technologies such as wearable devices in all sectors, it is logical to consider teaching/training the associated knowledge and skills as a stand-alone field or combining it with additional curricula areas to prepare for the modern work environment. Through the demonstration of learning using a digital portfolio, the accomplishments can then be certified through digital badging. The badging system offers an innovative medium for showcasing cutting edges experiences not currently being recorded by traditional means.

Even though the badging system is in its early stages of development and adoption, it provides a tremendous opportunity for leading universities to have a major say in how to implement the digital badges approach. Understanding that digital badges do the same thing as credentials with one critical difference: badges are transparent, meaning the criteria, learning, artifacts, and assessments are directly linkable to the student earning the badge and to the issuing institution. Examining the use of digital badges through assessments and for student learning, while acknowledging the possibilities of digital badges is an endeavor important to study, especially in the realm of higher education.

As higher education continues to move forward in creating and providing well-designed learning experiences for students, linked to authentic real-world experiences, work competencies and skills, it is practical to examine other possibilities for assessing student learning and institutional outcomes, as found with digital badges. These steps demonstrate a commitment to increased accountability for student learning and the institution's teaching. The adoption of digital badges is another solid step forward towards meeting that goal.

REFERENCES

Abramovich, S., Schunn, C., & Higashi, R. (2013). Are badges useful in education? It depends upon the type of badge and expertise of learner. *Education Technology Research Development.*

Bruckman, A. (2004). Co-evolution of technological design and pedagogy in an online learning community. In S. Barab (Ed.), Designing for Virtual Communities in the Service of Learning (pp. 239-255). Cambridge University Press. doi:10.1017/CBO9780511805080.013

Catalano, F., & Doucet, K. (2013). *Digital badges emerge as part of credentialing's future*. New York, NY: Professional Examination Service.

Dede, C., Brown-L'Bahy, T., Ketelhut, D., & Whitehouse, P. (2004). Distance learning (virtual learning). In H. Bidgoli (Ed.), *The Internet encyclopedia* (pp. 549–560). New York: Wiley. doi:10.1002/047148296X.tie047

ECAR Study of Undergraduate Students and Information Technology. (2014). *Educause Center for Analysis and Research.* Retrieved from http://net.educause.edu/ir/library/pdf/ss14/ERS1406.pdf

Gee, J. (2011). Human action and social groups as the natural home of assessment: Thoughts on 21st century learning and assessment. In V.J. Shute, B.J. Becker (Eds.), Innovative Assessments for the 21st Century: Supporting Educational Needs. NY: Springer.

Gibson, D., Ostashewski, N., Flintoff, K., Grant, S., & Knight, E. (2013). Digital badges in education. *Education and Information Technologies*, 2013, 1–8.

Grant, S. (2013). 5 buckets for badge system design: You are here. *HASTAC*. Retrieved from http://www.hastac.org/blogs/slgrant/2013/10/23/5-buckets-badge-system-design-you-are-here

Grant, S. L. (2014). *What counts as learning: Open digital badges for new opportunities*. Irvine, CA: Digital Media and Learning Research Hub.

Mozilla Foundation and Peer 2 Peer University, in collaboration with The MacArthur Foundation (2012). *Open Badges for Lifelong Learning*. Retrieved from https://wiki.mozilla.org/images/b/b1/OpenBadges-Working-Paper_092011.pdf

Murray, J. H. (1997). *Hamlet on the Holodeck*. Cambridge, MA: MIT Press.

Schmidt, J., Geith, C., Haklev, S., & Thierstein, J. (2009). Peer-to-peer recognition of learning in open education. *International Review of Research in Open and Distance Learning*, *10*, 5. Retrieved from http://www.irrodl.org/index.php/irrodl/article/view/641

Tally, S. (2012). *Digital badges show students' skills along with degree*. Purdue University News.

The 10 Skills Employers Most Want In 2015 Graduates. (2015). *Forbes*. Retrieved from http://www.forbes.com/sites/susanadams/2014/11/12/the-10-skills-employers-most-want-in-2015-graduates/

KEY TERMS AND DEFINITIONS

Assessments: Assessments are used to evaluate learning through evidence.

Digital Badges: Digital badges consist of the student earner, instructor issuer, and web-based displayer, to show the acquisition of knowledge and skills.

HoloLens: A holographic computer running Windows 10.

Mozilla Open Badge Infrastructure (OBI): The OBI provides the underlying open technology and standardization to support badge issuers, badge displayers, a cloud-based repository for badge collection, and management for students.

Passport and Passport Profile: Instructors and advisors provide students digital badges for mastery of skills.

The School of Webcraft: A free online community for social learning about open web development.

Wearable Technology: Technology worn on, near, or in the body.

Chapter 6
Wearable Cameras

Alessio Drivet
University of Turin, Italy

ABSTRACT

Wearable Technologies represent an emerging theme. Probably the next emerging market where companies will focus. These devices are replacing entire categories of electronic objects in everyday life and affect the way we live, work and socialize. Among the many applications available, the author limits attention to the field of the smart video cameras. This chapter examines some of the most interesting applications of wearable cameras, with special reference to the Italian situation. In particular, the text traces a summary of the main applications in sports, spying, police, army, education, health, disabilities, and lifelogging. A part is devoted to "wearable extensions" and the concept of augmented reality.

INTRODUCTION

In this chapter, we talk about wearable camcorders, a segment of the broader world of wearable technologies. Among the many significant aspects, the focus is on several technological solutions and the wide range of applications. This will be done by examining six application areas. Because the technology is not neutral, we must consider the social aspects of the phenomenon. The part concerning issues, controversies and problems outline the main themes.

Wearable technologies are a significant part of the current technological frontier (O'Driscoll, 2014). The viewer's perspective can be captured through glasses, watches, clothing, and other smart devices (Shum, Liao & Zhang, 2001).Those interested in an overview of all Wearable Technologies, with particular attention to the main products and companies of the sector may see the publication of K Toh Peng (Toh, 2013). Regarding Italy, a point of reference could be the project "We Are Able To" focus on the study and development of solutions, services and technologies; the results are published on an e-book constantly updated (Lalli & Prunesti, 2014).

A Reuter's article titled "Research and Markets: Wearable Cameras Market Report 2015: Global Market Analysis and Forecasts 2014-2020" (Apr 22, 2015) summarizes some themes to be discussed in the text: "The wearable camera market is still in the early stages of development and is experiencing rapid growth as use cases for wearable cameras expand. The consumer segment continues to grow beyond

DOI: 10.4018/978-1-5225-0069-8.ch006

its core base of sports and adventure enthusiasts and into lifelogging and capturing photos and videos from life moments and special events. At the same time, wearable cameras are beginning to proliferate more widely in the public safety and enterprise sectors. Body-worn cameras for police officers are used with increasing frequency, and enterprise users are experimenting with a wide variety of applications from retail to hospitality to user experience research".

In Italy, a remarkable expansion of wearable technology has occurred in the field of cameras.

In this segment, we concentrate our attention on examining the areas of applications including sports, spying, police, army, education, health, disabilities, and lifelogging.

The Google Glass, and its competitors, can be placed between the wearable cameras, but in reality it is more correct to define it as a wearable computer because the features are different: photo / video shooting, Internet browsing, voice control, GPS navigation, etc. In other words, it is a miniaturized device usable for multiple purposes, a true extension of our physical and mental potential. The chapter's primary focus is on wearable cameras therefore will not discuss other possible uses.

In Italy, most wearable cameras refer to the use in sports. They have been on the market here for several years, with the light first "GoPro", a brand that even today remains the industry leader. The most famous action camera was created by Nick Woodman, an American surfer who created their own version of the camera dedicated to underwater shooting. The name Go-Pro comes from the fact that surfers needed a camera operator for shooting. With the commercialization of the action camera it was no longer necessary, and its name is precisely the indication "go as a professional."

The difference of diffusion with the other wearables can be interpreted in different ways: a) the costs of a wearable camera are comparable to that of a traditional, b) there is a fair number of people interested in sports activities, c) many users wish to document their experiences increasing visibility on social networks.

A particular market concerns "invisible" cameras / camcorders. The camcorder concealment is a solution for shooting hidden videos, used by private citizens and investigation agencies, widely used in the creation of popular television programs of investigation. In this case, the market is less extensive because, aside from personal situations, is intended for professionals.

An application generating some debate is the use of wearable cameras by police, because it broadens the scope of control of the activities of the population. A mobile and individual instrument combined with the enormous spread of fixed surveillance cameras. The dissemination of this tool in some countries is beginning to be significant, while in Italy it is at the experimental stage. A completely different speech is made for military use, here we can distinguish between simple tools for recording of wartime events and equipment intended to improve the technological level of resources available.

An environment beneficial to the development of these technologies is related to education and training at all levels. The use of equipment made of or containing a camera has been widely tested and documented in many areas of training. The Italian experience seems very partial; just think of how little is developed in the literature on the subject. Unfortunately, as often happens, economic and cultural reasons cause a brake.

One area of particular interest, especially for social implications, involves health and disabilities. The growth of social attention on the issue of disability develops the need for new instruments to promote integration. Using wearable tools can facilitate both operators in the sector of the disadvantaged. Some applications are proving to be very promising.

The dissemination of digital equipment like smart phones and tablets and the use of social message boards has changed the standards of everyday life for millions of people. To describe the process by which a huge number of people capture fragments of life to share with others was coined by the term Life Log: the diary of life, in the form of pictures, publishable on Internet. For some it is a new frontier, for others it is just one example of technology in the service of individual vanity.

Among the many wearable devices available, one must count "action cameras". These cameras are small by definition, waterproof and impact resistant, can be worn, and mounted on the handlebars of a bicycle, or on a car. Some cameras do not belong to the world of wearable technologies but being connected to a car or a bike can be seen as an extension of our body, and as a "wearable extension". In the text, it will be made a small mention of this topic.

In order to complete this introduction is important to mention two other terms: "augmented reality" and "smart devices". These technologies have opened the way for great progress in many different fields.

Often we talk about "augmented reality". This expression first appeared in the literature in the forties, but really spread the late eighties. The concept had to wait, however, during the XXI century it matured, and the first products in common use overlooked the market at the end of this decade.

The technologies related to the augmented reality represent an evolution compared to consultation of the information, the use of multimedia content and the interaction with the user's data, especially regarding the vision of the data superimposable to the world surrounding the user. Augmented reality has been introduced, at least so far, in very specific areas such as medical or military. We should not confuse augmented reality with virtual reality, these are two different approaches. Although both are based on the display of information in the field of view, show a substantial difference: the first superimposed images and text to portray what the user sees around them, the second recreates a virtual environment giving the sensation of being in another place.

MIT researchers have been able to develop a camera-shaped ring designed to aid the blind in the task of identifying objects and reading the text. The camera can point to an object, take a picture, and receive feedback through a digital voice about what has just been focused on. The ring in augmented reality also has a small camera, a processor and Bluetooth connectivity. With a simple double-click on a button at the side of the ring a voice command starts the process.

Often the word "smart device" is associated with these products. This term is not only an emphasis on the concept as it is now available as a small camera worn with a clip using a series of internal sensors and a microphone to understand when it is time to record images or not. The camera is activated when it is stable and is capable of framing faces.

A wearable camera taking pictures automatically overcomes many problems with users who do not want to draw attention to the technical device. One of the principles underlying the smart cameras is to make easier use of the instrument. Farewell to cameras engaging eyes and hands and room for creativity. An interesting project is Air Clicker, able to capture every moment with a simple click and transfer images directly to your smartphone using Bluetooth technology. The device consists of two modules in the shape of a ring (Figure 1), the first to be worn in the thumb (CCD) and the second in the index (shutter). If you want to take a picture, you have to simulate the act of clicking. To shoot a video, you need to place your hand as if you were to take a video camera in hand (Figure 2).

BACKGROUND

The points of interest are two: a) the size of the phenomenon, b) the sociological and technological impact. According to several popular media sources, the global market for wearables could reach sales

Figure 1. Wearable ring (Yanko Design, 2011)

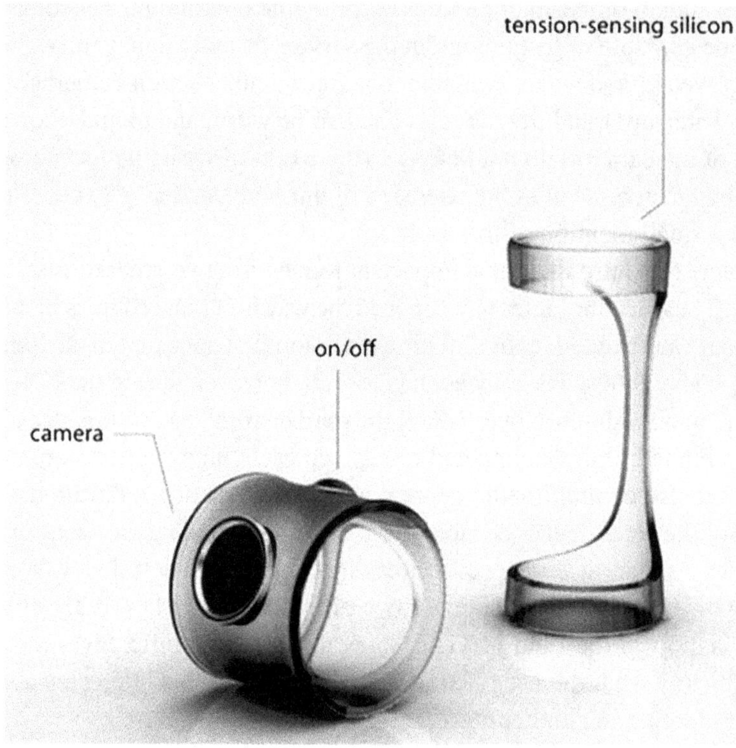

Figure 2. Wearable ring gestures (Yanko Design, 2011)

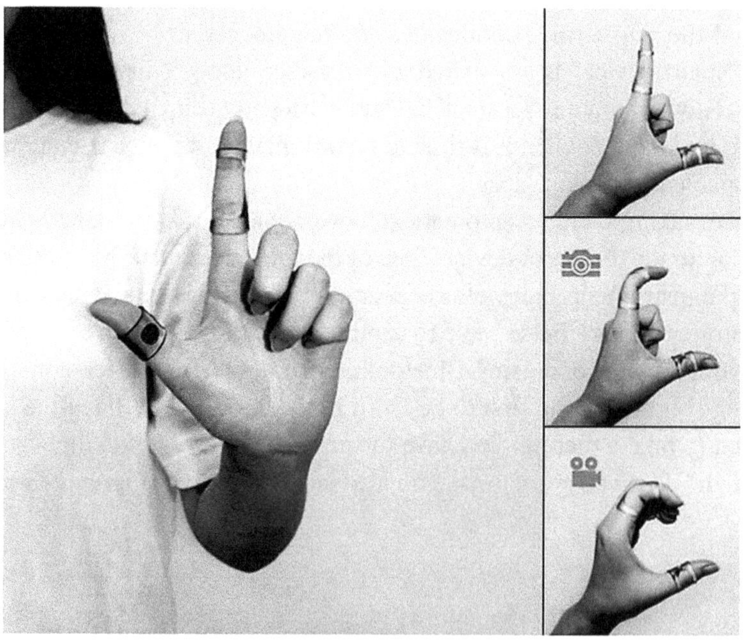

of 125 million units in 2018 with revenue of nearly $ 39 billion. According to other market analysts, the annual deliveries of wearable cameras will reach more than 30 million units in 2020. In the 2014-2020, it is expected to increase in the volume of deliveries by 5.6 million units to 30.6 million units, equivalent to a compound annual growth rate of 32%.

Although it is the category of actioncam to drive the market, over the next year devices intended for consumer use, for enterprise applications and public safety will have strong growth. However, this market is still at an early stage, experiencing rapid growth thanks to the expansion and the discovery of new uses for wearable devices. Of course, these forecasts will be strongly influenced by the novelty of the products, market prices, competition with other devices like smart watches and, of course, changes in the world economy. In Italy, the data shows a growth trend comparable to the world market. Is interesting to note the data results from the corporate (Figure 3) and business sectors (Figure 4). Of course, this concerns information referring to the entire field of wearables and not the more limited area of wearable cameras.

Speaking of tools such as smart cameras or, more generally, of smart devices one can refer to the expression Augmented Reality (Wassom, 2014), as mentioned in the introduction. For augmented reality we mean the enrichment of human sensory perception through information, generally handled and conveyed electronically. Items "increasing" the reality can be added through a myriad of vision devices adding multimedia information to reality already normally perceived. Sensors help to optimize performance by offering individualized feedback to wearers. In the age of appearance, where the image reigns, the cameras become an integral part of the human body to film every moment of the day.

Of course, this leads to a change in perspective, technological innovations have an impact on society and people's psychological attitude. This process, however, is also bidirectional, meaning the introduction of new tools become operational in terms of the market when new demands are growing in population.

Figure 3. Sharing of the market for corporate form in Italy (Wearable.to, 2014)

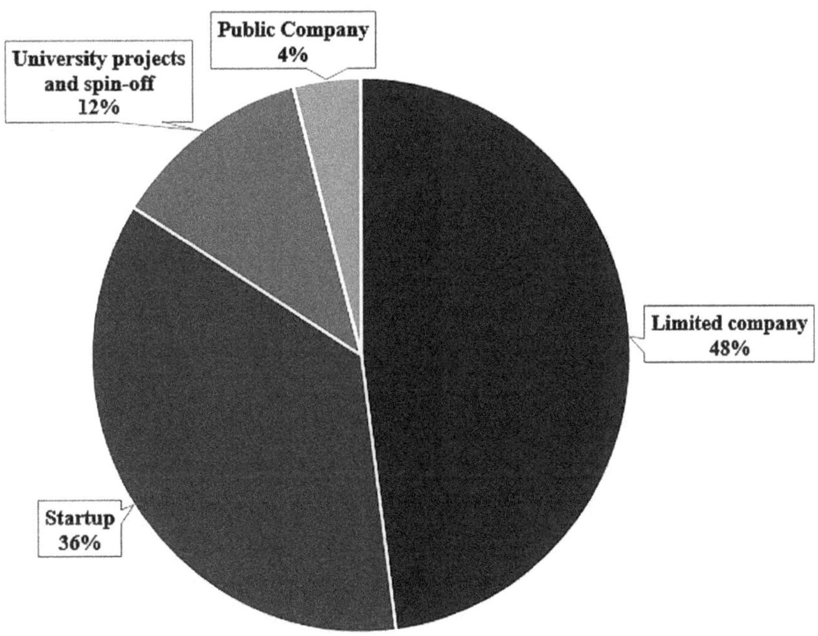

Figure 4. Sharing of the market for business sectors in Italy (Wearable.to, 2014)

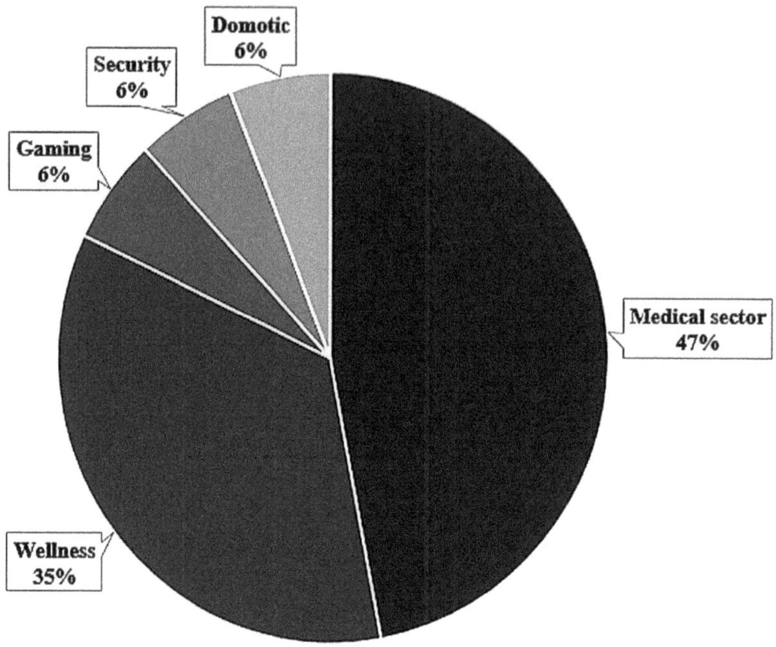

Many years ago an interesting research, which used a WearCam (a wearable multimedia computer equipped with camera, and wireless Internet connection), placed, experimentally, a series of interesting questions: "Once we accept that images can be easily falsified, we will someday soon (some have already) began regarding images as having no more truth than verbal or written accounts. The difference between whether images themselves are presented to a jury from someone's WearCam or a verbal description of incredible detail is provided by that same person through their looking into their own WearCam's memory, will matter less and less as technology evolves" (Mann, 1997, p. 173). It is a point of view very interesting, although there are some critics who underline the costs and technological divide can increase the differences between users.

SIX AREAS OF APPLICATION

Sports

Even more frequently, we see small or even miniaturized cameras used to record a journey by motorbike or a bicycle tour. In Figure 5, you can see an example of the author's personal experience, engaged in rafting. Now the cameras for sports, men and women must take account of their wishes: multisport tools easy to use and with high quality HD video. The latest cameras combine HD video recording, Wi-Fi and improved GPS functions.

Almost all cameras are equipped with various accessories: waterproof enclosures for underwater shooting or tools for mounting. Generally, we can distinguish two types of support. Support for helmets: with this accessory, you can dock the camera to various types of helmets, like those used in cycling,

Figure 5. Raft ride (Drivet, 2015)

motorcycling and skiing, where you normally wear head protection. Support belt: with this accessory, you can use the camera for performance in extreme sports such as parachuting, where there is high stress related to free fall.

There are several models of wearable cameras, addressed for the mass market, within the last decade; they differ in size, weight, cost, and accessories. The manufacturer best known is certainly the GoPro. The new frontier now begins to move towards products tending to resemble "smart glasses". In figure 6, it is possible to observe a pair of glasses with a careful design, in its simplicity, capable of shooting with an angle of 170 ° and a definition of 5 megapixels. The video, recorded on the card, can be downloaded and viewed on a PC or TV.

Figure 6. Smart glasses (Drivet, 2015)

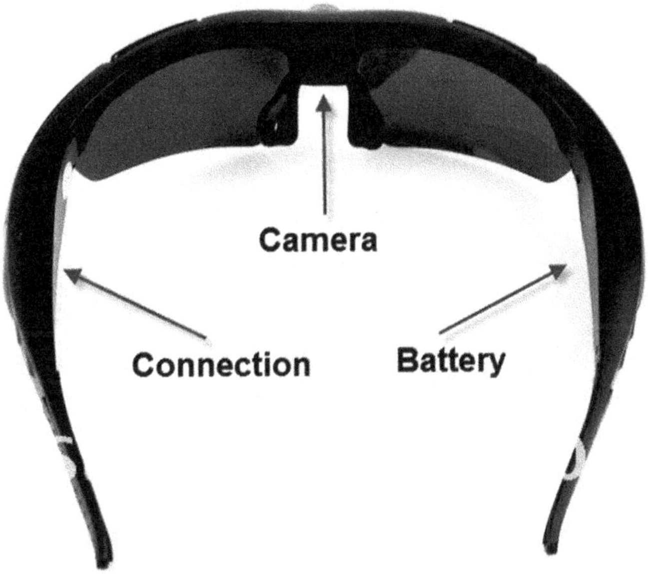

A particular use, agonistic or recreational (Miller, 2015) is linked to ski goggles (Figure 7). In the first case, the data on the movements of the athlete go directly to the tablet of the coach who can analyze the performance. The system is capable of recording all the parameters of skiing, distance covered, acceleration, speed, biological parameters such as heartbeat, breathing and then allow for analysis of the performance in the points of strengths and weaknesses. In the second case, the user can view the video to improve their skiing performance and share it on social networks. In this case, social content is more relevant than the sporty appearance.

The last device mentioned allows divers to shoot video hands free (Figure 8). Of course, we refer to amateurs because professionals use a very different kind of underwater cameras.

For best results, you can also add two spotlights. These lights are essential under 30 ft. to add colour to everything, even during the daytime, since the water filters out all the warm light as you go deeper.

With all these cameras you can review, in subjective, the practice of an extreme sport. In this way the subject is free to exercise the favorite sport with the strong emotions lived and then calmly analyze the

Figure 7. Ski goggles (Drivet, 2015)

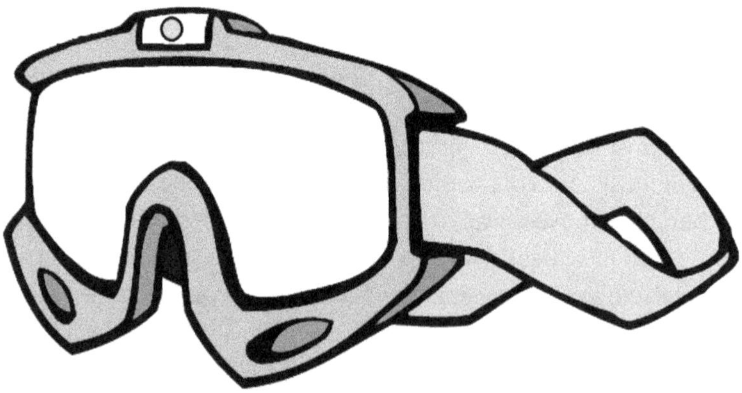

Figure 8. Dive Mask Camera (Drivet, 2015)

experience. One aspect not to be underestimated is the possible change of perspective, in theory anyone can become beneficiary of the experience: you can watch the world through the eyes of the protagonist comfortably seated in front of the television.

Is it restrictive to think about these tools only in a sense of play or, at best, useful for monitoring sporting performance. The tool can be used to make most spectacular sporting events. An example is the shirt with a built-in camera allowing you to follow the soccer game from the perspective of the player, a project developed by a Spanish company.

The integrated cameras have appeared for the first time in an official game of basketball in 2015. It took place in Madrid, during the meeting between the Real and the Lithuanian team Zalgiris Kaunas. Two players of the visiting team have worn, under the official T-shirt, a small camera. These images have allowed fans to experience the action in real speed, lowering oneself into the role of the players and experiencing their point of view.

Spying

The camera for concealment (Figure 9) has been specifically designed and manufactured to be easily hidden.

All forms of concealment devices rely on the use of camouflage in the most general sense of the term. A miniature spy camera is almost completely essential as compared to a normal camcorder, for example has no viewfinder.

Figure 9. Mini Recorder Conceal Video Camera (Drivet, 2015)

The module can be installed, given its small size, in items of clothing as tie, button, watch, jacket, bag, baby carrier. Superior quality, lightweight and small size make it the ideal product for creating professional candid camera shots. Often the recording starts automatically as the device perceives a sound that exceeds 65 db. The cameras for concealment may be available to any private person, but represent a solution increasingly used by law enforcement. In Italy, the law does not allow the use of such equipment during tests, university admission, conjugal control, etc., these cases constitute a crime. Naturally, to fight the use of hidden cameras have built equipment for the contrast based on a combination of laser and scanning wireless. They are also widely used in the production of television programs where a subject is recorded without knowledge. In Italy, there are several cases where hidden cameras are used for television investigations. Italian broadcasts as "Report" and "Le Iene" (the Hyenas) often use a hidden camera to expose fraud or perform journalistic investigations. The use of traditional methods often do not reach the same result. A journalist, following narrow ethical principles in the exercise of his professional function, should always qualify as such and declare the presence of tools of image or voice capture. The Guarantor of privacy has frequently stressed professional information cannot resort to "tricks" to conduct its business and must always make known his profession, unless there are real risks to their safety or cannot fulfill the function. The problem with hidden cameras are the legal concerns. A recent ruling by the European Court of Human Rights (ECHR) affirms the principle allowing the use of hidden cameras only when the movie protects the privacy and identity of the subjects.

Although not closely related to the problem of "spying", it is interesting to point out another aspect involving technology development and sociological aspects. The widespread diffusion of technology, cost reduction of technology products, increasing computer literacy, increasing capacity and sophistication of high-tech tools, and their facility of use have made possible not only the production of information and documents from the citizen journalists, but also their easy-dissemination through the Internet, which has become a worldwide showcase to access potentially anyone. With computer networks, citizen journalism has taken it a step further, becoming netizen journalism" (Bravo, 2012).

One last consideration: each time a new media technology has been made available, it was often "abused". Examples include movies or concerts which are more easily recorded to get an illegal copy. The process of making and distributing such recordings is known as bootlegging, a term initially attributed to alcohol smuggling. Bootlegs can consist of recordings of live performances or material created in private or professional sessions. The anti-piracy laws prohibit the use of cameras in concerts, cinemas and theaters, so these locations do not allow the use of the devices. Of course, it is easier to identify pirates with smartphones rather than with micro cameras. They spell out what it means (MPAA) and the National Association of Theatre Owners in 2014 have officially banned Google Glass and other wearables from movie theatres'. Those who want to enjoy a movie have to take their glasses off, as these can be used to illegally record movies. It is an interesting change, especially considering how they currently overheat in a very short amount of time and must be turned off. I would imagine over time this issue will be resolved.

Police and Army

Recently, in several countries, the police use small wearable cameras during public events.

In Italy, an experimental program provides for the assignment of wearable cameras to police agents. As a result, it was implemented as an experiment in four cities. During this study, the cameras and memory cards were equipped with a serial number recorded in a register with details of day, time, service performed, position and name of the agent who signs the return forms. The cameras will not always be in operation, but will be activated by agents only against a specific order of the officer, who will determine "the first sign of criticality". After the service, the images are downloaded on special secure servers and acquired by the Scientific Police. Later, if necessary, the images will be made available to the judicial authorities in accordance with the guarantees of the legislation. The camera is applied onto individual tactical vest (Figure 10). As you can see in the image, the camera is placed on the chest, the shot provides a wide-angle view of the scene, but the device could be damaged during a physical confrontation. Mounted on the glasses could provide a more correct view but could have other contraindications especially with the use of the helmet. The Italian Guarantor of privacy provided a positive opinion on the use of the new camera system. As the system is subject to the principles of the Privacy Code must be guaranteed the right of access to the images. This is a critical issue because the exercise of this right is not easy for the resistance of organisms involved.

These projects open wide spaces for companies working in the security sector with the supply of cameras offering the chance to expand applications. Police Corps of U.S.A. will allocate large sums in the coming years to have the cameras. Only in recent times cloud computing has made it feasible to start thinking about saving and maintaining, on a server managed by companies like Microsoft the mass of video data from police-worn cameras. The problem is most cloud storage platforms don't meet the security policy standards. But the drawbacks of local storage are numerous, requiring police departments to hire IT personnel for maintenance and security. Locally stored police data is also a huge target of hackers, with departments regularly fending off malicious cyberattacks. Hardware is only a crowbar to get their hands on ancillary services. A good part of the money will be spent on cloud storage enabling departments to organize, catalog and store all photos and videos produced during work. Some issues still need to be addressed such as what is the purpose for the recordings? When should the data be deleted? The

Figure 10. Camera onto tactical vest (Corriere della Sera, 2014)

definition of rules concerning the retention of data is not clear. The risk is the movie not only serves as evidence, but can also become a monitoring tool used for completely different purposes. This is a very important aspect (Mateescu, Rosenblat & Boyd, 2015).

If augmented reality is still limited in the consumer market, a very different level of development and optimization is in the Army. The Armed Forces of today are completely different from those of just a few years ago. For example, the US military uses rifles combined with Daylight Video Scopes, a digital video camera with zoom. The weapons also have an integrated laser rangefinder with digital compass. The sensors can transmit images directly on the computer display and on the Head-Mounted Display positioned above the eye of the soldier.

British Aerospace Systems (BAE) has designed and manufactured the helmet Striker II, to interact in a new way with the pilot. The real highlight is the transparent visor, a real display capable of showing images at a resolution of 1280x1060 pixels. In this way, the pilots of fighter jets are able to see clearly critical information such as speed, altitude, compass, targets and weapons available.

The aspect perhaps most striking is the use of the wearable cameras to document war actions. The war in Syria and Afghanistan can be seen through the eyes of the soldiers (Figure 11): A foray into the world of "combat footage". Now, available online video filmed by the US Army, from Syria, and even by the ISIL (ISIS or IS). Military videos have become popular recently. Sites like "Funker350" share the video, circulating quickly in the military community online.

These type of films should be banned by the US military; the document called General Order Number 1C prohibits among others the following activities:

h. Photography and Videotaping.

(1) Except as authorized for official use and purposes described below, this Order prohibits the taking, possession, reproduction, or transfer (to include uploading) of photographs, videos, depictions, and audio-visual recordings of the following: Detainees or former detainees; detention facilities; active combat operations (e.g., firefights); flight-line operations or equipment, subject to written, local exceptions, etc.

Figure 11. War scene (YouTube, 2015)

Education

The use of video in teaching dates back at least to the 60's and now we can speak of many examples of field research and an established theory. Even in Italy, there are many examples of the use of video recordings, especially regarding the primary school. For example, professors of mathematics education at the University of Turin often use this technology to understand the processes of formation of mathematical knowledge. Taking care to minimize the intrusive presence of the medium and researcher, you can collect behavioral codes of the children and, later, analyze these codes with the teachers of the school and the university students. "The use of the video camera to capture teaching and learning has the advantage of encapsulating and preserving a detailed representation of the dynamics of complex interactions making them available for in-depth analysis" (Otrel-Cass, Cowie & Maguire, 2010).

This approach has been valid for the use of conventional cameras, but a question arises: is it possible to assume the impact of new wearable technologies? For a brief analysis of the problems related to the use of the video, you can see a paper by Jewitt introducing the theoretical and practical issues related to the use of the camera. (Jewitt, 2012). Lahlou describes a new family of methods for social science research based on: first-person audio-visual recording with a miniature video camera worn at eye-level ('subcam'); confronting subjects with these first-person recordings to collect their subjective experience; formulating the findings and discussing the final interpretation with the subjects. These procedures enable subjects to reconstruct and describe their psychological state at the moment of action, especially their goals, by reviewing films of their own activity recorded from their own perspective with subcams. (Lahlou, 2011).

A premise: in education and training, the use of wearables remains largely experimental.

One can attempt to provide some indication, effective for the future. "Inconspicuous wearable devices could change the landscape of educational computing. Wearable cameras, for instance, allow a learner to engage simultaneously as observer, reporter, and participant, enabling more detailed life-blogging and provide a subjective point of view for digital storytelling. As the things we own and wear acquire sensors, process data, and connect us beyond ourselves, they open new opportunities for us to see further, hear differently, and touch things we have never been able to reach, enabling a new self-awareness and an enhanced perception of the world around us" (Educause, 2013, p. 3).

There are several possible uses for these new technologies in the field of education. For example, students can document their activities in the classroom or outside the classroom; teachers can register their protocols teaching with a subjective view. One of the most interesting aspects of wearable technology in education could be related to an increase in productivity. The possibility for the user to send information, with voice commands or gestures, via (SMS), email and social networks could help students and educators to communicate and keep track of updates.

In the field of education, at least in Italy, the development of wearable technology still lags behind. For example, the use of devices such as the Wearable Multitouch Projector (Figure 12), working through a laser projector and a video camera allows you to use hand gestures and touch to interact with the graphics projected onto any wall, is still distant.

Health and Disabilities

The eye-tracking technique has been used for about thirty years for surveys on language proficiency of individuals through analysis of ocular paths. There are several technologies for eye-trackers. The cat-

Figure 12. Wearable multitouch projector (Microsoft Research, 2012)

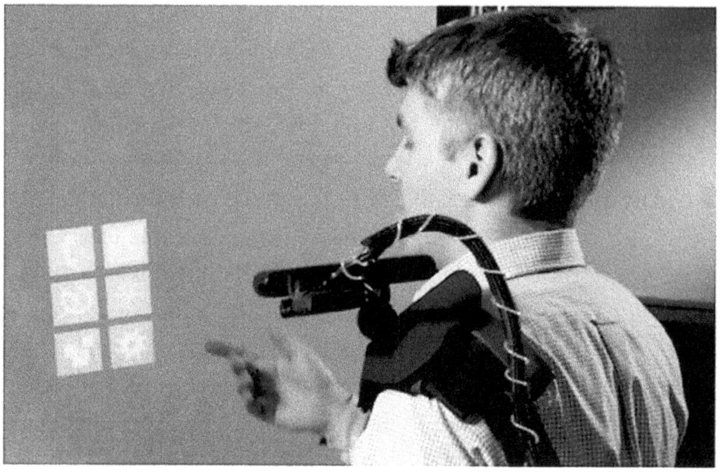

egory of wearables comes in the form of a hat with a visor using a fixed emitter infrared camera and, more recently, using special glasses. Eye-trackers of this type allow for observation and measuring eye movements with reference to the position of the head. Eye tracking is a great space in different fields of application, the most important aspects are: a) the connection between different measurements of motor activity of the eye and algorithms of text comprehension of written and spoken; b) use as active interface to control your computer through the eye. Being able to control a computer through the eyes is particularly useful in the case of diseases severely limiting and progressively worsening motor skills of a person. (Porta & Rastelli, 2013)

The most promising developments are in the field of clinical education (Doherty et al., 2013).

Wu concludes in an interesting clinical research study: "Wearable technologies such as Google Glass can be successfully integrated into simulation-based training exercises without disrupting the learners' experience. Data obtained from this integration can be utilized to improve debriefing sessions and self-reflection. Future research is underway and required to evaluate other potential uses for wearable technology in simulation-based training" providing excellent insights (Wu, Dameff, & Tully, 2014).

The possibility of integrated biometric sensors placed into wearables provides a significant boost to the scope of medicine. An example is in a video of the operating room, available on YouTube. It is an experimentation carried out by Philips in collaboration with Accenture. In Italy, in several experiments, doctors (surgeons and dentists) wear Google Glass to allow students to follow the operations from the doctor's point of view, while he has his hands free to work. It is necessary to reflect on many problematic issues. For example, Kelly and al. have defined ethical guidelines for the use of automated, wearable cameras in observational health behavior research (Kelly, 2013).

A topic of great interest is related to the disabilities. A research group at MIT (Massachusetts Institute of Technology) has created a special wearable device, Finger Reader (Figure 13), which can be very useful to visually impaired or people who are blind. It is a ring with a small video camera and a special software allowing one to synthesize vocally the written words in a book. Just wear it and slide your finger along each line of text in the book. A small sensor emits a vibration feedback to guide the finger in its path. Through a software for handwriting recognition, ring scans and returns text as audible audio

Figure 13. Finger Reader (Fluid Interface, 2014)

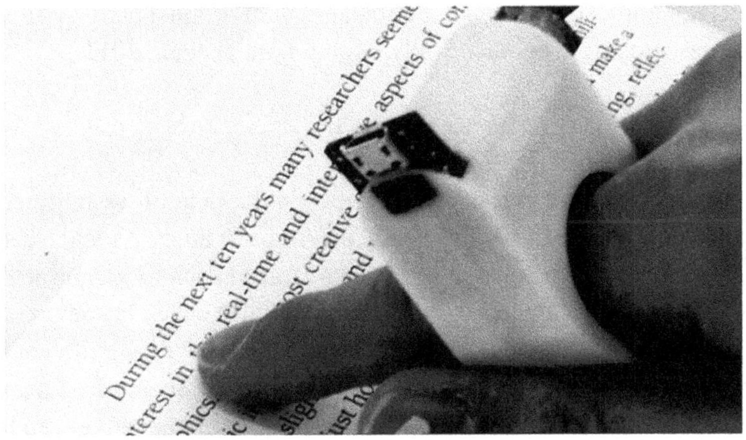

through the built-in speaker. This can be extremely useful in education, just think about the possibility of assisting blind students.

In this line of research is a project of an Italian start-up that, thanks to a process of crowdfunding, has developed Horus, a wearable electronic device developed to assist blind and visually impaired.

The device can be mounted on any pair of glasses, Horus observes reality, includes and describes it to the person, providing useful information using bone conduction. The device, in addition to advanced software, has a video camera able to recognize what is framed: a text or the face of a person. An external battery, carried in a pocket or bag, powers the device and guarantees a day of independence.

The user can interact with buttons, voice interactions or performing simple actions allowing the device to understand the context in which it is located, and then act accordingly.

A group of researchers at Second Sight was able to project patterns directly onto the retina of a blind patient. The retinal prosthesis system (Figure 14) includes a grid of sixty electrodes implanted on the innermost layer of the retina, a small video camera mounted on a pair of glasses, and a wearable computer, which processes video images and determines the stimulation current of each electrode in real time. These instructions are transmitted wirelessly to an antenna in the implant. The signals are then sent to the electrode array, which emits small pulses of electricity. These pulses can bypass the damaged photoreceptors and stimulate the retina's remaining cells, which transmit the visual information along the optic nerve to the brain. This process is intended to create the perception of patterns of light in which

Figure 14. Argus system (Second Sight, 2013)

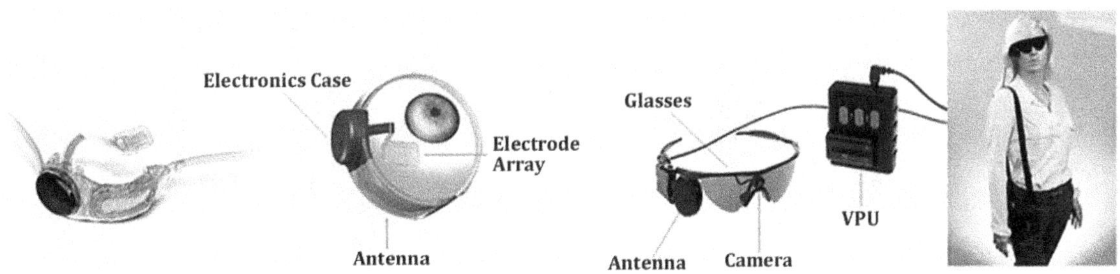

patients can learn to interpret as visual patterns. (da Cruz et al., 2013). To determine what extent subjects implanted with Argus can improve performance compared with residual native vision in a spatial-motor task you can read the research from another Study Group. (Ahuja et al., 2011).

Lifelogging

Wearable cameras can also be found in research activities in the field of lifelogging.

Lifelogging means digital acquisition of the daily activities of an individual, seen from the point of view of the person. (Gurrin et al., 2014). The term can be traced back to Vannevar Bush who, in 1945, imagined scientists with wearable cameras to record lab work. He called his device a memory extender (memex), a term some researchers also use today. Lifelogging can have a significant impact on our lives because it fixes, in a simple way, the memories. Memories can be shared in a narrow circle or made available on the network. The opportunity for people to meet again in a movie can help you remember events occurring during the day with more precision and reflect later on various aspects related to what has been done. These micro cameras (about the size of a few centimeters), is able to create a continuous life log, also called "narrative cameras" precisely for this characteristic. The photos, taken every thirty seconds and georeferenced using GPS, are organized in a time line in order to represent a video story of the life of the user. It is interesting to observe the evolution of the species, i.e., the increasing miniaturization of the object (Figure 15).

Lifelogging can also be used in preventive medicine. The SenseCam developed by Microsoft (Figure 16) is able to help you better understand the correlation between lifestyles and health. Digital eyes, hanging from the neck, register user behaviors in real time. After connecting the device to a computer, software is able to outline what lifestyles present a danger to the health of the subject. Another proof of the effectiveness of the "Big Brother" comes from research on neuropsychological performance (Silva, 2013).

With the camera for lifelogging, we can collect the behaviors of students in order to analyze processes and performance with statistical information. The analyses relating to learning could be crossed with

Figure 15. Evolution of narrative cameras (Glogger, 2015)

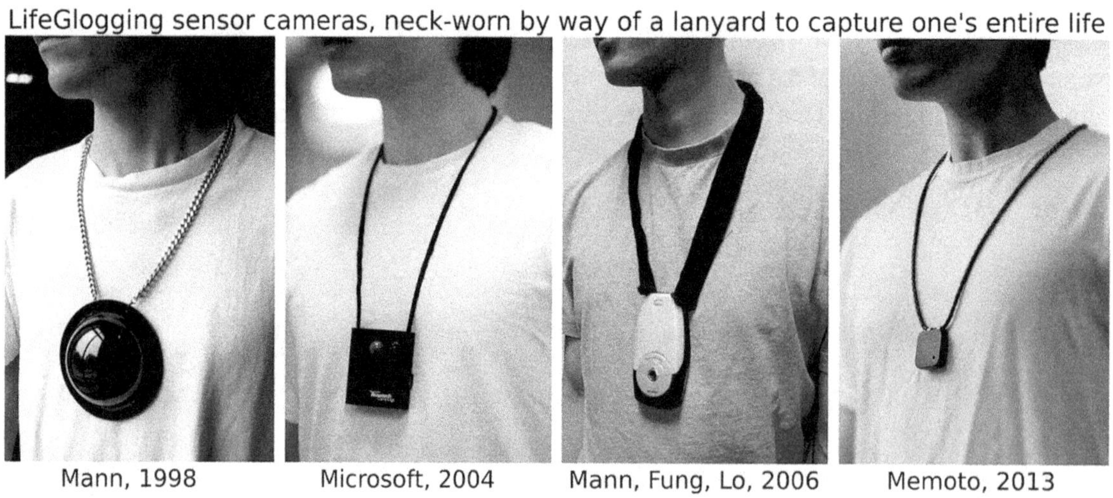

Figure 16. SenseCam (Microsoft Research, 2013)

the monitoring of lifestyles. In addition, these devices can offer assistance to those who have physical disabilities or help in monitoring distance learning.

Those using lifelogging tools and social networks can share content with those who have common interests. Any action or event of a person is, at least in theory, perpetually available in an archive. This is an interesting aspect, but also disturbing. The information, however, is at safety risk and not all users are aware of how this can be dangerous. If the data is accessible to third parties, there is a risk of profiling by external entities. As written by Anita Allen "We must hope that the changes in the quality of life affected by the proliferation of lifelogs will not result in a further deterioration of the taste for privacy or fewer legal privacy protections" (Allen, 2008). Optimists think we will not need the memory assist to remember our lives, they save the digital image archive that flows like a "second life", a true reflection of existence. For pessimists this digital information overload can become a threat to privacy when it is shared with other users on the web. Social scientists see an unprecedented opportunity. Wearable cameras have a great potential to understand the lifestyles of people in context. This is definitely useful to suggest how to modify behavior errors. We could define the users of these tools as voluntary guinea pigs of a great experiment. Indeed several studies have shown to revise (visually) the facts of the day, leading to improvements in autobiographical memory. "This study provides evidence that at least some kinds of cues captured by life-logging technologies, in this case SenseCam images, can be shown to provide effective links to events in people's personal past. Further, it suggests the automatic way in which SenseCam captures these images results in cues effectively triggering memory as images which people capture on their own initiative. In fact, with regard to the recollection of past events, these passively captured images may even cause people to remember more events than they would with their own actively-captured images". (Sellen et al., 2007).

A very interesting aspect, relating to the use of SenseCam, is the use for monitoring travel behavior. (Kelly et al. 2011). The mini camera is used to make a report of these trips, on foot or by bicycle. Journey duration derived from direct observation of travel behavior from time-stamped images appears to suggest over-reporting of self-reported. Therefore journey duration is much more reliable. The use of digital images can be useful both for researchers who want to study the behaviors of travellers and for those who plan interventions of active transport. Active transportation is an important behaviour in the fields of public health, environmental sustainability and transport planning.

Two Examples of Indirect "Wearability"

In the first case, the video camera can be used to automatically record and save video files in high definition directly on SD / MMC. The built-in microphone allows for audio recording as well. The use represents a particular case of "wearability" because, normally, the camera is placed on a car and, thanks to its own battery, or through connection with the car's cigarette lighter, it allows video recording. In this way, the driver can concentrate on driving and vibrations of the camera are greatly reduced.

Even a cheap video camera (Figure 17) allows stable and exciting images of driving in the desert (Figure 18).

Figure 17. Car video camera (Drivet, 2015)

Figure 18. Desert trip (Drivet, 2015)

The second example (Figure 19) refers to a tool called Connected Cycle Pedal (Kelly, 2013). As the pedals move, you activate the sensors to monitor activity in the saddle. You can record not only the location but also speed, slope, calories burned. The data can then be transmitted to a device and stored in an app.

A cyclist always runs the risk of having an accident. A video camera connected to an accelerometer can record what happens. In the event of an accident, they are able to preserve evidence of the claim. Science fiction? No, a U.S. company has built a sort of black box with these characteristics (BlackVue, 2013).

Issues

All the gadgets that can be defined as wearable, unlike the technological tools now part of our existence (computer, smartphone, tablet, etc.) tend to introduce a new paradigm. All digital objects require a high

Figure 19. Connected cycle pedal (Connected Cycle, 2015)

level of attention but the wearable can save a part of mental and physical resources. A wearable camera allows the subject to focus all his attention on the task and not on the management process. A second aspect is the wide range of applications. Potentially wearable cameras could integrate, at least in part, conventional camcorders and smartphones used for shooting. In many cases these are experimental prototypes. An example is Neurocam, the first camcorder functioning on the basis of the analysis of brain waves, presented in 2013 in Japan. It consists of a headset with a sensor to detect and analyze brain waves and connect to the iPhone, the functioning relies on brain parameters that start recording automatically as soon as the analysis suggests interest in something in our field of vision. Recorded content is then saved in an album freely available. It is also a predisposition an application to the analysis of the brain waves whose task is to quantify the level of user interest. As soon as initial scores exceed 60% recording starts automatically.

Two issues seem attractive: the importance of the video in education and the use of wearable devices for disabilities. The first theme interesting to report is the conclusions on the use of wearable cameras in learning research "In our discussion of specific cases of using wearable cameras and analyzing first-person-perspective video, we have concluded that using these cameras is important and valuable to research in learning and education. We encourage other scholars to consider wearable cameras and first-person-perspective video as valuable tools and possible solutions to the concerns in the field about the creation versus the collection of video data" (Lee, 2014). It might seem obvious the importance of the use of video in the school, in reality the impact of new technologies (especially those wearable) has, at least in Italy, a picture with few lights and many shadows. The images do not lack interesting experiences, but involves the use of conventional cameras or camera phones. The focus is centered on content and videos are used to develop a research idea and not to document a collective experience. For the second theme, one can remember an experiment using wearable devices in a museum. The Egyptian Museum in Turin, the second largest in the world after Cairo, has experienced the Google Glass 4 Lis. The goal is to translate the information on the most important artifacts in the Italian Sign Language (ISL) for the hearing impaired, thus ensuring an extraordinarily inclusive experience thanks to an avatar appearing on the small screen embedded in the glasses. The Google Glass has been used inside the museum to achieve three main objectives: technological innovation in museums, integration for the deaf, and a complete and addictive experience for the visit.

Controversies

A simple search on the Internet reveals the most controversial points related to wearable technologies. In summary, the debate is centered on Google Glass and the role of video cameras worn by members of the police. What about these two aspects? Leaving aside economic issues related to the cost, according to some researchers, Google Glass could reduce vision and cause blind spots. Moreover, these devices could pose serious privacy issues. One thing the boom of social media has taught us is that privacy is a valuable asset which can escape our control too easily. From this point of view, to place a small video camera on a pair of glasses may be a miscalculated risk due to security concerns.

In regards the use of cameras worn by police, this control method seems to be quite effective, at least in the opinion of many police officers and experts. The decision to permit this use may be a good idea if it decreases the use of force and increases the possibility, by the courts, to have the documentary material to be analyzed. On this issue, it is useful to focus on one of the empirical studies available. A search of the University of Cambridge in 2011 showed a decrease in the use of force by 59 percent and of grievances against officials by 87 percent. However, some criminologists argue the videos are partial because they are susceptible to different interpretations. A police officer, a lawyer or a simple citizen interpret a movie based on his experience. Regardless of objective limits as the choice of the angle of recovery, the technology does not allow us to analyze the context.

Problems

Technological changes always involve changes of a social nature. To understand these changes we coined new words: surveillance, coveillance, sousveillance. The first term is well-known, all cities are filled with video cameras in order to monitor the territory. Video surveillance uses a VCA (Video Content Analysis) software that automatically analyzes the images to extract relevant information to the user. This program can play a number of different functions: automatically detect the entry of a person into a given perimeter, suspicious objects, obstruction of emergency exits etc. VCA software, however, is not only used for safety; video analysis can also be used for monitoring the behaviour of people. The second term, coveillance, is the result of the development of social networking sites, people have adapted to practices of mutual observation or towards public entities. The third term, sousveillance, sometimes confused with the second, refers to a process in which the observed becomes observer; a sort of control from below to use against those who hold power. The difference between surveillance and sousveillance is that the few watching the many and the many watching the few. Sousveillance means "watching from below [...] Sousveillance, by its nature, tends to be distributed, so its efficacy ('swollag') would naturally be enhanced by widespread adoption, where its swollag is distributed through a large network, rather than smaller groups from the 'likes' of hacktivism" (Mann & Ferenbok, 2013).

This raises a number of issues summarized in the following quote: "The proliferation of small, easily concealed and readily networked digital cameras can be a headache for those trying to retain some degree of privacy, but they're a nightmare for those trying to keep hold of some degree of secrecy. But the world of the participatory panopticon is not as interested in privacy, or even secrecy, as it is in lies. A police officer lying about hitting a protestor, a politician lying about human rights abuses, a potential new partner lying about past indiscretions -- all of these are harder in a world where everything might be on the record." (Cascio, 2005).

One of the main problems raised by politicians, sociologists, journalists, etc. is related to privacy.

It is reasonable to assume a person would not later see an interception of their previous actions or words? What is the limit when technology offers us a more sophisticated and easily camouflaged tool?

An emblematic example of the contradictions related to this subject is given by the above-mentioned judgment of the European Court of Human Rights. It gave a ruling on the use of a hidden camera by journalists, giving reason to these and condemning Switzerland whose national courts had sanctioned a reporter for the dissemination of recordings without permission of the person concerned.

One final problem: the impact of connections from mobile to mobile (and wearable devices). M2M refers to applications enabling wireless systems to communicate with similar devices. The wearable devices are considered a sub-segment of the category of M2M connections. In 2014, the wearable devices have generated an average monthly traffic of 6 times greater than mobile base. To give an example of using of wearable high-end devices, the live streaming of the GoPro camera could generate about 5MB of mobile data traffic per minute.

SOLUTIONS AND RECOMMENDATIONS

What has been written, relative to the problems associated with the technologies presented, deserves a look by "technicians" but also needs a broad debate involving all. We see a kind of pendulum in which, from time to time, the population is divided between privacy and security, uncertain between confidentiality and the need for total transparency. The increasing use of the Internet is showing us we are inconsistent: we all want more privacy, but in fact, we leave every day more and more tracks and data in the network, ignoring how the data may be used. New technologies can increase this dilemma and a solution is not easy. Only a collective discussion can enact laws to ensure the right balance.

This problem is present in devices such as Google Glass. The reasons suggest a reflection linked to "disruptive technology", a prosthetic technology not only interfacing with the outside world but something much bigger in terms of knowledge and understanding. Video and data collected by millions of people can be stored and archived by a company like Google a great power. In Italy, a special commission launched a reflection on the code of conduct on such delicate matters. The Group has initiated a discussion on fingerprinting devices (use of elements of information aimed at identifying and tracking users), and wearable computing (the case of Google Glass). In a 2013 letter, Google was invited to engage in a dialogue with data protection authorities to clarify the many profiles of privacy. The Guarantors specifically requested explanations on the measures pro privacy adopted by the developers of the product, the types of data collected by Google through Google Glass and shared with third parties, the purpose of treatment in place, and any risk assessments put in action by Google. The controversy raised, related to aesthetic reasons, privacy issues, contradictory aspects on the topic of usability, have perhaps contributed to the fact the original Google Glass Explorer Program has ended. There is division among analysts as to whether this is a step back, even if some say Google is working on a new version with the Italian company Luxottica. Probably Google, but other competitors will return to the charge with similar products in a near future. Several companies have already begun to market products following this track. Probably it is risky to draw conclusions or to hazard solutions, however, it is certainly possible for some reflection. The first is in the area of new technology and how it is still evolving and consumers are probably not willing to pay for expensive leaps forward; perhaps this is why the action camera is experiencing positive results. The second is we need to distinguish the role of the early adopters from the consumer mass. The third is in any case the wearables are the future and experimentation in the areas highlighted in the text can only lead to great results.

CONCLUSION

This is a classic case in which it is difficult to draw conclusions because it is a growing sector, not only for the evolution of technology, but also because a specific paradigm has not been established.

Retracing the application areas examined in the text it is possible to note that, in most cases, the most interesting developments relate to professional applications of wearable cameras. It is no coincidence the examples given in the text refer, in large part, to sports professionals, reporters, researchers and police forces. Wearable camera adoption will be the strongest in areas where the context in which the camera is used is clear and specific. Currently Google Glass is expensive and is intended for a limited audience, at least in Italy. This wearable technology is not for everyone: most users have been developers who need to explore its functionality. The Google Glass explorers program was halted while working with industry partners to increase its usability. There are rumors of a new and improved version in the works. At this time large retailers and consumers must still wait. There are numerous other smart glasses in the works by competitors with a variety of functions.

We have a different speech for the "action camera" now making up a significant share of the Italian market. Recall that on YouTube videos are uploaded every minute with approximately three videos tagged action cam and the GoPro channel scored about one million subscriber and 260 million views. The size of the phenomenon is likely to lead us to a consideration: there is no contradiction between this data and what has been said about the role of professionals? To answer this question we must keep the two worlds separate. In the first we have a huge audience of people who, in a more or less technically satisfying way, want to express personal creativity or, more simply, tell the world their own existence. In the second case we are dealing with a more restricted world but using professional media. For example world-renowned filmmakers have used un'action GoPro cam to shoot some scenes of their movies.

The spread of wearable cameras seem to open interesting perspectives to the companies involved, but also to some startups able to provide applications and services. However this process must take into account several factors: 1) Cost: the items must be within the reach of customers with budgets not excessive; 2) Value: the wearable must have added value, provide the user with functionality not otherwise available; 3) Fashion: the product must not exceed the critical threshold separating the technology enthusiast from the mass audience; Innovation: the technological development has to solve some problems such as battery life and improve the quality of products in terms of apps available.

As we wrote in the previous sections we provide a broad picture of the topic but, obviously, it is a theme evolving. Time will tell which applications will develop and which will be abandoned.

Certainly we can assume that the constant miniaturization and demands of stakeholders will lead to a growth in the sector.

Supplemental Website Resources

Introduction

- http://www.reuters.com/article/2015/04/22/research-and-markets-idUSnBw226530a+100+BSW20150422
- http://ieeessit.org/technology_and_society/recent-contents/
- http://www.cs.indiana.edu/~kapadia/papers/sensitive-lifelogs-chi15.pdf
- http://www.wareable.com/headgear/how-to-choose-the-best-wearable-camera
- http://www.educatorstechnology.com/2013/12/augmented-reality-transforming-education.html
- http://ilpiccolo.gelocal.it/sport/2015/03/12/news/i-iesta-il-primo-calciatore-telecamera-1.11033906
- http://www.yankodesign.com/2011/11/18/two-finger-camera/#sP8ei5A9YfmYTKIx.99

Background

- http://www.idc.com/getdoc.jsp?containerId=prUS25519615
- http://www.fotografidigitali.it/news/videocamere-indossabili-30-milioni-di-pezzi-nel-2020_56928.html

Sports

- http://www.wareable.com/cameras/best-wearable-action-cameras
- http://corriereinnovazione.corriere.it/coverstory/2015/20-gennaio-2015/pista-sci-diventa-smart-230868714052.shtml

Spying

- https://epic.org/privacy/google/glass/
- http://www.articolo21.org/2015/03/fino-a-che-punto-e-lecito-pubblicare-filmati-girati-con-telecamere-nascoste/

Police and Army

- http://www.vievu.com/
- http://www.dailymail.co.uk/news/article-2304978/From-people-brought-Taser-headcam-turns-policeman-Robocop.html
- http://www.npr.org/sections/alltechconsidered/2015/01/22/379095338/how-police-body-camera-videos-are-perceived-can-be-complicated
- https://consaproma.wordpress.com/2014/10/02/telecamere-per-luniforme-delle-forze-di-polizia-via-libera-allutilizzo/
- http://www.rivistastudio.com/standard/guerra-pov-isis-afghanistan/
- http://mikeshouts.com/striker-ii-helmet-mounted-display-by-bae-systems/
- https://www.youtube.com/user/FUNKER530

Education

- https://net.educause.edu/ir/library/pdf/eli7102.pdf
- http://www.teachthought.com/technology/36-ways-to-use-wearable-technology-in-the-classroom/
- http://youtu.be/WLoMecZ80BQ

Health and Disabilities

- https://www.youtube.com/watch?v=H9_piiD_j0M
- http://www.wired.it/attualita/tech/2014/04/16/google-glass-sala-operatoria/
- http://www.sordionline.com/index.php/2014/05/i-google-glass-in-italia-alcuni-casi-duso/
- http://www.sociale.it/2014/08/01/finger-reader-lanello-che-legge-libri/
- http://horus.technology/it/
- http://www.secondsight.com/

Lifelogging

- http://www.elearning-conf.eu/docs/cp14/paper-9.pdf
- http://en.wikipedia.org/wiki/Lifelog

- http://publik.tuwien.ac.at/files/PubDat_192440.pdf
- http://research.microsoft.com/en-us/um/cambridge/projects/sensecam/publications.htm

Indirect "Wearability"

- http://cameras.reviewed.com/features/6-reasons-you-should-own-a-dashcam-2
- http://www.macitynet.it/ces-2015-con-connected-cycle-pedal-la-bicicletta-e-sempre-connessa/

REFERENCES

Ahuja, A. K., Dorn, J. D., Caspi, A., McMahon, M. J., Dagnelie, G., Dacruz, L., & Greenberg, R. J. et al. (2011). Blind subjects implanted with the Argus II retinal prosthesis are able to improve performance in a spatial-motor task. *The British Journal of Ophthalmology*, *95*(4), 539–543. doi:10.1136/bjo.2010.179622 PMID:20881025

Allen, A. L. (2008). Dredging up the past: Lifelogging, memory, and surveillance. *The University of Chicago Law Review. University of Chicago. Law School*, 2008, 47–74.

Bravo, F. (2012). Giornalismo investigativo, vittime e diritto di cronaca *Rivista di Criminologia, Vittimologia e Sicurezza*, 6(2). Bologna: IT.

Cascio, J. (2005). The rise of the participatory panopticon. *World Changing*. Retrieved from http://www.worldchanging.com/archives/002651.html

da Cruz, L., Coley, B.F., Dorn, J., Merlini, F., Filley, E., Christopher, P., ... Dagnelie, G.(2013). The Argus II epiretinal prosthesis system allows letter and word reading and long-term function in patients with profound vision loss. Retrieved from http://bjo.bmj.com/content/early/2013/02/19/bjophthalmol-2012-301525.full

Doherty, A. R., Hodges, S. E., King, A. C., Smeaton, A. F., Berry, E., Moulin, C. J., & Foster, C. et al. (2013). Wearable cameras in health: The state of the art and future possibilities. *American Journal of Preventive Medicine*, *44*(3), 320–323. doi:10.1016/j.amepre.2012.11.008 PMID:23415132

7 things you should know about wearable technology. (2013). *EDUCAUSE*. Retrieved from https://net.educause.edu/ir/library/pdf/eli7102.pdf

Gurrin, C., Hopfgartner, F., Hurst, W., Johansen, H., Lee, H., & O'Connor, N. (2014). MultiMedia Modeling. *Proceedings of the 20th Anniversary International Conference MMM '14* (Part 1). Berlin, GE: Springer

Jewitt, C. (2012). An Introduction to Using Video for Research. *NCRM Working paper*. London

Kelly, P. (2013). *Assessing the utility of wearable cameras in the measurement of walking and cycling* [Doctoral dissertation]. University of Oxford.

Kelly, P., Doherty, A., Berry, E., Hodges, S., Batterham, A. M., & Foster, C. (2011). Can we use digital life-log images to investigate active and sedentary travel behaviour? Results from a pilot study. *The International Journal of Behavioral Nutrition and Physical Activity*, *8*(1), 44. doi:10.1186/1479-5868-8-44 PMID:21599935

Kelly, P., Marshall, S. J., Badland, H., Kerr, J., Oliver, M., Doherty, A. R., & Foster, C. (2013). An ethical framework for automated, wearable cameras in health behavior research. *American Journal of Preventive Medicine*, *44*(3), 314–319. PMID:23415131

Lahlou, S. (2011). How can we capture the subject's perspective?: An evidence-based approach for the social scientist. *Social Sciences Information. Information Sur les Sciences Sociales*, *50*(4), 607–655. doi:10.1177/0539018411411033

Lalli, F., & Prunesti, A. (2013). *Wearable: Connect you with everything*. IQUII Social & Mobile Factory.

Lee, V. (2014). Learning Technologies and the Body: Integration and Implementation. In *Formal and Informal Learning Environments*. New York, NY: Routledge.

Mann, S. (1997). "Smart clothing": wearable multimedia computing and "personal imaging" to restore the technological balance between people and their environments. *Proceedings of the fourth ACM international conference on Multimedia* (pp. 163-174). ACM.

Mann, S., & Ferenbok, J. (2013). New Media and the Power Politics of Sousveillance in a Surveillance-Dominated World. *Surveillance & Society*, *11*(1/2), 18–34.

Mateescu, A., Rosenblat, A., & Boyd, D. (2015). Police Body-Worn Cameras (Working Paper). NYC: Data & Society Research Institute

Miller, M. (2015). *The Internet of Things: How Smart TVs, Smart Cars, Smart Homes, and Smart Cities Are Changing the World*. Indianapolis, IN: Que Publishing.

O'Driscoll, G. (2014). Your Definitive Guide to Wearables. Dublin, IR: HomeMentors.

Otrel-Cass, K., Cowie, B., & Maguire, M. (2010). Taking Video Cameras Into The Classroom. *Waikato Journal of Education, 15(2)*. doi:10.15663/wje.v15i2.117

Porta, M., & Rastelli, S. (2013). Lo studio dei tracciati oculari (eye-tracking) nella ricerca sul linguaggio In La ricerca sperimentale sul linguaggio: acquisizione, uso, perdita (pp. 165-197). Pavia University Press.

Sellen, A. J., Fogg, A., Aitken, M., Hodges, S., Rother, C., & Wood, K. (2007). Do life-logging technologies support memory for the past?: an experimental study using sensecam. *Proceedings of the SIGCHI conference on Human factors in computing systems* (pp. 81-90). ACM. doi:10.1145/1240624.1240636

Shum, H. Y., Liao, M., & Zhang, S. (Eds.), (2001, October 24-26). Advances in Multimedia Information Processing-Pcm 2001. *Proceedings of the Second IEEE Pacific Rim Conference on Multimedia*, Beijing, China (*Vol. 2*). Springer Science & Business Media.

Silva, A. R., Pinho, S., Macedo, L. M., & Moulin, C. J. (2013). Benefits of sensecam review on neuropsychological test performance. *American Journal of Preventive Medicine*, *44*(3), 302–307. doi:10.1016/j.amepre.2012.11.005 PMID:23415129

Toh, P. (2013). *The New Age of Consumer Wearables*. CreateSpace Independent Publishing Platform.

Wassom, B. (2014). *Augmented Reality Law, Privacy, and Ethics: Law, Society, and Emerging AR Technologies*. Waltham, MA: Elsevier.

Wu, T., Dameff, C., & Tully, J. (2014). Integrating Google Glass into simulation-based training: Experiences and future directions. *Journal of Biomedical Graphics and Computing*, *4*(2), 49. doi:10.5430/jbgc.v4n2p49

KEY TERMS AND DEFINITIONS

Action Cameras: Small cameras with high standards of quality in recording, built to be "worn" in various contexts for recording images.

Augmented Reality: The enrichment of human sensory perception through information, generally handled and conveyed electronically.

Hidden Camera: Small camera normally used for concealment of images recorded.

Life Log: The diary of own lives with images, often posted on the Internet.

Narrative Camera: A device of lifelogging, its primary task is capturing random moments.

Multi-Touch Projector: An in depth-sensing and projection system using multi-touch applications on the display surface.

Second Life: An online virtual world.

Ski Goggles: A system capable of recording all the parameters of skiing.

Smart Glasses: Newer technological glasses incorporating a bright high-definition display.

Wearable Camera: Cameras placed on the body.

Wearable Technologies: Electronic devices which can be worn.

Chapter 7
Wearable Technologies in Academic Information Search

Robert Gibson
Emporia State University, USA

ABSTRACT

Wearable technologies and appliances are making inroads into a variety of consumer and commercial applications, including leisure and entertainment, health care, and gaming. A natural evolution of the technology is in academe where faculty and students have begun exploring the possibilities of the technology in a variety of settings, most visibly in libraries where the process of seeking information using such devices holds significant promise. This chapter provides an inventory of the state of wearable technology, its challenges, its possibilities, and how it might be used in academe, including a study regarding the ability to access common library indexes using two wearable appliances: Google Glass and a smart watch.

...wearable technology is not limited to just the wrist. There's a whole lot more. The whole sensor field is going to explode. It's a little all over the place right now. With the arc of time, it will become clearer.

- Tim Cook, Apple CEO

INTRODUCTION

As wearable technologies become more commonplace in society, the capabilities of the devices is likely to improve. While generally limited to basic utility, including checking email, calendaring, messaging, checking weather conditions, etc., the functionality of these devices is expected to drastically improve in the near future. Already we've seen the potential of wearable technology with the now ill-fated introduction of Google Glass Explorer which was introduced without a clear market segmentation – a hallmark of many Google product launches. However, it's within reason to expect these devices to have many

DOI: 10.4018/978-1-5225-0069-8.ch007

of the same capabilities and characteristics as Glass and other mobile technologies – including built in video, access to search engines, etc. Wearable technologies will also become more discreet. The form factor will likely be smaller and less obtrusive than glasses and headgear when worn – sometimes directly embedded into jewelry, clothing, and other fashion. The use of these discreet devices will range from recreation to fitness to travel to health care. Some health screening prototypes are being developed that take the form of pills and other investigable components. Other systems will be permanently implanted chips that provide real-time sensory feedback no matter the environmental context. Still others are being developed that can be used in conjunction with mass transit systems.

This chapter explores the current state of wearable technologies, where it's being used, and how academia might be able to leverage the technology in the context of information search – a common activity conducted by both students and faculty. An analysis of common library search engines is provided in the context of using wearable technology. It's safe to say that the technology is not yet mature enough to be used effectively in most common information search processes. The indexes themselves have not yet built out their infrastructure to support these types of devices – likely because there isn't yet a critical user mass. However, there is considerable promise. Whereas current search processes are relatively two dimensional in nature, future wearable devices will likely provide information search to be conducted in a third dimension using embedded RFDI sensors within the library that provide the researcher related resources, author biographical information, and even media that supports the query.

BACKGROUND

Wearable technologies are broadly defined as unobtrusive, miniaturized sensors attached directly to the body or garments. These sensors can even be embedded directly into the fabric of garments (Bonato, 2005). In recent years wearable technologies have become part of the larger ontology called *The Internet of Things* (IoT) which is broadly defined as a world of connected devices, objects, vehicles, machines, consumer durables, clothing, and other components all hooked to a network (Kranenburg, Anzelmo, Caprio & Dodson, 2011).

Increasingly, more people are acquiring and using wearable technologies and appliances; however exact adoption metrics remain elusive given its relatively recent introduction into the consumer market. In fact, EDUCAUSE (2013) still considers wearable technology "experimental". Perhaps the best evidence regarding adoption metrics are provided by Mary Meeker from Columbia University who provides a macro snapshot of growth rates as of 2013 (Figure 1). Cisco, Inc. forecasts there will be more than 50 billion such devices connected to the Internet in just five years (Evans, 2011). ABI Research has indicated the wearable computing device market will grow to an astounding 485 million annual device shipments by 2018 (OPC, 2014). Forbes magazine projects by 2017 the revenue from these devices will exceed $20B (Sabhlok, 2013) making this a very lucrative growth market.

As wearable devices become more commonplace within the framework of the *Internet of Things*, their use in a variety of academic disciplines will likely begin to challenge traditional instructional technologies and methods of research. In particular, ready access to information - a hallmark of academe - will likely be influenced, if not fundamentally shaped using these types of devices and appliances. This shift may also influence libraries and other traditional curators of information. The ability to instantly gain access to information based on geo-positioning, voice-activated commands, body/Radio Frequency Data Identi-

Figure 1. The internet of things growth trajectory (Columbia Business School, 2015)

Mary Meeker (KPCB), Internet Trends 2014 Report: "Biggest Re-Imagination of All = People Enabled with Mobile Devices + Sensors Uploading Troves of Findable and Sharable Data"

Cisco: By 2020, there will be **50bn** internet connected devices in the world.

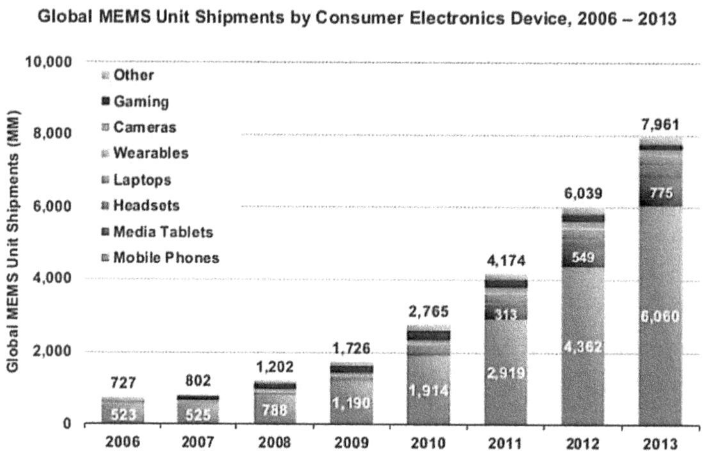

fier (RFDI) sensors, tactile input, gestures, heads-up displays, and other discretely transportable input/output sensors allow users to access a variety of academic resources in locations never before considered.

Although most academic databases and other information repositories are commonly available online, and although mobile technologies capable of accessing those resources are nearly ubiquitous, imagine an ecological field trip where students can conduct research activities using devices and sensors attached to their person. These "always on" devices and sensors can instantly access resources and information in any number of situational locations. Wind, humidity, temperature, barometric pressure, precipitation, and other atmospheric conditions can be captured and exchanged with cloud based storage systems in a continuous state. Smart phones, while arguably powerful, are generally not well-suited for all types of environmental elements and are generally not suited to capturing data in a continuous, always-on state. Certain wearable technologies, on the other hand, can easily be used underwater, in extreme weather, and in other conditions beyond that of traditional mobile media. One company, for example, has designed and developed a wearable technology used by swimmers to measure how they move through water (sporttechie.com, 2014). Still other wearable technologies are being introduced in health care environments where handling and operating traditional handheld devices can lead to the spread of bacteria (Luckowitz, 2014).

Aside from the momentum in consumer and commercial market applications, most wearable technologies and appliances are deeply rooted in academic research laboratories. Somewhere beyond the flashy storefronts and tech blogs, early prototypes were borne in labs and testing facilities of leading technology research labs and university testing facilities. In fact, one of the first digital touch prototypes was developed by a Japanese academic researcher years before Apple introduced the iPhone (Wagstaff, 2015). Surprisingly, some of the earliest mobile and wearable prototypes actually appeared in the 1980s – long before today's browsers and commonplace Internet access (Racoma, 2013).

What distinguishes wearable technologies and appliances from other forms of mobile, wireless, and handheld technology is its relative discreetness, portability, and functionality. These technologies are often difficult to detect when worn by an individual. Wearable technologies also allow the user to interact with the Internet using input methods not necessarily requiring the user to hold or tactilely control the device - which might be beneficial in certain circumstances. An archeologist or geologist, for example, might be able to use wearable technologies while at an excavation site to retrieve data samples. The conditions of dirt, wind, rain, and heat may not lend themselves to using a smart phone or laptop in the field, but a lens or heads-up based technology built directly into specialized apparel or headwear might be employed to access research resources using voice, head control, or gesture-based commands. Another distinguishing characteristic is the reduced time between the user's intention to perform a research task and the ability of the device to retrieve the desired information. A common Internet search using a wearable device, for example might take a user a second or two to formulate a conditional search string, while simultaneously performing another task. Wired magazine refers to these as *microinteractions* (Starner, 2013).

While not necessarily research-oriented, some commercial companies are introducing wearable technologies such as ski goggles with built-in Internet-based displays. These always-on displays can provide information on everything from snow conditions, to the skier's location on the mountain or slope, to speed and elevation - all available through a built-in display embedded directly inside the lens (Wasik, 2013). This type of technology provides hands-free access to real-time data. Future uses for lens-based technologies may appear in any number of occupational or recreational devices.

Gamers and game-makers are also getting into the wearable act. Researchers Christiaan Ribbens and Patrick Kersten (2015) developed a prototypical persuasive game called *Spooky* that wraps the players into a digital environment where their physical person becomes a game participant. Using specially designed clothing with embedded sensors, players can embark on a series of paranormal adventures.

A more utilitarian application of wearables can be realized in the field of disability services and support. Starner (1998) demonstrated that wearable and desk-based video can be used to help recognize sentence-level continuous American Sign Language with a rather astonishing 98% degree of accuracy. In this study, the wearable camera was able to decipher sign language into digital speech and speech recognition systems. Other research into wearable technology in the area of disability services has resulted in some remarkable findings. For example, Ehrenkranz (2014) found that paraplegic and quadriplegic patients, and patients with severe forms of muscular dystrophy were able to effectively engage in Internet-based activities using wearable technologies. The patients were able to use head (nodding) and eye motions (blinking), and voice activation to control the device without the use of their hands, arms, or fingers. Given that 15% of the world's population identifies as having a physical disability, this is a significant and important development. Even patients suffering from the effects of Parkinson's Disease were able to effectively leverage the technology.

Through a Google, Inc. grant, a New York based dance instructor named David Leventhal commissioned an app called *Moving Through Glass* that assisted Parkinson's suffers by providing a series of lens-based extension exercises. These exercises were rooted in ballet and modern dance movements and are common to dancers who work themselves through a series of progressive warmup steps and routines. Patients wearing the technology are exposed to a variety of warm-ups, balance routines, walking routines, and other systematic steps that provide visual assistance to Parkinson's sufferers (CNN, 2015).

Seemingly futuristic at the time it was launched in 2013, Google Glass provided a glimpse as to the possibilities of wearable lens technology. While Glass is no longer manufactured as of this writing, its intriguing potential was that the wearable technology can 'be there when you need it, and gone when you don't need it.' The ultimate desire of wearable technology is it actually helps 'get technology out of the user's way' (Starner, 2013). Glass provided ready access to the Internet, apps, web conferencing, and other utilities through a glass projection lens. A particularly fascinating use of Glass was demonstrated by Andrew Vanden Heuvel, a virtual high school physics teacher and community college faculty member, who wore his Google Glass while touring the CERN laboratory in Switzerland. Using wearable technology, he was able to host a live webinar with his students who were located in the United States while he rode a bicycle through the corridors of the research facility and demonstrated the nuclear super collider in real time (Scott, 2013).

To encourage broader adoption of wearable technologies, the challenge ahead will be to find market differentiation from smart phones and other traditional mobile technologies. What would entice someone to purchase and utilize wearable technologies when they may already own a smart phone, tablet, or other handheld/mobile product? In the general commercial market the differentiating factor may be fashion. An Internet-aware device disguised as a bracelet or ring may provide both form and function. In fact, these fashionable devices can be quite powerful. Smart rings, for example, provide gesture-based commands controlling smart phone applications, check email, and even pre-loaded with money to pay for mass transit fares. The user simply holds the ring up to a fare sensor to tabulate the transaction fee (Gonzalez, 2014). The latest smart watches also provide telephony capabilities. A user does not necessarily need to carry a phone at all in order to make a call. Fitness bands provide access with health and wellness data, including steps, heart rate, speed, etc. using a relatively discreet wrist band. Sensors are becoming extremely small and discreet; in fact, they will be included in all manner of consumer products – even household appliances.

Already, researchers are considering the future of embedded sensors; that is, microprocessors surgically implanted under the skin or directly into the cornea, or ingested in the form of a pill (Figure 2). These types of technologies would be able to transmit patient data to a health care provider in real time for remote patient monitoring (Niewolny, 2013). Microsoft, for example, has partnered with the University of Washington to create a type of wearable contact lens that can serve as a prosthetic device for sight-impaired individuals. The lens will be able to monitor glucose levels in the body and transmit that information back to the patient's health care provider (Woda, n. d.)

Figure 2. Ingestible wearable technology of the future (Quora, 2014)

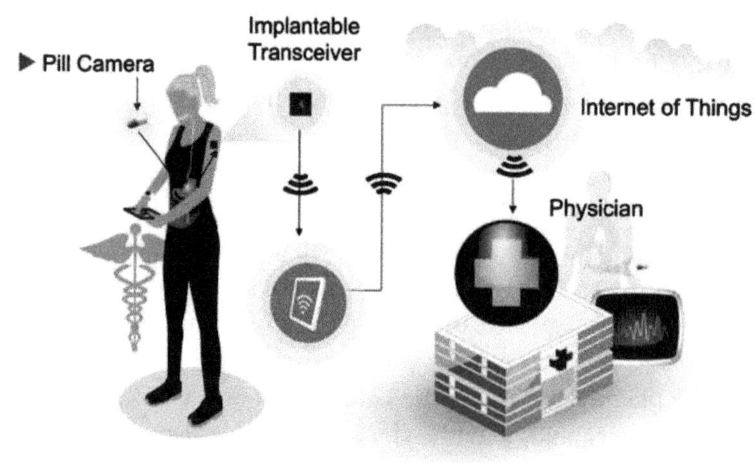

From gaming, to health care, to leisure and recreation, to biometrics, to academia, wearable appliances are making several inroads. As with any new technology, applications and utilities often follow the product introduction. What might be initially intended as a leisure or entertainment-based product often is later found to be quite effective for other activities, including academe or health care. This chapter will consider the use of wearable appliances in academic information seeking and inter-library functions. This survey includes the current state of the technology, innovative uses, and the future of wearables in academic libraries.

Characteristics of Wearable Technology

Wearables share certain characteristics distinguishing them from traditional handheld mobile media and other forms of portable technology. As illustrated in Figure 3, wearables are typically voice and/or gesture controlled, are always on, are typically connected to the Internet at all times, and are environmentally aware. That is, they can be used for geo-positioning and geo-spatial awareness. These unique features provide users with the ability to easily and unobtrusively access the Internet from nearly any location. As we'll explore later, there can be unintended consequences from wearing a device accessing data in a continuous state.

One might argue smart phones share many of the same characteristics. The primary difference being that wearable technologies and appliances are typically less obtrusive and distracting than a phone, tablet, glasses, or a small form factor computer. Hence, the technology is normally difficult to detect. Researchers at MIT's Media Lab (1997) provided a brief overview of wearable technology years before the term became commonplace. According to the MIT researchers, the following characteristics separate wearables from smart phones, smart watches and the like; although the lines are beginning to blur as the technologies begin to include similar functionality:

Figure 3. Characteristics of wearable technology (Columbia Business School, 2015)

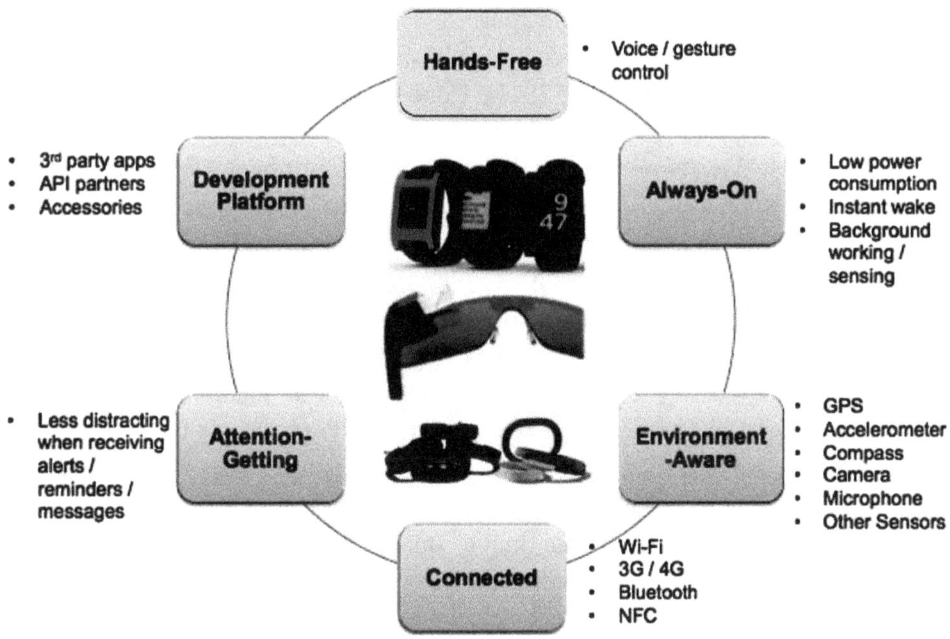

1. **Portable While Operational:** Meaning, of course, the technology is functional and operational in a variety of transportation modes.
2. **Hands-Free Operation:** Keyboards, screen gestures, tactile input, etc. are normally required for most mobile technologies. However, wearable technologies can be operated realtively hands-free.
3. **Embedded Sensors:** The sensors can provide spatial-recognition data, microphones and video, and other forms of real-time information input-output.
4. **Immediate, Unobtrusive Access:** Wearable technologies should be able to convey information and communications immediately and in nearly any condition. Email, for example, might be readily accessible using a wearable device – as opposed to opening an email client on a computer.
5. **Always On:** Wearable technology is not normally something that is turned on/off or requires the users to access a secondary device.

Wearable technologies are exemplified by being much more sophisticated than traditional handheld mobile technology. As indicated earlier, this technology can provide sensory and scanning features not typcially a characteristic of smart phones. This provides biofeedback and real-time tracking of physiological functions (Wearable Devices, 2014).

Researchers also envision a variety of wearable technology that can be worn on or embedded into the body. For example, head-based technology can be used to record brain activity; microscopic technology can be embedded into such things as contact lens' and fashion eyewear; apparel-based technology can be used for health monitoring; bio-monitoring devices can track vital signs and other diagnostics can be surgically implanted; wristbands can provide data regarding bodily functions; foot or ankle bands can provide power, speed, and motion data; and aural sensors can provide data via implanted earbuds. It's foreseeable that users will wear multiple appliances simultaneously – each with a different characteristic and each used for transmitting or collecting different types of information.

Challenges of Wearable Technology

In order to support this new generation of wearable devices, a completely new wireless transmission standard called Wi-Bo (wireless body) has been proposed specifically for the anticipated growth of wearable technologies and appliances (Pultarova, 2015). Conceivably people will need to support multiple connected devices at the physical layer. The problem arises in how all these devices will be supported on a single network, or if those devices will even interconnect with traditional wireless networks. Whereas, Wi-Fi can be used to interconnect a variety of traditional devices over short range, Wi-Bo would instead use a new standard used to interconnect devices that are either mounted on or embedded into the body (Figure 4). Wi-Bo would not interfere with a user's traditional Wi-Fi network if used in the same proximity (Sheehy, 2015). Hence, a user could potentially use a variety of wireless devices without the problem of saturating a single network or taxing the available bandwidth.

A bigger challenge for wearable technology industry will not be the connectivity constraints, but rather the privacy and copyright issues surrounding their use. Transmitting or collecting information from a variety of sources can lead to serious privacy and data control issues. Unlike smart phones and other forms of mobile technologies that typically control access to resources and utilize encrypted passwords, wearable technologies might be in a continuous state of information transmission and collection. As it relates to academe, for example, how will a library control who has access to its data without any sort of authentication? Would those resources be available to anyone who is using a wearable technol-

Figure 4. Proposed wireless standard – Wi-Bo (Gartner Research, 2014)

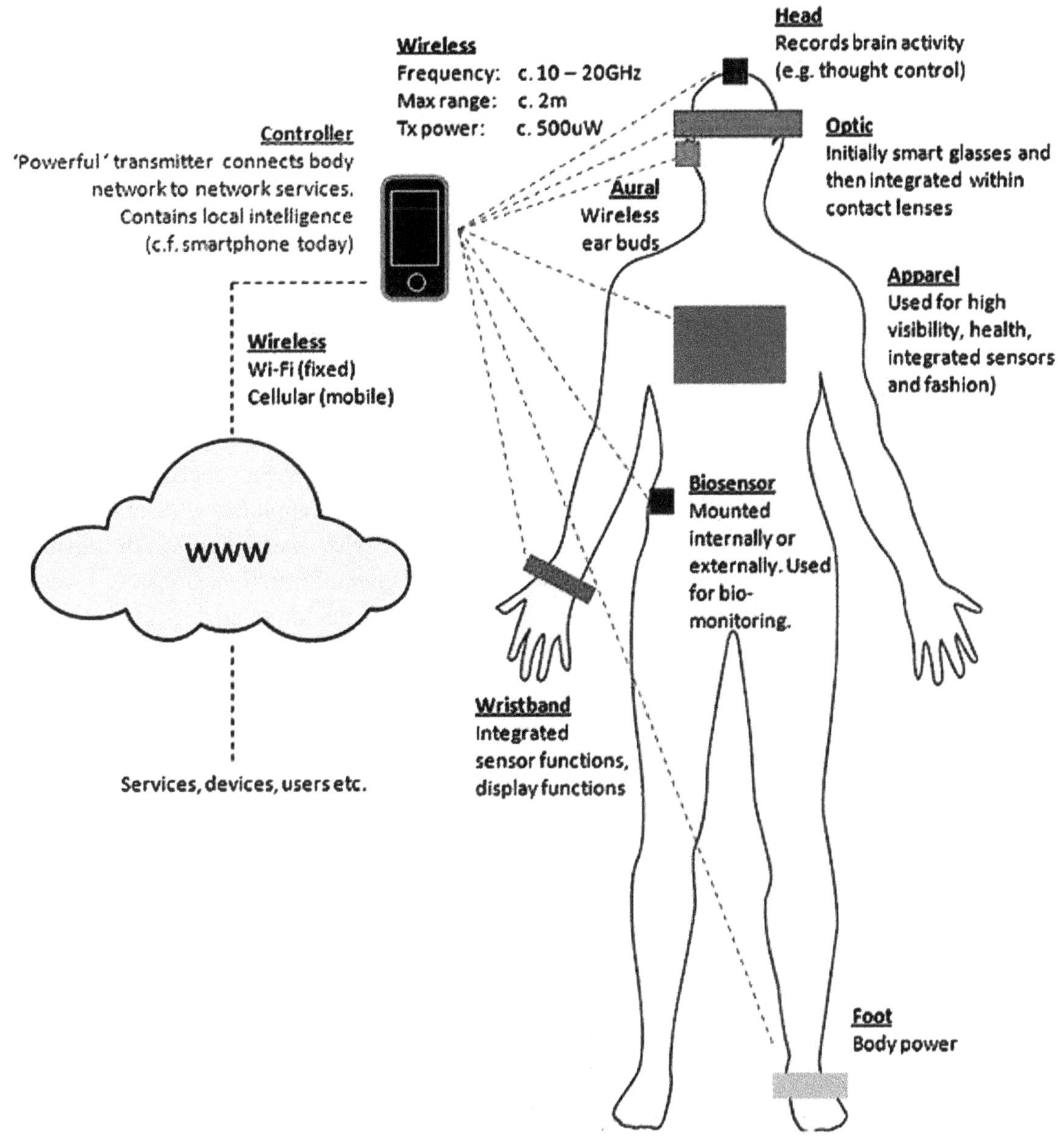

ogy? Keep in mind libraries typically contract with data providers by headcount for many of its indexes and academic journal subscriptions. Should information be available from any type of wearable device and for any patron? What are the consequences of transmitting the user's profile data back to a storage warehouse in order to collect user profile information? These are only some of the questions beginning to emerge as these devices appear in the commercial market.

These same concerns are magnified in health care. Will data security be controllable in the health care spectrum where stringent laws govern the transmission and collection of patient health care information? Will transmission of vital health care data be secure? Wearables expose a world of possible privacy and data security concerns, including surveillance, privacy, data control, and access. Already, research is underway to combine data generated by sensors within the current generation of smart phones so as to identify and authenticate individuals on a network (OPC, 2014).

Finally, product fallout is also a concern. Forrester Research's J.P. Gownder (CNN, 2014) indicates that between 80% and 90% of wearable technologies will likely fail within two years following product launch. This is primarily because the appliance is unusable, bulky, and has an unattractive form factor; there is no clear or identifiable advantage over other forms of mobile technology; the social stigma is still high; the technology is still reserved for the early adopters and hasn't yet reached mass appeal; the device has high power consumption and short battery life; and the technology is still at a price point beyond most consumers. Google's abandonment of the Glass technology is a case study in consumer market failure discussed in more detail later in this chapter.

Wearable Technologies and the Information Search Process

As search technologies have migrated from passive devices, such as laptops and Personal Digital Assistants, to small form factor mobile devices, and soon to wearables devices and appliances, several challenges emerge. Not all of these challenges involve the mechanics, size, security, or efficiency of the device, but rather how to collect, manage, curate, sort, rank, qualify, and decipher information. Consider a wearable device in the context of an academic library where a variety of information related to a topic might be emerge in the process of conducting a search. How the information seeker manages and organizes this content becomes an added challenge. The American Library Association indicates "one of the potential challenges for a growing Internet of Things is the variety of data collection and reporting methods used. Without standards for this data's collection and transmission, the benefits of the Internet of Things may not be as fully or easily realized" (ALA, 2015).

The size and discreetness of the appliance is not necessarily a characteristic in assisting to locate information easier than traditional media – nor assist the information seeker in locating quality information. Rather, a wearable appliance might help the information seeker by both sending and receiving information relative to a query. For example, a sensor collecting and storing queries or providing predictive search algorithms and analytics might provide the information seeker feedback along more than one dimension. Body worn appliances, like smart watches or activity trackers, will become particularly important as they will allow individuals to not only monitor and track research activities, but also provide data to other objects using the results to improve user experiences (temperature, acoustic, and lighting adjustments, for example) or to control other devices and physical spaces, such as unlocking doors and accessing archival information (ALA, 2015).

For an academic information seeker the distinguishing characteristic would likely be function over form. The wearable device needs to satisfy the requirements of providing research and information retrieval functionality that is different, if not better than other forms of information search. What type of functionality might an information seeker be interested in using? Clearly, being able to easily access library indexes and data warehouses using a wearable appliance would be important; the speed of information access; the ability to quickly compose complex queries; the ability to access data in a variety of conditions and constraints, and perhaps the accuracy and relevancy of the search results. The discrete characteristics of wearable technology might not afford any particular advantage to an information seeker. Rather, the ability to access information using a technology activated in unique, unanticipated, or unorthodox conditions would provide utility over other types of mobile technology.

Forrester Research, a leading technology think tank, recently asked potential wearable technology consumers the question "....would you be interested in wearing/using a sensor device, assuming it was from a brand you trust, offering a service that interests you?" The results of that research indicated the majority of individuals surveyed prefer a watch or wrist-based technology (42%), followed by technology that is clipped onto clothing (35%), and earbuds or headphone (21%). Glasses and glass-based wearable technology, ironically, ranked relatively low (18%) (Figure 5). According to this study, glass or head wear is not as appealing to consumers as technology developers once envisioned. It might be that the

Figure 5. Popularity of various wearable devices on the body (Forrester Research, 2014)

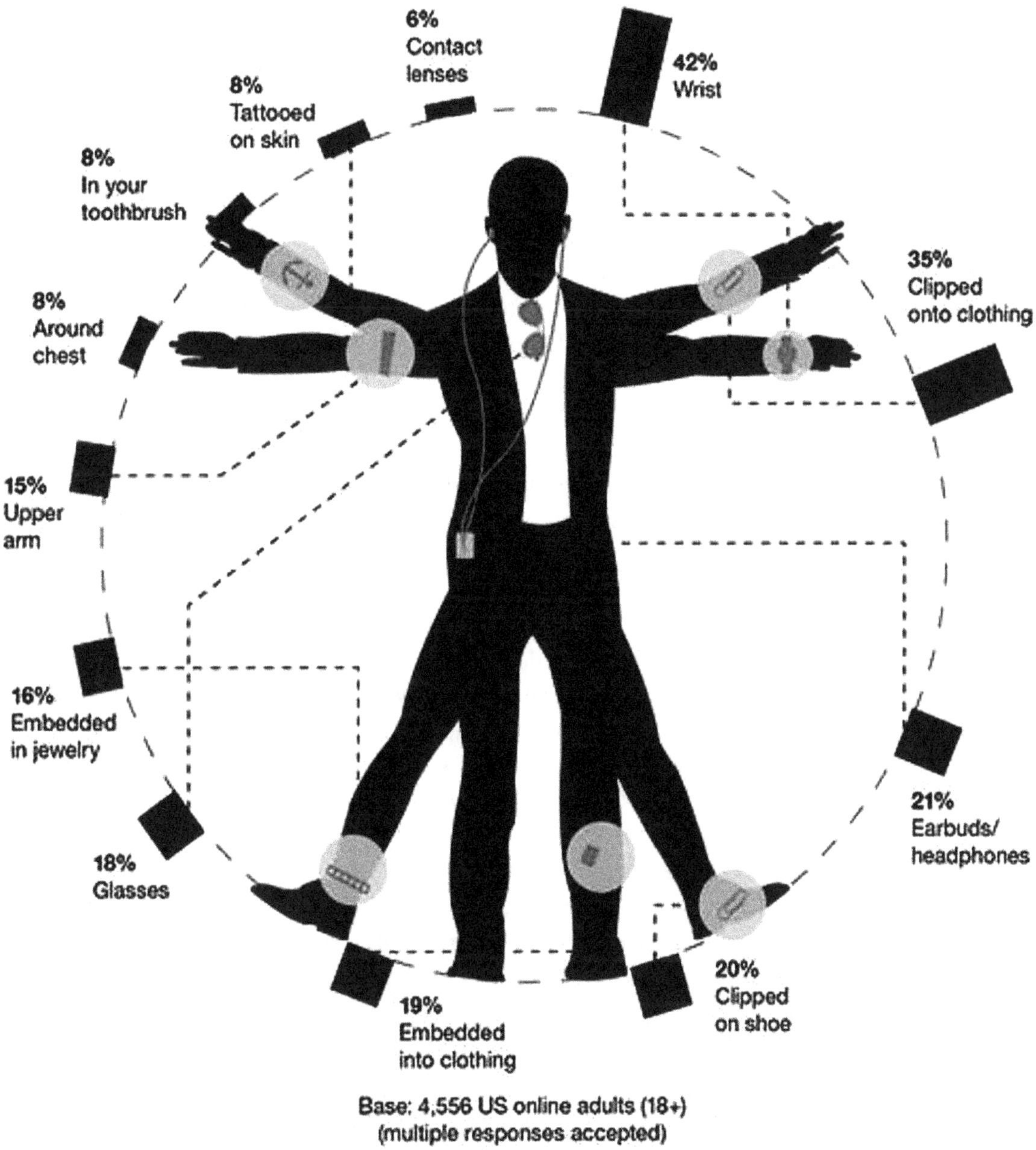

entire Google Glass technology was still a bit ahead of its time. Perhaps the key will be designing the technology to be less obtrusive, more practical, and offering functionality not available in other portable technologies.

Regarding how information seekers locate academic resources using mobile technologies, voice-based search still appears to be one of the most commonly used input methods for accessing information from libraries and data warehouses. Three of the leading voice-based search engines are, of course, Apple's Siri technology, Google Search, and Microsoft's new Cortana. The promise of these technologies is not simply for conducting one-dimensional search queries, but for the technology to actually include predictive algorithms aiding the information searcher to locate a variety of content. For example, a wearable device might allow the researcher to locate interconnected information tangential to the original search. Think of this as a triangulated search query (Alford, 2015). A query for a particular medical condition might yield interconnected resources and research results referenced within the search target. Or, videos, audio files, and other media content that supports and enhances the reference.

Although still relatively new – and primarily available with smart watches – haptic input/output is another, albeit less sophisticated type of information feedback system. Haptic technologies provide kinesthetic feedback in the form of micro-vibrations. For example, a smart watch wearer can easily discern a phone call from an email from a text message simply through the type of haptic vibration. Haptic technology is already embedded into video gaming technology. The next frontier appears to be integrating haptics into wearable appliances. In fact, Google incorporated a haptic sensor into its Glass technology transmitting audio through a bone located above the wearer's ear and into the ear canal. Apple uses haptics in its watch to provide turn-by-turn directions, whereas researchers are considering the technology in such things are auto steering wheels and public maps for more natural interactions (Moss, n. d.). It is foreseeable that haptic enabled appliances might also be used in the information search process. Consider a device transmitting micro-vibrations to guide users in locating library resources.

An emerging, if not interesting characteristic of information search in academic libraries now involves gaming and game-based activities. The Association of College and Research Libraries has found that gamification is very effective in higher education as a way to promote collections, and engage and motivate students to learn more about the library's services and resources (Kim, 2015). Multiple academic libraries have developed mobile-based games, such as the University of Huddersfiled's *Lemontree* allowing patrons to earn points and digital badges based on their reengagement with library resources (Broussard, 2011). It is possible the game-based metaphor can extend to wearable technologies as well. Consider patrons engaging with library services using a wearable appliance, such as goggles, an auditory search system, or a haptic system while they earn points or badges toward specific outcomes. This might be in the form of augmented reality, web quests, or other interactive information searches.

Glass and Head-Based Wearables

Google's Glass, perhaps the most lauded technology in the wearable spectrum to date, was a recent entrant into the consumer market. Conceived in 2009, Glass was the brainchild of Stanford researchers Sebastian Thrun, Astro Teller, and Babak Parviz (who is with Amazon as of this writing). These researchers were at the forefront of wearable technology several years earlier, along with fellow researcher Isabelle Olsson who designed several of the early, albeit crude apparatus prototypes (Bilton, 2015). Sponsored under Google's secretive "X Project", the earliest technology was developed in 2012 and made generally available to consumers in 2014 under the heralded "Google Explorer Program" which made Glass technology available to consumers based on selected research applications. Early Glass prototypes included everything from a scuba mask to industrial glasses (Figure 6). Eventually, the technology was refined and reduced in size, but remained painfully obvious when worn in the general public.

Users interacted with the technology using a series of swiping motions – up, down, forward, backward - that progressed through a series on-screen "cards" that appeared inside a small lens located on the right side of the frame. A microphone allowed voice-activated commands, such as "Ok Glass" followed by a basic voice command, such as "take a photo." Audio was available through an ear-based haptic system that was engineered directly into the temple and earpiece of the frame. Basic functionality included capturing photos, videos, conducting a search, browsing maps, and browsing web sites via crude on-screen browser. Secondary applications can be downloaded and installed using a phone-based application called "My Glass", or side-loaded through a direct USB connection to a computer. Applications for Glass were specifically designed for the appliance.

The rather sudden demise of the technology has been linked to a variety of corporate missteps and rollout blunders. Although it held significant promise, the recording capabilities raised concerns regarding privacy and personal information sharing. It should be noted these recording concerns, which will eventually be even more discrete and un-detectable with future appliances, has not abated. Bars, theaters, casinos and other public venues generally banned the devices altogether given its potential to record people and events surreptitiously. Additionally, it simply wasn't a well-conceived technology. Without a clear understanding of how Glass might be used beyond a novelty, there was very little compelling customers to pay a rather hefty $1,600 for a pair of glasses with a video overlay. It was also prone to short battery life and a bit temperamental. For example, the technology would often crash when certain apps were launched, modified, or side-loaded. The lens optics was also very limiting. For example the field of view is a mere 55 degrees horizontally and 43 degrees vertically, making the technology very restrictive for sensitive applications. As a result, researchers Barbara Chapparro and Jibo He from Wichita State University developed a lens application retrofitted to Glass to increase the field of vision to 110

Figure 6. Early Google Glass Prototype (Mashable, 2014)

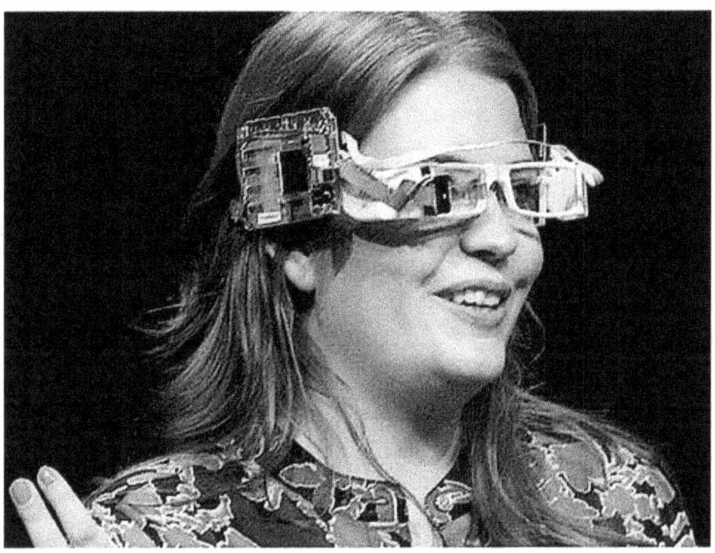

degrees and 58 degrees respectively, providing more adaptability in medical and information retrieval applications (Rusco, 2015).

Another serious flaw in the technology was it was still too obvious when worn. Even though the technology had been reduced in size and complexity through multiple prototype iterations, it was still quite apparent when someone was wearing them, leading to a host of privacy concerns. Indeed, the designer Olsson herself believes the future of wearable technology is in smaller, less detectable form factors (Davies, 2014).

Arguably, Google's mistake was releasing the technology too soon – well before it had been thoroughly tested. A Google executive admitted the Glass consumers saw and purchased was not the final product, but rather an early prototype. Without a clear vision for the technology and somewhat flawed hardware, it was doomed. There was also the issue of "glassholes" – the unending haranguing experienced by some users who wore the technology in public. As it turned out, social stigma was much harsher than Google researchers anticipated. However, the technology behind Glass still remains relevant and will likely resurface again in an improved wearable form factor (Metz, 2014).

Applications for Glass Technologies in Information Search

The New Media Center (NMC) in conjunction with the EDUCAUSE Learning Initiative (ELI), which together produces the annual *Horizon Report – Library Edition*, posed the following question to readers: *How might wearable technology be relevant to academic and research libraries?* Several readers provided a variety of interesting ideas and projects that are germinating in academic libraries. The Claremont Colleges Library (CCL) in Claremont, California, for example, was an early adopter of Glass technology. Their staff began the implementation by conducting a series of demonstrations among faculty, staff, and students. Users were then invited to submit proposals for how they intended to use the devices in support of their academic pursuits. Everything from mathematics application development, to environmental fieldwork, to tutoring, to language learning and academic productivity were among some of the proposals submitted (Booth & Brecher, 2015). According to the researchers, response to the device was extremely favorable. The biggest issue, they noted, was a disparity between the availability of the devices and the high user demand.

The New Media Center indicated in its most recent *2014 Horizon Report* that the *Internet of Things* will make a substantial impact in academic libraries within the next 5 years (NMC, 2014). Futurists envision a plethora of research activities conducted in academic libraries using wearable devices, including the ability to control the physical surroundings, provide information regarding current and previous searches, and promote content in real time. Some of this is already being conducted in museums where room conditions are tracked in real time using a series of internet-connected devices.

Librarians envision databases and indexes being made widely available to users by way of the Internet of Things. Using a wearable device, library patrons may be able to download directions, maps, and secondary apps to support the information search process. According to the 2014 Horizon Report, we're beginning to see some early affordances of this technology advancement. For example, Archeologists from the University of Bristol are embedding sensors directly into historical objects from the transatlantic slave trade for their Reflector Project (Skelly, 2015).

Sarah Roullard, a librarian and library blogger (2013) indicated the Internet of Things holds great promise for academic libraries. According to Roullard, "The Internet of Things can also be applied to library collection management. By affixing radio frequency identification (RFID) tags to collection items, they can be identified and inventoried using computers and RFID readers. In effect, each item with an RFID tag is assigned their own virtual representations in the system. RFID tags can be integrated into library cards to streamline checkout. Patrons can bring their items to the circulation counter for checkout and place the items on a checkout counter along with their library card. The circulation counter, equipped with an RFID reader, scans the items' tags and the library card, then delivers the data to the library's servers and updates the catalog on the Internet."

With such promise regarding wearables, many academics and university researchers have been quick to adopt the technology and explore its potential. Google Glass was one of the first commonly available wearable technologies academicians could readily obtain given its broad release and encouragement from Google to apply the technology in new and innovative research environments. One of many fascinating applications was introduced by a Cornell University Professor of Music Cynthia Johnston, who used the technology while she conducted an orchestra in real time. The image was simultaneously recorded and displayed on a large screen for the orchestra patrons. Her goal was to provide the audience an augmented representation of what it is like to conduct an orchestra from the vantage of the conductor (Otani, 2013).

The technology also found its way into academic libraries under the Explorer Program. According to College and Research Libraries News (2015) "To date, most uses of Glass in academic libraries are represented by one or more of the following three categories: pedagogy and research, community lending, and application development." Early academic libraries to adopt the technology include the University of Colorado at Boulder, Nevada State Colleges, Miami University, Virginia Tech, and North Carolina State University. Ernesto Hernandez from the Marydean Martin Library in the Nevada State Colleges successfully engaged Glass to record freshman information literacy courses from a first-person perspective (Hernandez, 2014). Jan Heller, a librarian from Miami University found the technology particularly useful for certain types of literacy instruction in the library. As an early Google Glass Explorer, she was intrigued about the possibilities the technology promised for library support services (Hawkins, 2014).

However, beyond general technology awareness and basic library instruction, the primary limitation of Glass in an academic library setting is the ability to share content. Indeed, the Glass technology is one of being very intimate and personal, whereas librarians are in the business of being very collaborative with their patrons. Logistically speaking, account restrictions, licensing, and personal profiles made the devices difficult to circulate in a library environment. Ultimately it proved to be difficult technology to check in-and-out of a library for general patron use.

Additional problems with the technology included the ability to easily connect to an academic library network. To use the technology effectively, Google Glass relies on an Android Phone application called 'My Glass' that passes data back and forth over a personal wireless network - which is normally less secure than a university network. In short, there was no easy way to access a secure, private network within an academic environment outside of the patron using his or her own personal account. That, in turn required each patron have a smart phone, a Gmail account, the My Glass application installed, and be able to connect to his or her own data provider or personal wireless network.

Secondly, the physical swiping motion required of Glass takes some getting used to. It can be very easy to access the wrong application, accidently initiate a recording, or inadvertently share content with your social network. Hence, training users to quickly understand and adapt to the technology proved to be very problematic.

Thirdly, the Google Glass technology relied heavily on proprietary applications developed by Google. While users could connect to Facebook, Twitter and other applications, the primary engine behind Glass was Google Apps – including G+. Without a G+ account, the user was limited in what they could do with the device. Other applications, such as Hangouts, were initially offered as part of Glass, but performance was so poor that it was quickly dropped.

Finally, in order to use the technology with corrective glasses the user needed to purchase a special part of frames only sold through Google. Along with custom made lens, the cost for the technology increased substantially for a nascent technology with limited functionality.

Application of Smart Watches in Information Search

Another relatively new entrant into the consumable wearable appliance market is the smart watch technology. Companies in this spectrum include Pebble, Sony, Samsung, Meta, Omate, Trendy, and multiple others (Figure 7). IBM was actually one of the first companies to introduce the technology in 2001 with its WatchPad (Ranger, 2015). In 2012 Sony introduced its smart watch which many consider to be the template for today's designs. It goes without saying, however, that Apple's iWatch, which was only recently introduced following years of speculation, has garnered the most publicity (Apple, 2014). Some of these smart watches now include telephony capabilities, while others are simply Internet-aware devices.

Still relatively new, the wearable technology provides affordances that allow the user to access the Internet from nearly any location where there is a data connection. Samsung's Gear smart watch relies on the user's Android-based phone to connect to the Internet via Bluetooth. The user simply installs an app on their phone providing connectivity capability with the watch. The phone serves as a reader for email and text messages; performs Internet searches; provides calendar reminders; provide alarm capa-

Figure 7. Smart watch market share 2013 (Smartwatch Group, 2014)

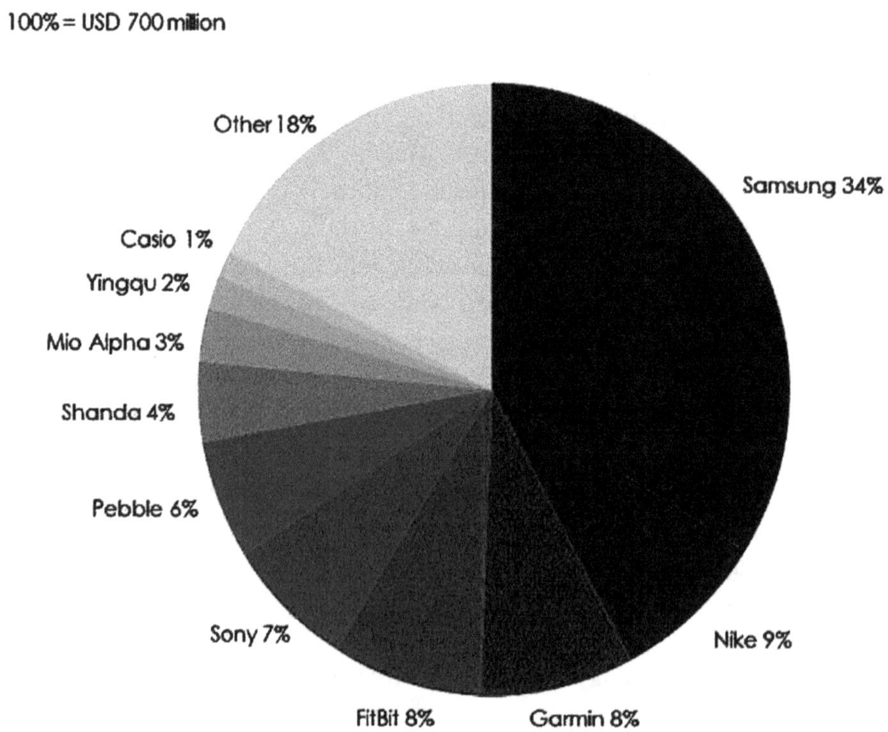

bilities; serves as a health monitor, and allows the user to install apps from the Google Play Store. Using voice input capabilities, the watch can perform basic Internet searches using the built in Google search engine. Most watches include both a microphone and a speaker. The newest versions of the watch, called the Gear S, provide telephone capabilities so the user doesn't require a phone to make and receive calls.

Another variant of wrist-based wearables are considered 'fitness bands', such as the popular FitBit and Garmin's Vivosmart that can also be used for basic communication capabilities, such as checking email, text massages, and phone calls, in addition to monitoring heart rate, steps, and other fitness metrics. However, these devices generally lack the sophistication to perform any sort of information search beyond general queries.

Libraries are at least beginning to entertain smart watches as a potential information search utility. The American Association of Law Libraries, for example, found certain native watch applications provided at least some search utility. For example, certain Gear series watches include a camera useful for note taking or even scanning. (Although the scan quality is still very poor.) Most watches also include native Google searching to provide for quick fact checking (Bell, 2014). The technology is still evolving, but holds some promise.

A potential concern for all of these wearable devices in academe is test security and cheating. Wearable devices are basically a search engine, camera, and communication tool worn on the body, allowing students to retrieve factual information, communicate with one another, and share information. While relatively easy to ban smart phones while taking a test, it might be more difficult trying to ban a device that's difficult to detect or doubles as another device or apparel item.

Potential Impact of Wearable Technology on Library Services

Library and information seeking strategies have been studied for several decades. In particular, the psychology of human interaction involved in seeking information. Carol Kuhlthau's seminal research regarding the theoretical *Information Search Process* (ISP) dates back more nearly three decades (Kuhlthau, 1991). Kuhlthau provides a multi-stage model user's pass through in the information seeking process (Figure 8).

Nick Belkin followed Kuhlthau with his *Anomalous States of Knowledge* (ASK) model. Belkin's research provided much of the foundation for how users interacted with information retrieval systems and digital libraries (Belkin, 1990). Belkin maintained that in many cases the user was unable to precisely formulate the type of information they were seeking. An ASK is a situation where "the user realizes that there is an anomaly in [their] state of knowledge with respect to the problem faced." The person may address the anomaly by seeking information. After obtaining information, the person will evaluate again whether the anomaly still exists. If it does, and the person is still motivated to resolve it, more information may be sought.[16] according to Belkin every search begins with a problem and a need to solve the problem, the gap between this refers to information need which leads to information seeking.

Gary Marchionini from the University of North Carolina's School of Library and Information Science later provided additional context as search relates to digital information. Marchionini maintained while information is largely available from anywhere in the world thanks to the Internet and digital databases, this same information is difficult to express without the aid of machines allowing us to transpose information in a format we can perceive (Marchionini, 1995). Therefore, regardless the device used to access this information and despite the ability to access information in new and unforeseen ways, how the user interprets and uses the information remains the goal of any information search process.

Figure 8. The Information Search Process (ISP) Model

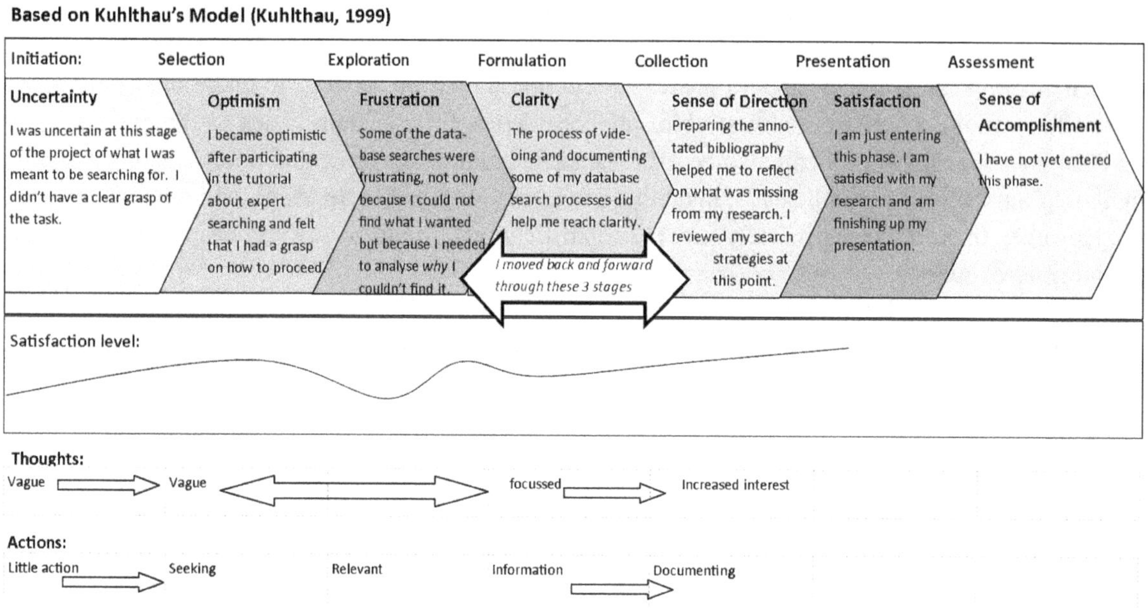

So why use these wearable devices to access library resources? Is there an inherent advantage afforded to the information seeker? Will they transform the library information search process? Not yet, but the future is bright. It's still too early in the evolution of these devices to provide evidence they are somehow better or different than a traditional search using any other type of search technology. However, as these devices evolve and the applications improve it is well within the realm of possibility such appliances will provide a wealth of library resources when physically attached to the information seeker. Imagine, for example, a user with a wearable device who can instantly access the Library of Congress – from anywhere - by simply speaking a search string into eyewear, or a watch, or even Internet-aware apparel not only always on, but context sensitive.

A Survey of Library Databases: Access Using Wearables

The following is a brief survey of where the library index services industry stands regarding access to common indexes using wearable appliances. Suffice it to say that we're probably several years away from being able to conduct sophisticated information searches using anything more than Google on one of these devices. Common academic library indexes sometimes sponsor a mobile app or mobile version of their web site to access their search engine, but until those apps are developed expressly for the wearable appliance it's likely that the library patron will not realize any noticeable affordances.

So, the question might be "why?" Is there an advantage in using a wearable device in tandem with a smart phone to conduct a search? In short, it still may be too early. In experiments conducted at Emporia State University there wasn't any intrinsic advantage using a wearable device over any other Internet aware technology. However, that shortcoming was due, in large measure, to the lack of a native app designed expressly for the wearable technology. If such an app were available and downloadable through one of the popular app stores or app sharing services the results may have been much different.

WorldCat

Worldcat doesn't offer a native mobile application. Rather, WorldCat provides a mobile web site for access to their database. Essentially, any Internet aware device – including wearables – can access WorldCat using a search engine. Searches are relatively simple. Using a smart watch it was relatively easy to invoke an academic library search by speaking into the device's microphone. To begin, simply initiate a Google search from the smart watch which, in turn, launches a mobile browser on a nearby smart phone connected via Bluetooth. From there, access to the WorldCat mobile search interface was accessible where a search string can be entered. Note that a phone was required in order to conduct the search, however.

The same was true using Google Glass. Using the voice-activated commands and the on-board device microphone, simply open the WorldCat index using the Glass native browser. However, browsing a web site on the device is not easy. Native swiping motions are only moderately effective. The user can move up and down a web page, but selecting options and navigating to linked pages was nearly impossible.

EBSCO Host

EBSCO, including Academic Search Complete and ArticlesPlus, provides a free downloadable app for iOS and Android mobile devices. However, the app is not yet installable or usable on most wearable appliances as of this writing. In order to make this client useful on a wearable appliance, a specifically designed app will need to be developed. Hence, specific access to EBSCO-based content is limited to smart phones and other traditional forms of mobile media. Until EBSCO develops an application specifically for wearable appliances, this particularly library index will be relatively inaccessible on a smart watch or other wearable appliance.

LexisNexis

LexisNexis offers the most replete catalog of mobile applications for the Apple ecosystem, including a mobile version of their database for iPads and iPhones. However, they do not currently offer an Android version of their product. Installing the app on an iPad was quite easy. However, without an Android version of the product, testing the product on a Samsung Gear smart watch or on Google Glass was impossible. It's likely without an app developed specifically for a wearable appliance the search engine will not currently function with most contemporary wearable products – certainly nothing in the Android ecosystem.

ProQuest

As with the other popular indexes, ProQuest provides options for a variety of mobile devices. In fact, ProQuest arguably provides the most options regarding mobile product support, including tablet support for iPads, Kindle Fire, Samsung Galaxy products, Google Nexus, and Windows Surface devices. All manner of iOS, Android, and Windows smart phone devices are also supported. However, each of these options is supported using a web site adapted for mobility rather than native applications. Hence, the patron must use a mobile browser to access resources. This is not a problem on a smart phone, but testing this product was unsuccessful using both a smart watch and Google Glass since neither product includes a true "browser". The Glass browser, for example, is a proprietary application written specifically for the product. Swiping motions do not work well for many web sites. Opening the ProQuest web site on Google Glass was relatively easy. Swiping forward and backward moves the user through the web site. However, selecting an option or inputting information via the touch pad was nearly impossible. In short, the wearable device is not designed for the interactive components built into a traditional web site.

JSTOR

Like ProQuest, JSTOR provides a mobile web site rather than a native application. Therefore the mobile version of their services are available using most types of hardware. However, testing using both a Samsung smart watch and Google Glass was again very limited. A smart watch can be used to open mobile browser on a nearby smart phone. However, opening the site or conducting a search on the watch itself was not yet a possibility. Using Google Glass, the JSTOR web site will load, however, there was little or no ability to navigate and control the web site.

ARTstor

As with other library indexes, ARTstor, a large image database with nearly 2 million photos and art images, provides both an Android and iOS version of its mobile application. In limited testing, however, the wearable devices were not able to access the index through the mobile app – primarily because the mobile app has not been developed specifically for the devices. Access to the web-based site was possible, but browsing and navigation was very limited, rendering the site basically unusable through Google Glass or a Gear smart watch.

In summary, most of the leading academic indexes simply haven't yet developed products and services for the wearable market. While some queries were available via a Bluetooth connected smart phone and a browser, this search method still requires a secondary device in order to complete the search. Other companies had native mobile apps, but these typically won't work for the wearable device unless it has been expressly designed for the product. This lack of wearable appliance applications is likely due to economies of scale. Without a critical mass of users, it's probably not the best use of programming time to develop apps for such a small market segment. Additionally, the wearable appliances still have not evolved to the point where they can easily access those services through voice, gesture, or kinesthetic control.

SOLUTIONS AND RECOMMENDATIONS

The use of wearable technologies is gaining considerable momentum in a variety of areas, including health care, biometrics, disability services, sports, leisure and recreation, gaming, shopping, and even academics. In fact, many of the earliest wearable prototypes were conceived in academic research laboratories. Many of those prototypes have gradually made their way into the commercial market where they are being adapted to work with a variety of clothing, glasses, goggles, watches, and other apparel. One of the earliest wearable technologies to reach mass appeal was Google's Glass, launched in 2013. This technology was a bit of a paradox, however. It gained mass appeal in a very short period of time. Hundreds of applications were developed for a variety of purposes, including entertainment, video and photo sharing, email, and even web conferencing. Applications were also developed to support for individuals with physical disabilities and other motor impairments. However, the technology was simultaneously lauded due to its premature launch and lack of clear market differentiation. Lacking a clear market direction and clouded with a variety technological limitations, Glass Explorers version was short lived. The technology was unceremoniously discontinued in 2015. This sudden and somewhat surprising demise of Glass Explorer was a temporary setback for wearable technology, but it's likely to reappear in a more discreet form factor in the near future.

Around the same time, smart watches and other forms of wearable technologies began to appear on the market. First as health monitoring bands and later as fully functioning phones and Internet aware devices. While smart watches have been gaining in popularity and market share, they are still too new to predict their eventual influence in the market. It is likely watches will continue to appeal to a niche consumer market. However, smart watches appear to have a completely different market appeal for consumers. First, they are much more discreet than Glass. Secondly, they have more pragmatic utility – serving as watches, email readers, search engines, and now even phones. As research has shown, watches are by far the most popular wearable device – fully three times more appealing than head-based technology or eyewear at this time.

Other forms of wearable technologies are beginning to appear in a variety of clothing and apparel. The difference between these later incarnations and products such as Glass and Smart watches are its relative discreetness, ability to be used in a variety of conditions, and its ability to remain in a continuous state of internet activity. The notion of 'there when you need it and gone when you don't' appeals to individuals who want the technology to be available when and where they desire, but hidden when not in use. The challenge for product designers will be to design wearable technologies that can be discreet, yet provide the affordances of a new generation of powerful information collection systems.

As these technologies begin to reach mass appeal it is only natural academicians have begun experimenting with their potential in a variety of teaching and learning environments. Everyone from researchers to traditional classroom faculty have explored the potential of wearable technologies and other *Internet of Things* devices to provide new dimensions for teaching, learning, and research. Despite the limitations of the current technology and despite the dearth of applications in this space, academicians should continue to pursue wearable technologies as a mid-to-long term solution strategy. It is very likely as these devices reach mass appeal the developer community will begin to realize the potential in academic and research library environments. It is quite possible that this interest will initially be driven by other factors, such as wearable technologies in museums and various public venues.

The wearable community appears to be ready for software developers to begin producing applications for this space. Prior to the discontinuance of Google Glass, hundreds of software applications were being developed specifically for the device. However, academic products normally trail commercial products in the mobile and wearable space. There has to be a critical mass of users in order to justify the cost of development. Librarians, in particular, are often among the earliest adopters of emergent technologies on campus. It is likely that a library- or information-search-based wearable application/appliance will be among the first type of academic product we'll see emerge in this space.

FUTURE RESEARCH DIRECTIONS

The natural instructional curiosity extends to the academic and research libraries where a number of wearable technologies have begun to make inroads. Starting with Glass in 2013, several libraries across the country have begun experimenting with the technology to understand its capabilities. A review of those early experiments revealed some promise, but still not a panacea. Glass, for example, was introduced in several libraries for its patrons to check out, test, and experiment. However, one for the first limitations encountered was Google requires a Google account to pair with the device. Given the nature of libraries and their equipment circulation, this proved to be an impediment. Secondly, users generally found the devices difficult to use without a least a moderate level of training. The goal of having devices that were easily circulated and easy to use for information search was somewhat dashed. Hence, many libraries resorted to demonstrations, general education, and technology awareness seminars rather pressing the technology into service as a functional apparatus.

Additional challenges include the limited availability of library index apps for wearable devices such as Samsung Gear, the iWatch, and Google Glass. Currently, there are very few mobile library and index apps available for the wearable spectrum. While some apps are available for smart phones, they've not yet made their way into the device specific app management utilities, such as Galaxy Apps, Glass Apps, and other application indexes. Therefore, users are generally limited to Google search-based queries for the foreseeable future. The ideal would be to directly access library-specific indexes using wearable technologies and wearable apps.

Attention is now turning to a new generation of wearable technologies and the promise of using these technologies in a variety of academic information search activities. These technologies will likely be much more discreet, always on, location and context sensitive, and provide the ability to retrieve a variety of information from embedded sensors located directly on the target. So, rather than using a wearable device to simply *locate* the resource within the library, the wearable device may be used to retrieve additional information regarding the source, including an abstract, reviews, links to secondary research associated with the information, the authors' biographical sketch, and other types of data associated with the target. Additionally, the patron will be able to store a variety of queries in a searchable index for easy retrieval, share queries with other researchers, and even directly embed the query into a paper or presentation.

It is likely that within the next 3-5 years we'll see a variety of wearable technologies emerge in the academic space. It is also likely that these devices will provide access to information in multiple dimensions e.g., geo-positioning/special recognition; multi-sensory input, including RFDI, audio, tactile, and gesture recognition; in/out access to information, including indexes that can both receive and respond to inputs; multi-dimensional search queries, including searches that can be executed using more than one type of simultaneous input, etc. As these products emerge, a wealth of research opportunities will follow. In addition, how information is indexed, structured, stored, curated, and accessed may also change to accommodate these devices, leading to further research and investigation regarding access to digital resources and the role of libraries to meet these challenges. The possibilities are both endless and quite exciting.

REFERENCES

Alford, E. (2105). Wearables to heighten demand for local search: prepare now! *Search Engine Watch*. Retrieved from http://searchenginewatch.com/sew/news/2396194/wearables-to-heighten-demand-for-local-search-prepare-now#

Belkin, N. J. (1980). Anomalous states of knowledge as a basis for information retrieval. *The Canadian Journal of Information Science*, *5*, 133–143.

Belkin, N. J. (1990). The cognitive viewpoint in information science. *Journal of Information Science.* Retrieved from http://jis.sagepub.com/content/16/1/11.abstract

Bell, D. (2014). Smart watches in the library? *CSSIS.org*. Retrieved from http://cssis.org/news/blog/2015/01/smart watches-library

Bilton, N. (2015) Why Google Glass broke. *The New York Times*. Retrieved from http://www.nytimes.com/2015/02/05/style/why-google-glass-broke.html?_r=0

Bonato, P. (2005). Advances in wearable technology and applications in physical medicine and rehabilitation. *Journal of NeuroEngineering and rehabilitation.* Retrieved from http://www.jneuroengrehab.com/content/2/1/2

Booth, C., & Bretcher, D. (2015). Ok, library: Implications and opportunities for Google Glass. *College and Research Libraries.* Retrieved from http://crln.acrl.org/content/75/5/234.full

Broussard, M. S. (2011). A spoonful of sugar: Instructional games in libraries. *Games in libraries.* Retrieved from http://gamesinlibraries.blogspot.com/2011/12/lemontree-game-at-university-of.html

Davies, C. (2014). Google's Glass designer talks less-is-more in wearables. *Slashgear.com.* Retrieved from http://www.slashgear.com/googles-glass-designer-talks-less-is-more-in-wearables-16325496/

7 Things You Should Know: Wearable Technology. (2013). *EDUCAUSE*. Retrieved from https://net.educause.edu/ir/library/pdf/ELI7102.pdf

Ehrenkranz, M. (2014) Wearable Tech for people with Disabilities: How a Tech Specialist is Modifying Google Glass to be More accessible. *Idigitaltimes.com.* Retrieved from http://www.idigitaltimes.com/wearable-tech-people-disabilities-how-tech-specialist-modifying-google-glass-be-more-accessible

Evans, D. (2011). The Internet of Things: How the next evolution of the Internet is changing everything. *Cisco.com*. Retrieved from https://www.cisco.com/web/about/ac79/docs/innov/IoT_IBSG_0411FINAL.pdf

Gonzalez, Z. D. (2014). Put a ring on it. *Wearable Technologies.com.* Retrieved from https://www.wearable-technologies.com/2014/04/put-a-ring-on-it/

Hawkins, D. (2014). Library services and Google Glass. *Libconf.com*. Retrieved from http://www.libconf.com/2014/04/09/library-services-google-glass/

Hernandez, E. (2014). Google Glass in the academic library. *CreativeLibraryPractice.org*. Retrieved from http://creativelibrarypractice.org/2014/03/25/google-glass-in-the-academic-library-the-exploration/

Kersten, P., & Ribbens, C. (2015). *Wearable games.* Retrieved from http://wearablegames.eu/

Kim, B. (2015). Keeping up with….gamification. *Association of College & Research Libraries*. Retrieved from http://www.ala.org/acrl/publications/keeping_up_with/gamification

Kuhlthau, C. C. (1991). Inside the search process: Information seeking from the user's perspective. *Journal of the American society for information science*, 42(5), 361-371. Retrieved from http://ptarpp2.uitm.edu.my/silibus/insidesearch2.pdf

Lukowitz, P., Kirtein, T., & Troster, G. (2014). Wearable systems for health care applications. *Methods of Information in Medicine*. Retrieved from http://www.researchgate.net/publication/8480044_Wearable_systems_for_health_care_applications

Marchionini, G. (1995). *Information seeking in electronic environments*. University of Maryland. Retrieved from http://catdir.loc.gov/catdir/samples/cam031/94044629.pdf

Metz, R. (2014). Google Glass is dead: Long live smart glasses. *MIT Technology Review*. Retrieved from http://www.technologyreview.com/featuredstory/532691/google-glass-is-dead-long-live-smart-glasses/

Wearable computing FAQ. (1997). *MIT*. Retrieved from: https://www.media.mit.edu/wearables/lizzy/FAQ/FAQ.txt

Smart phones are fading. Wearables are next. (2014). *CNN Money*. Retrieved from http://money.cnn.com/2014/03/19/technology/mobile/wearable-devices/

There's finally something awesome to do with Google Glass. (2015). *CNN Money*. Retrieved from http://money.cnn.com/2015/06/09/technology/google-glass-parkinsons/

Moss, R. (n. d.). Haptic technology: The next frontier in video games, wearables, virtual reality, and mobile electronics. *Gizmag.com*. Retrieved from http://www.gizmag.com/haptic-tech-vr-wearables-games-sightlence/35616/

Horizon Report – Library Edition. (2014). *New Media Center*. Retrieved from http://cdn.nmc.org/media/2014-nmc-horizon-report-library-EN.pdf

Niewolny, D. (2013) *How the internet of things is revolutionizing healthcare*. Retrieved from http://cache.freescale.com/files/corporate/doc/white_paper/IOTREVHEALCARWP.pdf

Wearable computing - Challenges and opportunities for privacy protection. (2014). *Office of the Privacy Commissioner*. Retrieved from https://www.priv.gc.ca/information/research-recherche/2014/wc_201401_e.asp

Otani, A. (2013). Cornell music professor becomes Google Glass explorer.

Pultarova, T. (2015). Researchers propose body WiFi for the wearable technology age. *Engineering and Technology Magazine*. Retrieved from http://eandt.theiet.org/news/2015/apr/body-wifi.cfm

Racoma, J. A. (2013). Wearable tech: A brief history and a look into the future. *Android Authority*. Retrieved from http://www.androidauthority.com/wearable-computing-history-238324/

Ranger, S. (2015). Before the Apple iWatch: A history of smart watches in pictures. *ZDNet*. Retrieved from [REMOVED HYPERLINK FIELD]http://www.zdnet.com/pictures/before-the-iwatch-a-history-of-smart-watches-in-pictures/8/

Roullard, S. (2013). *The Internet of Things in the library*. Retrieved from http://libserra.com/the-internet-of-things-in-the-library/

Rusco, L. (2015). Wichita State professors invent accessory lens for Google Glass. *Witchita State News*. Retrieved from http://www.wichita.edu/thisis/stories/story.asp?si=2743

Sabhlok, R. (2013). Smart watches, Google Glass and the wearable technology chocolate box. *Forbes.com*. Retrieved from http://www.forbes.com/sites/rajsabhlok/2013/12/05/smart watches-google-glass-and-the-wearable-technology-chocolate-box/

Scott, M. (2013). Google Glass: West Michigan online physics teacher takes students on virtual field trip to Switzerland. *Michigan Live*. Retrieved from http://www.mlive.com/news/grand-rapids/index.ssf/2013/05/google_glass_west_michigan_onl.html

Sheehy, A. (2015). Now that wi-fi is done it's time for wi-bo. *Generator Research.com*. Retrieved from: http://www.generatorresearch.com/tekcarta/analysis-insight/now-that-wi-fi-is-done-its-time-for-wi-bo-wireless-body/ July, 2015.

Starner, T. (1998). Real-time American Sign Language recognition using desk and wearable computer based video. *IEEE Transactions on Pattern Analysis and Machine Intelligence*, 20(12).

Starner, T. (2013). Google glass lead: How wearing tech on our bodies actually helps it get Out of our way. *Wired.com*. Retrieved from http://www.wired.com/2013/12/the-paradox-of-wearables-close-to-your-body-but-keeping-tech-far-away/

Van Kranenburg, R. (2011). *The Internet of Things*. Retrieved from http://www.researchgate.net/profile/Matt_Ratto/publication/228360933_The_Internet_of_Things/links/0912f513755ebd1e87000000.pdf

Wagstaff, J. (2015). The path to a wearable future lies in academia. *Reuters.com*. Retrieved from http://www.reuters.com/article/2015/05/04/us-tech-academia-idUSKBN0NP1YR20150504

Wasik, B. (2013). Why wearable tech will be as big as the smart phone. *Wired.com*. Retrieved from http://www.wired.com/2013/12/wearable-computers/

Introduction to wearable technology. (2014). *Wearable Devices*. Retrieved from http://www.wearabledevices.com/what-is-a-wearable-device/

Wearable technology dives into the pool with Xmetrics. (2014). *Sporttechie*. Retrieved from http://www.sporttechie.com/2014/11/10/wearable-technology-dives-pool-xmetrics/

Woda, T. (n. d.). *Privacy and wearable technology* (White Paper). Retrieved from http://www.uknow.com/wp-content/uploads/2014/03/Location-Services_White_Paper.pdf

Young, J. (2015). Wearable teaching? College to experiment with Apple Watch as a learning tool. *Chronicle.com*. Retrieved from http://chronicle.com/blogs/wiredcampus/wearable-teaching-college-to-experiment-with-apple-watch-as-learning-tool/56459?cid=wc&utm_source=wc&utm_medium=en

KEY TERMS AND DEFINITIONS

Google Glass: A wearable technology developed by Google that provides wireless connectivity to a variety of cloud-based services using a small computer and sensor that resemble glasses.

Smart Watches: The smart watches allow individuals to not only monitor and track activities, but also provide data to other objects using the results to improve the user experiences. Haptic technologies have been integrated into the smart watch to provide kinesthetic feedback in the form of micro-vibrations.

Wearables: Wearables are typically voice and/or gesture controlled, are always on, are typically connected to the Internet at all times, and are environmentally aware.

Wearable Appliances: Wearable appliances are being used in all fields such as gaming, health care, leisure, recreation, biometrics, academia, and more.

Wearable Technology: Wearable technologies provide the ability to instantly gain access to information based on geo-positioning, voice-activated commands, body/Radio Frequency Data Identifier (RFDI) sensors, tactile input, gestures, heads-up displays, and other discretely transportable input/output sensors allowing users to access a variety of resources.

Wi-Bo: To support the new generation of wearable devices, a completely new wireless transmission standard called Wi-Bo (wireless body) has been proposed specifically for the anticipated growth of wearable technologies and appliances.

Chapter 8
Design and Development of Educational Multimedia:
The Software Development Process for Mobile Learning

Ibrahim Arpaci
Gaziosmanpasa University, Turkey

ABSTRACT

The decision to implement new technologies such as smart glasses, smart phones, and tablets in an educational setting without determining optimal use scenarios is an evident universal problem as the adoption of such mobile platforms becomes widespread. Semi-structured interviews are conducted with two Science and Technology (S & T) and three Information Technology (IT) public school instructors to further investigate this significant problem. The constant comparative method was used to analyze the qualitative data resulting from these interviews. Preliminary results demonstrate the educational use of tablet computers has several advantages, along with a few limitations needing to be addressed. Specifically, one of the main limitations of these new instructional technologies is the lack of interactive content, embodied in the audio, video, and pictorial multimedia. The urgent need to address this limitation has motivated the development of multimedia software to work seamlessly on tablet computers.

INTRODUCTION

In this study, a design and development research methodology included the ADDIE model of analysis, design, development, implementation, and evaluation. During the implementation phase collaboration included students, experts, and teachers. During the evaluation phase a questionnaire was completed by 40 elementary students identified as target users' typical of the characteristics and readiness needed for adoption and utilization of tablet computers. In all, a triangulated approach, including semi-structured interviews, literature review, and, most importantly, software analysis was employed to identify both functional and non-functional requirements. In conclusion, an effective multimedia software solution for tablet computers was advanced for fifth grade elementary students, with the primary objective of

DOI: 10.4018/978-1-5225-0069-8.ch008

increasing overall success rates. Finally, a pilot testing was conducted with 17 elementary students to detect possible usability issues in the utilization of the software. In addition, implications for practice are provided with limitations of the study discussed. We begin with a review of the literature surrounding mobile technologies.

Mobile Technologies

The mobile telecommunications industry has experienced appreciable growth during the recent decade. The significant growth has led mobile device manufacturers and service providers to develop new smart technologies such as glasses, watches, phones, tablets and a new array of data-contingent applications such as the mobile web, cloud computing, and multimedia streaming. With the introduction of these user-centric devices and interactive services, it appears a new mode of learning has arisen through modern mobile devices.

Mobile learning can be defined as a new form of personal learning taking place anytime and anywhere with the aid of mobile devices such as notebooks, tablets, and smart phones (Kadirire, 2009). Mobile learning provides greater individualism, flexibility, and most importantly, ubiquity, contrasting sharply with traditional learning methods (Arpaci, 2015). Moreover, it enables an "always-on" connectivity for learners and provides mobility, interaction, and portability (Jeng, Wu, Huang, Tan, & Yang, 2010; Kukulska-Hulme, Sharples, Milrad, Arnedillo-Sánchez, & Vavoula, 2009; Sharples, Milrad, Arnedillo, & Vavoula, 2009; Traxler, 2009).

The advantages provided by mobile technologies have prompted instructional technologists and researchers to take a pedagogical view toward developing educational multimedia applications for mobile devices to promote teaching and learning (Wu, Wu, Chen, Kao, Lin, & Huang, 2012; Kukulska-Hulme & Traxler, 2007). For example, Güler, Kılıç, and Çavuş (2014) compared the difficulties novice instructional designers experience during the instructional design processes using mobile devices and desktop computers for learning content. Sixty-eight sophomore students developed learning content for mobile devices and desktop computers using the instructional design model of; Analysis, Design, Development, Implementation, and Evaluation (ADDIE). The results showed both development processes had similar difficulties, while the development process for mobile content requires some specific adjustments.

Doolittle and Mariano (2008) examined the effects of individual differences in Working Memory Capacity (WMC) on learning from a historical inquiry using multimedia tutorials in a stationary environment versus mobile learning environments using a portable digital media player. The results showed students in the stationary instructional environment performed better, while interaction effects demonstrated the low-WMC students performed poorly in a mobile instructional environment. In a similar study, Gedik, Hanci-Karademirci, Kursun, and Cagiltay (2012) investigated critical issues in designing a cellular phone-based mobile learning environment for 11th graders to examine students' perceptions on reasons for satisfaction, participation, and implementation processes. In the analysis phase, researchers gathered information about students' cellular phone usage habits, brands of phones, mobile network operators' names, and students' mobile learning preferences with regards to content, time, and methods used. In the design and development period, researchers designed the content using examples, explanations, and interpretive multiple-choice questions. The content and open-ended questions were based on the high school curriculum, textbooks, with input from the biology teachers. They found the limitations of regular cellular phones (i.e., storage, screen size, bandwidth) made it difficult to provide full instruction using them. However, the majority of these limitations have been addressed today by emergence of larger display devices and new mobile services such as cloud storage.

Fernández-López, Rodríguez-Fórtiz, Rodríguez-Almendros, and Martínez-Segura (2013) developed a mobile platform to cover the main phases of the learning process such as preparation, use, and evaluation. The platform includes four kinds of educational activities, including association, exploration, puzzles, and sorting. The researchers found the use of mobile platforms were associated with positive effects in the development of learning skills for children with special education needs. They observed increased results for basic math, language, environmental awareness, and personal autonomy. In another study, Ullrich, Shen, Tong, and Tan (2010) developed a Mobile Live Video Learning System (MLVLS) for a computer science course and found mobile devices are more widely used than desktop or laptop computers. Similarly, De-Marcos et al. (2010) designed an application for mobile phones to reinforce students' knowledge by means of self-assessment and found it improves student achievement, especially amongst younger learners, with a relatively low impact on current teaching activities and methodologies.

Evaluating the effectiveness of mobile learning has been a frequent theme of research (Chu, Hwang, & Tsai, 2010; Hwang, Yang, Tsai, & Yang, 2009; Chen, & Chung, 2008; Chen, & Hsu, 2008; Kukulska-Hulme & Traxler, 2007; Kukulska-Hulme, 2005; Schwabe, & Göth, 2005). For example, Evans (2008) investigated the effectiveness of mobile learning in the form of podcasting in a business course for university students. The results suggest students find podcasts to be preferable to their textbooks as a learning aid. In another study, Baya'a and Daher (2009) explored the effectiveness of mobile learning using mobile phones and found students respond positively to the use of mobile phones in learning mathematics.

Educational use of tablet computers has also been a common line of inquiry amidst researchers (Alvarez, Brown, & Nussbaum, 2011; Young, Trudeau, Odell, Marinelli, & Dennerlein, 2012; Pereira, Miller, Huang, Odell, & Rempel, 2013; Binboğa & Korhan, 2014; Boyce, Mishra, Halverson, & Thomas, 2014; Chen, Falkmer, Parsons, Buzzard, & Ciccarelli, 2014; Dündar & Akçayır, 2014; Rehmat & Bailey, 2014; Schneps, Ruel, Sonnert, Dussault, Griffin, & Sadler, 2014). For example, Couse and Chen (2010) investigated preschool children's ability to acclimate to tablet computers based on data obtained from a sample group of 40 preschoolers aged 3-6. Their results suggest differences in the level of tablet use between course sessions and overall engagement increase in proportion to age. They also reported children quickly learned to explore and navigate tablet computers and kept interested for much longer periods of time.

Ferrer, Belvis, and Pamies (2011) suggest there are significant differences in students' academic results according to their first-order social variables (i.e. gender, birthplace, etc.), socioeconomic and cultural backgrounds. They claim the incidence of lower test scores increased with the introduction of tablet computers in a public school in Spain. On the other hand, Richardson (2012) argued tablet computers have increased the benefits of educational games by making such games available to a wider audience. In another study, Galligan, Hobohm, and Loch (2012) found the use of tablet computers in a distance math course could facilitate assignment submission, provide efficient instructor feedback, and increase communications between students and instructors.

More recently, researchers Nortcliffe and Middleton (2013) found tablet computers may foster autonomous and pervasive engagement among university students in the UK. Similarly, Romney (2011) conducted a longitudinal study of undergraduate students and found when using tablet technologies to learn mathematics students are more likely to continue their studies than non-tablet users. In another study, Lavery (2012) integrated tablet computers in an online English and Creative Writing class and discovered college students participated more frequently in class activities.

The estimated cost of approximately $7 billion USD, has been invested towards the Movement to Increase Opportunities and Technology, known as FATIH. The educational FATIH project is designed to promote effective use of information and communication technologies in public schools for learning. In total, 737,800 tablet computers were distributed to all teachers and 9th grade students by February 2014. However, 10 million 600 thousand-tablet computers will be distributed to all students and teachers in elementary, middle, and high schools by February, 2016 (MEB, 2015).

Effective use of tablet computers, put to use in public schools in Turkey to enhance teaching and learning is an important topic of research. However, the decision to introduce so many tablet computers within the educational system without determining how to use them effectively is a significant problem, and therefore, requires investigation. Thus, this study aims to investigate effective use of the tablet computers distributed within the scope of the FATIH project. Initial problem analysis shows one of the main advantages of the new technology is delivering interactive and multimedia-rich content to learners, and on the flip side one of the more significant problems. Driven by this motivation, the current study aims to advance multimedia software for fifth grade elementary students to increase success amongst S & T students.

The rest of the study is organized as follows: First, the methodology used is explained. Then, the multimedia software development processes, including analysis, design, development, implementation, and evaluation are presented in detail. Finally, discussions of the relevant research findings are provided along with the possible limitations.

METHODOLOGY

The study employed a design and development research methodology of analysis, design, development, and implementation followed by collaboration with target users. Seels and Richey (1994, p. 127) define design and development research as "the systematic study of designing, developing and evaluating instructional programs, processes, and products that must meet the criteria of internal consistency and effectiveness". Design and development research involves not only the design and development of instructional products or programs, but their use and evaluation (Richey & Klein, 2014; Tracey, 2009). Using a design-based process-oriented approach, researchers focus on understanding and improving the multimedia software (Van den Akker, Gravemeijer, McKenney, & Nieveen, 2006).

In the literature, it is argued educational research is far from implementation and may have limited effect in everyday practices and fall short of providing guidance on educational practices (Lagemann & Shulman, 1999). Design and development research, has great potential for designing e-learning environments, is considered the most suitable research method for this study, whose main motivation is to address the limitations identified in the study by developing multimedia software for tablet computers.

It is important to note both qualitative and quantitative research methods can be used in design and development research (Van den Akker, 1999). Thereby, strengthen the validity of the study, using a mixed method approach for data collection, analysis, and interpretation (Creswell, 2014). A series of semi-structured interviews, surveys, and observations with a group of elementary students, teachers, and experts were conducted throughout the study. Table 1 shows an overview of the methods used in the study.

Table 1. An overview of the research methodology (Arpaci, 2014)

Research	Method	Sample
Problem analysis	Semi-structured interviews	Two Science and Technology teachers and three Instructional Technology teachers
Target group analysis	Survey	40 elementary students (18 male, 22 female)
Requirement analysis	Semi-structured interviews, document analysis, software analysis	Five field experts
Pilot testing	Observation and semi-structured interviews	17 elementary students (8 male, 9 female)

Analysis

During the analysis phase, the following questions were considered: What are the current FATIH educational tablet implementation challenges? What are some possible solutions to the challenges? Are the suggested solutions feasible? What are the characteristics of the target users? What are the system requirements? Is the content suitable for the learners? Thus, the analyses performed in this phase included problem, feasibility, target group, requirements, and content analysis.

Problem Analysis

In the problem analysis stage, advantages and limitations of educational use of tablet computers were identified on the basis of extensive discussions with two S & T and three IT public school instructors. Interviews were conducted using open-ended and semi-structured questions. The constant comparative method was used to develop concepts from the data resulting from transcriptions of the interviews. Data analysis process, called coding, involves three levels of analyses: open coding, axial coding, and selective coding (Strauss & Corbin, 2008). Table 2 shows the codes, categories, and themes identified.

In data analysis, matrices were used to present information systematically to the reader, enabling the identification of coding procedures and reducing the categories of information. Stages of the coding process were; reading text recorded from logs, dividing the text into segments of information, coding segments, refining codes, and collapsing the codes into themes.

Results of the qualitative data analysis suggest educational use of tablet computers have several advantages such as bridging equity gaps in access to the devices, enabling quick and easy access to information, providing high motivation, responsibility, and individuality for students. On the other hand, it has a number of limitations needing to be addressed such as; lack of pre-installed learning content, complexity of devices, and issues of maintainability, privacy, and health. Although one of the main advantages of these technologies is delivering interactive and multimedia-enriched learning content to students, it is actually one of the main limitations of the FATIH project. The urgent need to address this limitation has motivated us to develop multimedia software working on tablet computers.

Feasibility Analysis

Sommerville (2015) suggest software specifications or requirements for engineering processes include four fundamental activities including a feasibility study, requirements elicitation and analysis, requirements specification, and requirements validation. A feasibility study focuses on answering three questions: 1.)

Table 2. Themes, Categories, and Codes (Arpaci, 2014)

Theme	Category	Statements and Codes
Advantages of the educational use of tablet computers	Equity	*Bridging* ***equity*** *gap in access to tablet computers...*
	Accessibility	*Quick and easy* ***access*** *to information...*
	Always-on connectivity	*Access to information* ***when and where*** *students need it...*
	Motivation	*Highly* ***motivated*** *students...*
	Responsibility	*Increased* ***responsibility*** *of students...*
	Individuality	*Students can learn* ***individually*** *without relying on instructor...*
	Mobility	*Students do not have to* ***carry*** *their books using a backpack...*
	Communicability	*Students have an ability to* ***communicate*** *effectively using these devices...*
Disadvantages of the educational use of tablet computers	Lack of pre-installed learning content	*There is a* ***lack of pre-installed content*** *on tablet computers...*
	Maintainability issues	*Battery should be* ***charged*** *regularly, these devices can be* ***broken****...*
	Privacy	***Privacy issues*** *may arise when devices were lost*
	Complexity	*Some students have* ***difficulty*** *using touch screen tablet computers...*
	Health issues	***Health issues*** *like eye care problems may occur...*

Does the system contribute to the overall objectives? 2.) Can the system be implemented within budget and schedule using current technology? 3.) Can the system be integrated with other systems used?

This phase aims to answer these questions. Thus, the specifications of the tablet computers were investigated to evaluate the compatibility of these devices with current technologies. Each tablet device distributed to the public schools in Turkey has a screen size of 8.9 inch with 1280 x 768 pixel resolution, allowing touch-screen based navigation through a virtual keyboard. It weighs about 550 grams, and its battery life is about 5 to 10 hours. The processors used in the devices are Cortex A8 type and have 1 GHz speed. It has an Android operating system, 16-32 GB storage capacity, 1.3 M Pixel camera, and 512 MB RAM. Wi-Fi and Bluetooth adaptors are available. It supports various types of multimedia file formats such as MPEG4, MP3, WAV, WMA, AAC, ASF, WMV, AVI, 3GP, MKV, PNG, JPG, BMP, and GIF (MEB, 2015). These specifications were considered during the design and development of the multimedia software.

Requirement Analysis

Sommerville (2015) argues after an initial feasibility study, the next activity should be requirements elicitation and analysis. Requirements elicitation and analysis process activities initiates with requirements discovery. In this activity, the system requirements are discovered and how they interact with stakeholders. The researcher conducts interviews and observations with stakeholders. Scenarios and prototypes are used to assist in understanding the system. A prototype is an initial version of a software system to help with the elicitation and validation of system requirements. Then, the requirements are classified and organized. After that, the requirements should be prioritized and the requirement conflicts should be resolved. Finally, the requirements should be documented.

A triangulated approach, including semi-structured interviews with field experts, document analysis, and software analysis was employed to identify functional and non-functional requirements. First, functional requirements of the multimedia software were identified. Functional requirements describe what the system should do and should not do. For example, functional requirements for the multimedia software include:

- The user shall be able to take notes using the software.
- The system shall generate feedback for each activity.

On the other hand, non-functional requirements define constraints on the functions or services offered by the system. Non-functional requirements can be categorized as products, organizational, and external requirements. Product requirements such as usability, efficiency, dependability, and security requirements constrain the behavior of the software. Organizational requirements include operational requirements, development requirements, and environmental requirements, while external requirements include regulatory requirements, legislative requirements, and ethical requirements (Sommerville, 2015). Here are examples of non-functional requirements for the multimedia software:

- **Portability:** The multimedia software shall work on any platform using Android operating system.
- **Usability:** 95% of the functionalities of the software shall be used one-hour after training.
- **Implementation:** The software shall be designed and coded using Adobe Flash CS6 and Action Script 3.0 (Adobe Systems Inc., San Jose, CA).

Target Group Analysis

During the target group analysis, a user profile questionnaire was conducted with 40 elementary students (18 male, 22 female) to identify their demographic characteristics such as age, gender, and level of technology use and ownership. Thus, students' readiness for adoption of tablet computers was investigated. The results demonstrated students with an average age of 11 have been using computers and the Internet for five and three years, respectively. They mainly use the technologies for homework/research and game-playing purposes. Finally, the results indicated 40% of the participants already had access to mobile devices such as tablet computers (besides the distributed devices) or smartphones.

Content Analysis

In the distribution of tablet computers, 5th and 9th grade students have priority (MEB, 2015). Thus, the first two chapters of the fifth grade S & T course book were chosen for the content of the multimedia software. The content of the related chapters (foods and balanced diet) was obtained from classroom textbooks. Accordingly, the content of the interactive and multimedia-enriched materials such as tests, problem solving, and educational games aligning to learning outcomes were collected.

Design

Instructional design is defined as the systematic development process involving translating principles of learning and instruction into plans to improve the quality of learning (Dooley, 2005; Reiser, 2001). In the literature, there are various instructional design models such as ADDIE (Molenda, 2003), Keller's ARCS Model (1987), Dick and Carey Model (Dick, Carey, & Carey, 2003), Gagne and Brigs Model (1979), and ASSURE Model (Smaldino, Lowther, & Russell, 2012). The well-known instructional design model ADDIE, emphasizes five basic steps of: analyze, design, develop, implement, and evaluate (Molenda, 2003). In the present study, these steps were followed and included involving students, teachers, and field experts.

Multimedia can be defined as an integration of multiple media forms such as audio, video, graphics, text, and animation to present information in a more attractive, effective, and efficient manner (Hall, 1996; Mayer, 2009). Table 3 shows the principles of multimedia design, consisting of principles for reducing extraneous processing, managing essential processing, and fostering generative processing. Following the principles identified by Mayer (2009), the graphical user interface, logo, splash-screen, buttons, and images were designed using Adobe Photoshop CS6 (Adobe Systems Inc., San Jose, CA). Meanwhile, motivation, feedback, and evaluation components of the multimedia software were designed. Finally, the sequence of the learning activities was constructed.

DEVELOPMENT

Sommerville (2015) defines software as computer programs and associated documentation such as requirements, design models, and user manuals. However, a software process can be defined as a set of related activities leading to the production of a software product. Software processes include four fundamental activities: 1.) Software specifications defining the functionality of the software and constraints on its operation, 2.) Software design and implementation including production of the software to meet the specifications, 3.) Software validation to ensure it does what the customer wants, 4.) Software evolution is used to meet changing customer needs (Sommerville, 2015).

Table 3. Principles of multimedia learning (Mayer, 2009)

Principles for reducing extraneous processing	
Signaling principle	Use underlining, bold or italics to focus attention to important content.
Coherence principle	Eliminate extraneous visuals, text, and audio.
Redundancy principle	Eliminate redundancy in content delivery modes.
Spatial contiguity principle	Integrate explanatory text close to related visuals to avoid split attention.
Temporal contiguity principle	Present text and related visuals simultaneously rather than sequentially.
Principles for managing essential processing	
Segmenting principle	Pare content down to essentials.
Pre-training principle	Train the learner about the names and characteristics of the key concepts.
Modality principle	Use animation and narration rather than animation and on-screen text.
Principles for fostering generative processing	
Multimedia principle	Integrate multiple media forms such as audio, video, graphics, text, and animation.
Personalization principle	Use words in conversational style rather than formal style.
Voice principle	Use human voice rather than a machine voice for the narration in multimedia.
Image principle	No need to add the speaker's image to the screen.

Several software development process models are simplified representations to guide the analysis, design, development, test, and maintenance of software products. Traditional software development process models and approaches such as Waterfall, Evolutionary Model, Incremental Model, Spiral Model, Prototyping, and Code and Fix were reviewed here to focus on advantages and limitations. The advantages and disadvantages identified by Sommerville (2015) are shown in Table 4.

The waterfall software development model was adopted for the development of the multimedia software. In guidance of the design report, the multimedia software was then developed and coded using Adobe Flash CS6 and Action Script 3.0 (Adobe Systems Inc., San Jose, CA). The waterfall model is a

Table 4. Software development process models and approaches (Sommerville, 2015)

Advantages	Disadvantages
Code and Fix	
Suitable for small-sized projects	After several fixes the code deteriorates
Suitable when used by the one who built it	Poor match to the user needs
Efforts directly contribute the product	Costly to fix
Waterfall	
Extended documentation	Difficult to respond to changes
Suitable when requirements are well understood	Costly to fix undetected requirements
Suitable when requirements are stable	Product may not meet user needs
Suitable for large-sized projects	Restricted feedback from learners
Evolutionary Model	
Evolves by user feedback	Process is not visible
Specification, development, and validation activities are carried out concurrently	Software deteriorates by changes
Enables users to better understand their needs, and thus, problems are detected earlier	Skilled personnel is required
Suitable for small and medium-sized projects	Expert management is required
Prototyping	
Missing requirements, misunderstandings, and inconsistencies are identified	Prototypes may not reflect real project
Suitable for projects involving high risk	Prototypes can bring some overheads (time, cost)
Suitable for demonstration of feasibility, interface design, and performance analysis	Prototypes might create a false sense of confidence
Incremental Model	
In each increment there is a waterfall	Difficult to partition software increments
Functionality grows with increments	Skilled personnel is needed
Most important increments are delivered first, most important increments are tested most, thus risk of overall project failure is reduced	High-risk activities cause schedule and cost overruns
Spiral Model	
Takes advantages of other models	Requires effective risk management
Risk driven approach might avoid potential difficulties	Rely on subjective risk assessment expertise
Enables early detection and eliminate errors	Expert management is required

well-known process model including the fundamental activities of software specifications, development, validation, and evolution. The researchers took them as separate process phases such as requirement specifications defining the system's requirements, constraints, services, and goals; software design identifying the system's abstractions and their relationships; implementation and unit testing involving coding and testing program units; integration and system testing involving integration of the program units and testing as a complete system, and operations and maintenance involving installation of the system and correction of errors (Sommerville, 2015).

The waterfall model is a document-driven process model and documentation is produced at each phase to make the process visible for managers. However, major problems in the model are the inflexible partitioning of the project into distinct stages, difficult to respond to changing customer requirements. Therefore, this model should only be used when requirements are well understood and unlikely to change radically during system development (Sommerville, 2015).

However, evolutionary model is more suitable when a correct and complete understanding of the user requirements cannot be defined from the very beginning. The evolutionary model divides the developmental cycle into smaller incremental waterfall models. End users provide feedback on the software for the planning stage of the next cycle and developers respond by changing the software. By breaking the project into smaller sections, the management risks can be minimized and early feedback from end users can be received.

Implementation

Sommerville (2015) suggests a software system has to go through three stages of testing: 1.) Development testing is where the software is tested during development to discover defects and bugs. Development testing is carried out at three levels: a.) Unit testing, where object classes or individual program units are tested by focusing on the functionality of the methods or objects, b.) Component testing, where individual units are integrated to create composite components by focusing on component interfaces, c.) System testing, where some or all of the components in the software are integrated and the system is tested as a whole by focusing on component interactions, 2.) Release testing, where a separate testing team tests a complete version of the software before it is released to users to ensure the software meets the requirements, 3.) User testing, involves the end users testing the software in their own environment.

During the implementation phase, target users tested the usability of the multimedia software. Shackel defines the usability of a system as the "capability in human functional terms to be used easily and effectively by the specified range of users, given specified training and user support, to fulfill the specified range of tasks, with the specified range of environmental scenarios" (Shackel, 1991, p. 24). Usability can be measured by both objective performance metrics and subjective user metrics reflecting participants' opinion about the system (Brinck, Gergle, & Wood, 2002; Nielsen, 1997). Usability can also be tested by means of a set of operational dimensions, including ease of learning, ease of use, easy to remember, performance effectiveness, flexibility, few errors, system integrity, and user satisfaction (Nielsen, 1990; Chapanis, 1991; Shackel, 1991; Preece, Rogers, Sharp, Benyon, Holland, & Carey, 1994; Guillemette, 1995).

During implementation, basic procedures of usability testing for effective multimedia development were followed. First, the purpose, test objectives, methods, user profile, task list, test environment, and test monitor's role in a test plan were defined (Rubin, 1994). Then, the software was installed into a tablet computer and thereby a testing environment was prepared. Before conducting the usability testing, the purposes of the test were explained to the participants. In total, 17 elementary students performed tasks on the software while thinking aloud. During debriefing participant questions and recommendations were recorded to improve the design and effectiveness of the software:

1. What is your overall impression of the multimedia software?
2. What are the weak aspects of the multimedia software?
3. Is there anything you feel is missing from the multimedia software?
4. Do you have any suggestions about the multimedia software?

Evaluation

In the usability testing, five different tasks were assigned to 17 participants with different characteristics. The participants completed 74 tasks successfully, whereas seven tasks could not be accomplished. Total time spent for the test was about four hours. Participants identified seven critical usability problems. These problems were evaluated and categorized as functionality problems, design problems, and content related problems.

Possible solutions addressing these problems were identified. It is important to note, corrective, adaptive, and perfective maintenance activities can be required for software maintenance (Schach, 2005). Corrective maintenance is performed to remove errors detected in software, while perfective maintenance is performed to improve the quality of software by adding new functionalities, speeding up the software, or improving traceability of documentation. However, adaptive maintenance is performed to adjust software to changes in the environment by improving portability of the software to a new operating system (Sommerville, 2015). Accordingly, the multimedia software was re-designed by eliminating detected problems to improve the design, effectiveness, and user satisfaction.

ISSUES, PROBLEMS, AND CONTROVERSIES

OECD (2015) report, which provides results of an internationally comparative analysis of the digital skills that students have acquired and of the learning environments designed to improve these skills, suggests that there is no appreciable improvements in student achievement in mathematics, reading or science in the countries that had invested heavily in information and communication technology (ICT) for education. Moreover, technology is of little help in bridging the skills divide between advantaged and disadvantaged students. Their results suggest that students who use computers moderately at school tend to have somewhat better learning outcomes than students who use computers rarely. On the other hand, students who use computers very frequently at school do a lot worse in most learning outcomes. Furthermore, students who spend more than six hours online per weekday outside of school are particularly at risk of reporting that they feel lonely at school, and that they arrived late for school or skipped days of school.

One interpretation of all these issues is that building higher-order thinking and deep, conceptual understanding requires intensive teacher-student interactions, and technology may distract from this valuable human engagement. Another interpretation is that integrating 21st-century technologies to 20th-century poor teaching practices just dilute the effectiveness of teaching (OECD, 2015).

OECD (2015) report confirms advantages of educational use of technology; technology allows teachers and students ubiquities to access up-to-date knowledge and specialized materials well beyond textbooks in multiple formats, it provides great platforms for collaboration in knowledge creation where teachers can share and enrich teaching materials, and support new pedagogies that focus on interactive tools for inquiry-based pedagogies and collaborative workspaces.

However, according to this report, in order to deliver on the promises technology; countries need a convincing strategy to build teachers' capacity, policy-makers need to become better at building support for this agenda, educational institutions should invest in change-management and capacity development skills and back all that up with sustainable financing, teachers should become active agents for change in both implementing technological innovations and designing them.

CONCLUSION

The decision to implement new technologies such as smart glasses, smart phones, and tablets in an educational setting without determining optimal use scenarios is an evident universal problem as the adoption of such mobile platforms becomes widespread. Semi-structured interviews are conducted with two Science and Technology (S & T) and three Information Technology (IT) public school instructors to further investigate this significant problem. The constant comparative method was used to analyze the qualitative data resulting from these interviews. Preliminary results demonstrated the educational use of tablet computers has several advantages, along with a few limitations needing to be addressed. Specifically, one of the main limitations of these new instructional technologies is the lack of interactive content, which is embodied by audio, video, and pictorial multimedia. The urgent need to address this limitation has motivated the development of multimedia software working seamlessly on tablet computers.

In the scope of the FATIH project, a total of 737,800 tablet computers were distributed to all teachers and 9th grade students. Moreover, 10 million 600 thousand-tablet computers will be distributed to all students and teachers by February 2016 (MEB, 2015). However, without a proper understanding of how to effectively integrate such devices into the traditional curriculum, the distribution of such a large volume of tablet computers has encountered numerous problems, and thus, pertinent investigations were carried out.

Problem analysis results suggested the educational use of tablet computers has several advantages, along with a few limitations needing to be addressed. Based on these findings, it can be said the educational use of tablet computers provides equal access and equity for all students, and thus, bridges the digital divide, to improve communication skills, support self-learning of individuals, and increase students' sense of responsibility. In addition, it facilitates quick and easy access to information anywhere and anytime, providing both a strong motivation to learn and greater mobility with lighter, less burdensome backpack loads.

On the other hand, the difficulty of typing with touch screens, privacy, maintenance, and repair issues in case of lost or broken devices were identified as major limitations. Specifically, one of the main limitations of the FATIH project was providing tablet computers without interactive learning content enhanced with multimedia formats, such as audio, video, graphics, and animation. Driven by this motivation, the present study aims to advance the use of multimedia software for fifth grade elementary students, increase achievement in S & T courses, and make learning with tablet computers more attractive, effective, and memorable.

Taken together, the findings have several implications for key figures of the FATIH project, including school principals, teachers, students, tablet computer manufacturers, content designers, developers, and the government.

- The most common reason for failure of large-scale projects is due to resistance to change (Laudon & Laudon, 2007); therefore, the attitudes of students and teachers toward tablet computers are critical for the success of this project. In addition, resistance by school principals may also affect the success of the project. Thus, IT and S & T teachers should act as change agents to facilitate the integration process and ensure all users involved accept changes created by the new technology. They may also work to provide user-training services, which can dramatically reduce user resistance.
- Privacy issues should be considered for the intangible property, which can be easily replicated or transmitted. Therefore, accountability, liability, and control issues must be considered (i.e. if a tablet computer is lost, who is responsible?).
- Software bugs, errors, hardware or facility failures, and inappropriate learning content may cause failure of the project. As aforementioned, one of the main limitations of the FATIH project was providing tablet computers without interactive learning content. Therefore, interactive learning content should be developed for all courses with active input from users. On the other hand, the findings suggest many students were mainly using the technology for game-playing purposes. This implies designers/developers should focus on designing serious games to better captivate the target audience(s) for learning purposes.
- Negative social consequences of the technologies should also be considered. For example, extensive use of the tablet computers and the Internet may impair social, educational or personal functioning. In addition, students can grow to be more dependent on tablet computers, or they may use such devices as instruments of crime. It is important to note potential health risks caused by excessive electronic use of this kind, such as computer vision syndrome, techno-stress, carpel tunnel, etc. should also be incorporated into initial planning.
- The results indicated 40% of the participants already had mobile devices, such as tablet computers (besides the distributed devices) and smartphones. This suggests students who already have these devices could be integrated into the project by just installing the relevant content into their devices.
- According to Moore's law, the power of microprocessors is projected to double every 18 months (Moore, 1965). However, the government may not have enough time or sufficient budgetary allocations to respond to this rapid change. Furthermore, hidden costs, such as the cost of developing interactive content, cost of adjusting to cultural differences, and cost of training should be considered.
- Finally, there are four strategies needed to implement a new system: direct cutover, parallel, pilot, and phased operations (Laudon & Laudon, 2007). It seems a pilot operation could be the most suitable strategy for the success of the FATIH project, for the devices and interactive learning content to be tested first and then deployed nationwide.

The following limitations are significant to the study. The sample size in this study is limited to make such significant generalizations. For example, a small sample of teachers and students were employed to draw general conclusions about the educational use of tablet computers for the entire Turkish schooling system. Moreover, educators' and students' readiness and willingness to use tablet computers may significantly affect the extent to which the realization of prospective learning objectives. Future research should therefore focus on students' and educators' readiness in adopting new and emerging technologies and their attitudes toward the integration of these technologies into the educational system.

REFERENCES

Alvarez, C., Brown, C., & Nussbaum, M. (2011). Comparative study of netbooks and tablet PCs for fostering face-to-face collaborative learning. *Computers in Human Behavior*, *27*(2), 834–844. doi:10.1016/j.chb.2010.11.008

Arpaci, I. (2015). A comparative study of the effects of cultural differences on the adoption of mobile learning. *British Journal of Educational Technology*, *46*(4), 699–712. doi:10.1111/bjet.12160

Baya'a, N., & Daher, W. (2009). Learning mathematics in an authentic mobile environment: The Perceptions of Students. *International Journal of Interactive Mobile Technologies*, *3*, 6–14.

Binboğa, E., & Korhan, O. (2014). Posture, Musculoskeletal Activities, and Possible Musculoskeletal Discomfort among Children Using Laptops or Tablet Computers for Educational Purposes: A Literature Review. *Journal of Science Education and Technology*, *23*(5), 605–616. doi:10.1007/s10956-014-9505-9

Boyce, C. J., Mishra, C., Halverson, K. L., & Thomas, A. K. (2014). Getting Students Outside: Using Technology as a Way to Stimulate Engagement. *Journal of Science Education and Technology*, *23*(6), 815–826. doi:10.1007/s10956-014-9514-8

Brinck, T., Gergle, D., & Wood, S. D. (2002). *Designing Web sites that work: Usability for the Web*. San Francisco: Morgan Kaufmann Publishing.

Chapanis, A. (1991). Evaluating usability. In B. Shackel & S. J. Richardson (Eds.), *Human factors for informatics usability* (pp. 359–395). Cambridge: Cambridge University.

Chen, C. M., & Chung, C. J. (2008). Personalized mobile English vocabulary learning system based on item response theory and learning memory cycle. *Computers & Education*, *51*(2), 624–645. doi:10.1016/j.compedu.2007.06.011

Chen, C. M., & Hsu, S. H. (2008). Personalized intelligent mobile learning system for supporting effective English learning. *Journal of Educational Technology & Society*, *11*(3), 153–180.

Chen, J. D., Falkmer, T., Parsons, R., Buzzard, J., & Ciccarelli, M. (2014). Impact of experience when using the Rapid Upper Limb Assessment to assess postural risk in children using information and communication technologies. *Applied Ergonomics*, *45*(3), 398–405. doi:10.1016/j.apergo.2013.05.004 PMID:23735750

Chu, H. C., Hwang, G. J., & Tsai, C. C. (2010). A knowledge engineering approach to developing mind tools for context-aware ubiquitous learning. *Computers & Education*, *54*(1), 289–297. doi:10.1016/j.compedu.2009.08.023

Couse, L. J., & Chen, D. W. (2010). A tablet computer for young children? Exploring its viability for early childhood education. *Journal of Research on Technology in Education*, *43*(1), 75–98. doi:10.1080/15391523.2010.10782562

Creswell, J. W. (2014). *Educational research: Planning, conducting and evaluating quantitative and qualitative research* (5th ed.). Upper Saddle River, NJ: Pearson Prentice Hall.

De-Marcos, L., Hilera, J. R., Barchino, R., Jiménez, L., Martínez, J. J., Gutiérrez, J. A., & Otón, S. et al. (2010). An experiment for improving students performance in secondary and tertiary education by means of m-learning auto-assessment. *Computers & Education*, *55*(3), 1069–1079. doi:10.1016/j.compedu.2010.05.003

Dick, W., Carey, L., & Carey, J. O. (2003). *Courseware Authoring*. Boston, MA: Pearson Custom Publishing.

Dooley, K. E. (2005). *Advanced methods in distance education: Applications and practices for educators, administrators and learners*. Hershey, PA, USA: IGI Global. doi:10.4018/978-1-59140-485-9

Doolittle, P., & Mariano, G. (2008). Working memory capacity and mobile multimedia learning environments: Individual differences in learning while mobile. *Journal of Educational Multimedia and Hypermedia*, *17*(4), 511–530.

Dündar, H., & Akçayır, M. (2014). Implementing tablet PCs in schools: Students' attitudes and opinions. *Computers in Human Behavior*, *32*(0), 40–46. doi:10.1016/j.chb.2013.11.020

Evans, C. (2008). The effectiveness of m-learning in the form of podcast revision lectures in higher education. *Computers & Education*, *50*(2), 491–498. doi:10.1016/j.compedu.2007.09.016

Fernández-López, Á., Rodríguez-Fórtiz, M. J., Rodríguez-Almendros, M. L., & Martínez-Segura, M. J. (2013). Mobile learning technology based on iOS devices to support students with special education needs. *Computers & Education*, *61*, 77–90. doi:10.1016/j.compedu.2012.09.014

Ferrer, F., Belvis, E., & Pamies, J. (2011). Tablet PCs, academic results and educational inequalities. *Computers & Education*, *56*(1), 280–288. doi:10.1016/j.compedu.2010.07.018

Gagne, R. M., & Brigs, L. J. (1979). *Principles of Instructional Design*. New York: Holt, Rinehart & Winston.

Galligan, L., Hobohm, C., & Loch, B. (2012). Tablet technology to facilitate improved interaction and communication with students studying mathematics at a distance. *Journal of Computers in Mathematics and Science Teaching*, *31*(4), 363–385.

Gedik, N., Hanci-Karademirci, A., Kursun, E., & Cagiltay, K. (2012). Key instructional design issues in a cellular phone-based mobile learning project. *Computers & Education*, *58*(4), 1149–1159. doi:10.1016/j.compedu.2011.12.002

Guillemette, R. A. (1995). The evaluation of usability in interactive information systems. In J. M. Carey (Ed.), *Human factors in information systems: Emerging theoretical bases* (pp. 207–221). Norwood, NJ: Ablex.

Güler, Ç., Kılıç, E., & Çavuş, H. (2014). A comparison of difficulties in instructional design processes: Mobile vs. desktop. *Computers in Human Behavior*, *39*, 128–135. doi:10.1016/j.chb.2014.07.008

Hall, T. L. (1996). *Utilizing multimedia toolbook 3.0*. Danvers, MA: Boud & Fraser.

Hwang, G. J., Yang, T. C., Tsai, C. C., & Yang, S. J. H. (2009). A context-aware ubiquitous learning environment for conducting complex science experiments. *Computers & Education*, *53*(2), 402–413. doi:10.1016/j.compedu.2009.02.016

Jeng, Y.-L., Wu, T.-T., Huang, Y.-M., Tan, Q., & Yang, S. J. H. (2010). The add-on impact of mobile applications in learning strategies: A review study. *Journal of Educational Technology & Society*, *13*(3), 3–11.

Kadirire, J. (2009). Mobile Learning Demystified. In R. Guy (Ed.), *The Evolution of Mobile Teaching and Learning* (pp. 15–56). California, USA: Informing Science Press.

Keller, J. M. (1987). Development and use of the ARCS model of motivational design. *Journal of Instructional Development*, *10*(3), 2–10. doi:10.1007/BF02905780

Kukulska-Hulme, A. (2005). Mobile usability and user experience. In A. Kukulska-Hulme & J. Traxler (Eds.), *Mobile learning: A handbook for educators and trainers* (pp. 45–56). London: Routledge.

Kukulska-Hulme, A., Sharples, M., Milrad, M., Arnedillo-Sánchez, I., & Vavoula, G. (2009). Innovation in mobile learning: A European perspective. *International Journal of Mobile and Blended Learning*, *1*(1), 13–35. doi:10.4018/jmbl.2009010102

Kukulska-Hulme, A., & Traxler, J. (2007). Learning design with mobile and wireless technologies. In H. Beetham & R. Sharpe (Eds.), *Rethinking pedagogy for the digital age: Designing and delivering e-learning* (pp. 180–192). London, UK: Routledge.

Lagemann, E. C., & Shulman, L. S. (1999). *Issues in education research: problems and possibilities*. San Francisco: Jossey-Bass Publishers.

Laudon, C. K., & Laudon, J. P. (2007). *Essentials of Business Information Systems*. Upper Saddle River, NJ: Pearson Education.

Lavery, H. (2012). Breaking down barriers with tablet technology for teaching. *Educause Review Online*.

Mayer, R. E. (2009). *Multimedia learning* (2nd ed.). New York: Cambridge University Press. doi:10.1017/CBO9780511811678

Ministry of Education, FATIH Project. (2015). *MEB.gov*. Retrieved from http://www.meb.gov.tr/press-release/haber/8098/en/

Molenda, M. (2003). In Search of the Elusive ADDIE Model. *Performance Improvement*, *42*(5), 34–36. doi:10.1002/pfi.4930420508

Moore, G. E. (1965). Cramming more components onto integrated circuits. *Electronics*, *38*(8), 114–117.

Nielsen, J. (1990). Evaluating hypertext usability. In D. H. Jonassen & H. Mandl (Eds.), *Designing hypermedia for learning* (pp. 147–168). Berlin: Springer-Verlag. doi:10.1007/978-3-642-75945-1_9

Nielsen, J. (1993). *Usability engineering*. Boston: Academic Press.

Nielsen, J. (1997). Usability testing. In G. Salvendy (Ed.), *Handbook of Human Factors and Ergonomics* (2nd ed., pp. 1543–1568). John Wiley & Sons.

Nortcliffe, A., & Middleton, A. (2013). The innovative use of personal smart devices by students to support their learning. In A. Wankel & P. Blessinger (Eds.), *Increasing student engagement and retention using mobile applications: Smartphones, Skype and texting technologies (cutting-edge technologies in higher education)*. Bingley, UK: Emerald Group Publishing Limited. doi:10.1108/S2044-9968(2013)000006D009

Pereira, A., Miller, T., Huang, Y. M., Odell, D., & Rempel, D. (2013). Holding a tablet computer with one hand: Effect of tablet design features on biomechanics and subjective usability among users with small hands. *Ergonomics*, *56*(9), 1363–1375. doi:10.1080/00140139.2013.820844 PMID:23909815

Preece, J., Rogers, Y., Sharp, H., Benyon, D., Holland, S., & Carey, T. (1994). *Human-computer interaction, Workingham*. England: Addison-Wesley.

Rehmat, A. P., & Bailey, J. M. (2014). Technology Integration in a Science Classroom: Preservice Teachers' Perceptions. *Journal of Science Education and Technology*, *23*(6), 744–755. doi:10.1007/s10956-014-9507-7

Reiser, R. A. (2001). A history of instructional design and technology: Part II: A history of instructional design. *Educational Technology Research and Development*, *49*(2), 57–67. doi:10.1007/BF02504928

Richardson, W. (2012). Gaming gains respect. *District Administration*, *48*(6), 50–51.

Richey, R. C., & Klein, J. D. (2014). *Design and development research: Methods, strategies, and issues*. New York, NY: Routledge. doi:10.1007/978-1-4614-3185-5_12

Romney, C. A. (2011). Tablet PC use in freshman mathematics classes promotes stem retention. *Proceedings of the 41st annual Frontiers in Education Conference* (pp. F1J-1 - F1J-7). doi:10.1109/FIE.2011.6142773

Rubin, J. (1994). *Handbook of usability testing: How to plan, design, and conduct effective tests*. New York, NY: John Wiley & Sons.

Schach, S. R. (2005). *Objected-oriented and classical software engineering* (6th ed.). New York: McGraw-Hill.

Schneps, M. H., Ruel, J., Sonnert, G., Dussault, M., Griffin, M., & Sadler, P. M. (2014). Conceptualizing astronomical scale: Virtual simulations on handheld tablet computers reverse misconceptions. *Computers & Education*, *70*, 269–280. doi:10.1016/j.compedu.2013.09.001

Schwabe, G., & Göth, C. (2005). Mobile learning with a mobile game: Design and motivational effects. *Journal of Computer Assisted Learning*, *21*(3), 204–216. doi:10.1111/j.1365-2729.2005.00128.x

Seels, B. B., & Richey, R. C. (1994). *Instructional technology: The definition and domains of the field Association for Educational Communications and Technology*. Washington, DC: Association for Educational Communications and Technology.

Shackel, B. (1991). Usability-Context, framework, definition, design and evaluation. In B. Shackel & S. J. Richardson (Eds.), *Human factors for informatics usability* (pp. 21–37). Cambridge: Cambridge University.

Sharples, M., Milrad, M., Arnedillo, S. I., & Vavoula, G. (2009). Mobile learning: Small devices, big issues. In N. Balacheff, S. Ludvigsen, T. De Jong, A. Lasonder, & S. Barnes (Eds.), *Technology enhanced learning: Principles and products* (pp. 233–249). Heidelberg: Springer. doi:10.1007/978-1-4020-9827-7_14

Smaldino, S. E., Lowther, D. L., & Russell, J. W. (2012). *Instructional technology and media for learning* (10th ed.). Boston, MA: Allyn & Bacon.

Sommerville, I. (2015). *Software Engineering* (10th ed.). NY: Pearson.

Strauss, A., & Corbin, J. (2008). *Basics of qualitative research: Grounded theory procedures and techniques* (3rd ed.). Newbury Park, CA: Sage.

Students, Computers and Learning. Making the Connection. (2015). *OECD*, Paris. doi:.10.1787/9789264239555-en

The definition of educational technology. (2004AECT, Association for Educational Communications and Technology. Washington, DC.

Tracey, M. (2009). Design and development research: A model validation case. *Educational Technology Research and Development*, *57*(4), 553–571. doi:10.1007/s11423-007-9075-0

Traxler, J. (2005). Case studies: introduction and overview. In A. Kukulska-Hulme & J. Traxler (Eds.), *Mobile learning: A handbook for educators and trainers* (pp. 173–188). London: Routledge.

Ullrich, C., Shen, R., Tong, R., & Tan, X. (2010). A mobile live video learning system for large-scale learning-system design and evaluation. *IEEE Transactions on Learning Technologies*, *30*(1), 6–17. doi:10.1109/TLT.2009.54

Van den Akker, J. (1999). Principles and methods of development research. In J. van den Akker, R. Branch, K. Gustafson, N. Nieveen, & T. Plomp (Eds.), *Design approaches and tools in education and training* (pp. 1–14). Dordrecht: Kluwer Academic Publishers. doi:10.1007/978-94-011-4255-7_1

Van den Akker, J., Gravemeijer, K., McKenney, S., & Nieveen, N. (2006). Introduction to educational design research. In J. Van den Akker, K. Gravemeijer, S. McKenney, & N. Nieveen (Eds.), *Educational design research* (pp. 3–7). London: Routledge.

Wu, W. H., Wu, Y. C. J., Chen, C. Y., Kao, H. Y., Lin, C. H., & Huang, S. H. (2012). Review of trends from mobile learning studies: A meta-analysis. *Computers & Education*, *59*(2), 817–827. doi:10.1016/j.compedu.2012.03.016

Young, J. G., Trudeau, M., Odell, D., Marinelli, K., & Dennerlein, J. T. (2012). Touch-screen tablet user configurations and case-supported tilt affect head and neck flexion angles. *Work (Reading, Mass.)*, *41*(1), 81–91. PMID:22246308

KEY TERMS AND DEFINITIONS

ADDIE: An instructional design model emphasizing five basic steps within the instructional design process of; analyze, design, develop, implement, and evaluate (Molenda, 2003).

Design and Development Research: "The systematic study of designing, developing, and evaluating instructional programs, processes, and products that must meet the criteria of internal consistency and effectiveness" (Seels & Richey, 1994, p. 127).

Educational Technology: "The study and ethical practice of facilitating learning and improving performance by creating, using, and managing appropriate technological processes and resources" (AECT, 2004, p. 3).

FATIH: A nationwide educational project designed to promote effective use of ICT in public schools in Turkey.

Instructional Technology: "The theory and practice of design, development, utilization, management, and evaluation of processes and resources for learning" (Seels & Richey, 1994, p. 1).

Mobile Technology: Refers to portable devices offering ubiquitous access to information using smart glasses, smart watches, smart phones, tablet computers, and other similar devices.

Multimedia: An integration of multiple forms of media such as audio, video, graphics, text, and animation to present information in a more attractive, effective, and efficient manner (Mayer, 2009).

Section 3

Mobile and Wearable Technology Applications

Chapter 9
Wearables and People with Disabilities:
Socio-Cultural and Vocational Implications

Damara Goff Paris
Emporia State University, USA

Katrina R. Miller
Emporia State University, USA

ABSTRACT

Assistive technology including wearables, have a positive impact towards changing the lives of people with disabilities. Advances in glass, electronics, sensors, and a wide range of digital devices are substantially improving how people with disabilities (PWD) navigate the world, and allow greater possibilities for PWD who are competing in today's job market. This chapter explores the following topics related to disabilities: Historical developments changing the lives of people with physical and sensory disabilities, laws pertaining to technology access, technological advances in wearables for individuals who are deaf, blind, paralyzed, or other mobility and speech impairments, socio-cultural and vocational implications of wearables for individuals who are deaf, deaf-blind or speech impaired, and the applications needed for wearables to meet the on-the-job demands.

INTRODUCTION

In the past fifty years, particularly since the advent of the computer and Internet, great technological strides have been made to increase social and vocational accessibility for people with disabilities. Such advances have demonstrated the importance of ensuring PWD can access every facet of society a non-disabled person has access to, from everyday activities promoting independent living skills to vocational and recreational activities. The concept of accessibility is addressed by federal mandates, including the Americans with Disabilities Act of 1990 (ADA) and the Telecommunications Act of 1998. This legislation states that it is the right of PWD to have access to technology in order to live full, productive lives.

DOI: 10.4018/978-1-5225-0069-8.ch009

This chapter begins by reviewing sociocultural and vocational issues faced by PWDs. A brief overview of federal mandates supporting technological accommodations will follow. Discussion of the historical contribution of essential assistive technology for people with sensory disabilities (visual impairments and people with hearing loss, including deaf, deaf-blind, hard of hearing and late-deafened individuals) and mobility disabilities will be explored. Current developments in wearable technologies for other disabilities will be reviewed. A section on the perceived impacts of technology and wearables on vocational and sociocultural aspects of living with disability will be provided, concluding with issues on the barriers to the use of wearables for PWD.

The objectives of this chapter are to:

1. Familiarize the reader with the historical impact of significant assistive technologies for people with disabilities;
2. Review some current wearables assisting people with disabilities on a day-to-day basis;
3. Discuss the potential sociocultural impact of the use of wearables for people with communication disabilities who desire increased interactions with other people;
4. Provision of an overview of the current job market for people with disabilities;
5. Review potential wearables still in developing stages; and
6. Review barriers obtaining and using wearables that PWD experience, and potential solutions offered.

SOCIOCULTURAL AND VOCATIONAL IMPLICATIONS OF DISABILITY

Nearly one in five people in the United States are reported to have a disability (Bertoni, 2012). While some people are born with a disability, almost everyone is at risk of experiencing a short-term or long-term disability, whether through accidents, employment-related injuries, illnesses, disease, or aging.

Despite the public's increased exposure to PWD in the media, laws that protect disabled people from discrimination, and federal mandates ensure functionally equivalent access to social and work environments, there remains a barrier towards acceptance of people with disabilities by mainstreamed society. In a book chapter entitled *On the origins of negative attitudes toward people with disabilities*, Livneh (2012) proposed several theories behind the misperceptions of society about PWD. One of these theories was sociocultural conditioning. According to Livneh, social and cultural norms, standards, and expectations create negativity towards the disabled population. Some examples are listed below.

- Emphasis on the "body beautiful", "body whole", youthfulness and physical prowess, creating high societal expectations which members of society are expected to conform.
- Judgment, especially in Western cultures, on personal productiveness and achievement. People with disabilities are perceived as unable to be socially and economically competitive, particularly in capitalist based nations
- A negative attitude towards people of certain socioeconomic levels, particularly those who are perceived to be abusing welfare. People with disabilities are considered largely at poverty level and highly unemployable.
- Negativity around members of society who are part of the "sick role" mentality. The perspective is one of playing sick to be exempt from normal societal obligations and responsibilities, particu-

larly those with chronic illnesses still largely unexplained, such as fibromyalgia. Society suspects that individuals with unfamiliar illnesses, who display variations in their disability (i.e., able to walk one day, and in a wheelchair the next day) of malingering in order to receive social service benefits or sympathy.

- Social stigma related to disability plays a large role in negative attitudes. People with disabilities are viewed as "outsiders", "offenders" or "different." This devaluation by society of people with disabilities, where society sees them as undesirable.
- Fear plays a large part in how non-disabled members of society react to PWD. In some situations where the disability may be severe and highly visible, there is concern about how to act or talk with the PWD. In other cases, fear that stems from superstition or lack of knowledge may cause avoidance when members of society are afraid that the disability is contagious. For example, until recently, there was a belief that pregnant women should avoid contact with individuals who were deaf or visually impaired in order to protect the fetus.

While many of the negative attitudes stem from non-disabled people, individuals with a disability are likely to have similar negative attitudes towards themselves or other people with disabilities. Often the root of this self-concept resides within the individual and depending on their background, and may be a result of observations from their social environment. As a result, PWD internalize oppression, and may then oppress himself or herself or other PWD.

Like non-disabled individuals, PWD are heterogeneous in terms of personal identity. The intersection of culture and disability may be shaped by socially and individually defined values and beliefs (Ripat & Woodgate, 2011). The meaning of disability varies greatly throughout different cultures. Even in Western societies, the definition of disability is not homogenous. For example, the biopsychosocial approach to disability is a defined as a negative state where impairment, activity limitation, or participation restriction exist (Ripat & Woodgate, 2011). On the flip side, in a social model, an individual is disabled as a result of oppressive societal practices and environments creating disabling conditions.

Although disability is often more accepted in Western cultures, particularly in a society where legal mandates prevent the discrimination of individuals with disabilities, there are patronizing perspectives. Some members of mainstream society see PWD as helpless victims who need protection or decisions made for them. Others view PWD as heroes, assigning positive attributes of "superdisabled" applauding the slightest accomplishment a disabled person does that would be considered normal activities undertaken by a non-disabled individual. Such viewpoints can impact the perception of people with disabilities have of themselves.

Ryan et al. (2005) suggested intergroup communications between non-disabled and disabled people can result in stigmatization of the disability and negative stereotypes ranging from "blaming the victim" for the disability, to the imposition of homogenous expectations of PWD. In other words, PWD may be pressured into expressing the same values; beliefs and behavioral characteristics based on the belief all people with a specific disability are the same.

Goffman (Hindhede, 2012) theorizes feelings of dependence other people violate western culture's idea of the isolated, self-contained individual. In the US, for example, independence is a highly prioritized trait, and dependency is not encouraged. However, for some PWD (such as those who are deaf or hard of hearing), there might be pride in their identity. The deaf community often do not identify themselves as disabled, but members of a culture bound together by a common language (American Sign Language), and historical and cultural norms and values (Padden & Humprhies, 1988). Self-identity and perception, therefore, plays an important part in how an individual with a disability might view himself or herself.

Overall, there are many factors impacting one's self-concept and acceptance of oneself as a person with a disability. These factors include: age of onset (whether the individual was disabled at a young age or much older and experienced most of life as a non-disabled individual), environmental and cultural influences (whether they were brought up in an environment in which the disability was accepted as opposed to being brought up in a culture employing a biopsyhchosocial (rather than a social model approach to disability), the degree the disability is visible (whether the disability is overtly physical such as involving daily use of a wheelchair as opposed to an occasional recurrences of arthritis).

For PWD, identity, culture and technology are often interconnected. Since the use of wearable or assistive technology (AT) may increase the likelihood of being identified as a person with a disability, self-identity and self-acceptance of a disability will play a large factor in whether the technology will be used, even if it is a necessary tool towards independent activity. For those using smartphones with applications to communicate, keep track of tasks, such as medicine dosages, it may actually be normalizing to use devices like everyone else.

Individuals who identify with a minority or ethnic group and use technology may hold differing viewpoints on the meaning of disability and the value of AT for those in mainstream society. At an individual and family level, how AT and wearables shape an individual's culturally defined identity, and the meaning AT holds to the person and family influence, the choice to use of wearables.

In essence, while wearables and assistive technology may increase and improve interactions between the social and physical environments and PWD, whether they are used by PWD is determined by a number of factors that relate to his or her self-identity and self-acceptance.

When it comes to employment, people with disabilities are severely under or unemployed. According to the 2010 census, approximately 4-in-10 individuals aged 21 to 64 with a disability were employed, compared with 8-in-10 individuals with no disability. Eleven percent of the adults aged 15 to 64 with severe disabilities experience persistent poverty. Of those with no disability, 4% experienced persistent poverty. An investigation by the U.S. Senate's Health, Education, Labor and Pensions Committee found that only 30% of working-age people with disabilities were participating in the workforce, and that one in three PWD live in poverty (Heasley, 2014).

Workers with disabilities are almost twice as likely as nondisabled workers to be in contingent and part-time jobs (Schur, 2002). Contingent work encompasses a variety of conditional and transitory employment arrangements, providing little to no job security for workers and generally provide low pay and few benefits. It normally includes temporary help agency employment, on-call, day labor, independent contracting, and other jobs with limited tenure. In combination with rising displacement rates, the growth of contingent work is often perceived as an indication of a general weakening of employee-employer attachments and the decline of long-term, stable jobs. Even though a number of anti-discrimination laws have been passed to reduce oppression for this population, getting hired for jobs remains an obstacle.

Title I of the ADA provides for the elimination of disability discrimination in the workforce, particularly through its mandate employers implement "reasonable accommodations"— modifications to or variations in workplace policies, schedules, and tasks—without eliminating the essential functions of the job or pose undue hardships (administratively or financially) on the employer. The ADA provides guidelines for implementing respectful personnel practices when it comes to disability-related issues. It frames itself as a tool for eradicating a long history of bias and stigma against people with disabilities, as well as serious underemployment and unemployment of the population (Basas, 2013).

Disability scholars attribute the underemployment of PWD to attitudinal barriers preventing them from obtaining positions commensurate with their skills. There are also issues of retention within the workplace, if employers fail to effectively accommodate employees with disabilities. One of the concerns of employers when it comes to hiring workers with disabilities—whether new to the workplace or existing workers—is how they will be accommodated and who will be responsible for paying for the accommodation. In addition, many employers become overwhelmed when reviewing ADA and the requirements for an accessible workplace. They fear expensive solutions and are not aware of simple ways to provide accessibility in the workplace. Not all workers with disabilities need or ask for accommodations. Approximately 10% of employees with disabilities ask for AT or equipment-based accommodations (Basas, 2013). Some PWD incur costs for accommodations that are very low compared to other kinds of accommodations. In addition accessible tools that are often used are personal devices or personal attendants, which is not the responsibility of employers.

Historical and Legal Developments Changing the Lives of People with Disabilities

Before discussing the impact of AT on people with disabilities, it is important to mention that the advent of personal computers, laptops and smart devices have greatly helped PWD equalize their interactions with non-disabled individuals, as well as federal mandates that have opened the door to making technologies tenable. The use of digital equipment and applications designed for the general public, such as e-mail, social media and Internet communication software (i.e., Skype, Facetime, and Zoom), have been successfully applied to people with disabilities, allowing for enhanced communication and interactions with non-disabled individuals. Use of computers and the Internet has allowed PWD to keep limitations and differences in functioning invisible and resulted globally in interactions without stigmatization. For example, through computer bulletin boards, emails, and chats about social media events, deaf individuals are able to learn and share their views on cultural events that thcy traditionally would have no means by which to participate. In addition to computers, Internet, and open or closed-captioned television programs, there are a number of inventions that have increased the functional equivalence of PWD.

One of these devices is the telephone. Initially, deaf people found this invention to be one of the worst inventions in terms of presenting obstacles to communication. Ironically, the creation of the telephone was originally credited to Alexander Graham Bell, a hearing teacher of deaf children and a child of a deaf parent. As the telephone became more commonplace among businesses, deaf individuals were often excluded from eligibility for jobs because use of the telephone was central to many workplace environments.

In 1964, Robert Weitbrecht, a deaf scientist, devised an acoustic coupler that converted a series of tones into text. The Teletype machines were later discarded after World War II. Weitbrecht and a team of scientists created the Teletypewriter for the Deaf (TTY). Using a standard phone placed on a coupler the TTY transmitted and received audible signals, which was by translated into text so it could be printed on paper similar to an adding machine (Baynton, Gannon, & Bergey, 2007). While it took almost a full decade to bring a portable TTY into the households of members of the deaf community, across the nation, the device became instantly popular. Deaf people were now able to call other deaf friends, family members or anyone who had access to a TTY. Hooking the TTY to a flashing light allowed a visual signal when a call was coming in. From this invention, a number of other technologies were established that served to alert deaf people to standard noises in their environments. Visual alert systems such as a flashing light or vibrating alarm clocks, doorbell alerts connected to lamps to notify household members someone was at the door, and strobe light smoke detectors became the norm in deaf households and some places of business by the end of the twentieth century.

While the TTY was a technological wonder when it first emerged, it soon became apparent accessibility over the phone was still problematic. People who were not deaf and did not have access to TTYs were still unable to communicate with deaf people over the phone. Partially due to federal mandates such as the Americans with Disabilities Act (ADA) of 1990, and the Telecommunications Act of 1996, functionally equivalent telecommunications access for people with hearing and speech disorders was in demand. In addition, The Tech Act of 1988 and the Assistance Technology Act of 1998 were enacted by Congress to provide a formal and legal definition of assistive technology. Within this definition, assistive technology can range from something as simple as a modified spoon to be used for eating and pencil grips and adaptive switches, to complex, computer-based, augmented-communication systems.

It is important to note that prior to the development of TTYs for deaf individuals, the reading levels among the deaf population were varied, with many not reading beyond 2nd grade level—due to controversy and dissension in education regarding how to educate deaf people. Prior to the 1990's, there was no early hearing detention laws for infant, and often children were not identified as deaf until well after the critical period for learning a language, creating delayed access to language.

In the late 1980's, several states began experimenting with organizations that provided the service of passing information from a phone to a TTY and back again so that people who were deaf, deaf-blind, hard of hearing or late-deafened could make calls. Beginning as volunteer services to connect deaf citizens to hearing people so that they could conduct everyday tasks such as setting up doctor appointments or communicating with the IRS, the FCC eventually mandated that each state set up what was called Telecommunications Relay Services (TRS). By 1993, every state provided TRS that was paid for by state and federal funds.

In 1996, Ed Bosson, a TRS administrator for the state of Texas, developed the concept of Video Relay Services (VRS). Through the use of videophones, deaf American Sign Language (ASL) users were able to connect visually to sign language interpreters who interpreted calls with non-signers, and then relay communication between the two parties. The service, which was formally funded by the FCC in 2001, instantly transformed the lives of ASL users. For the first time, they were able to communicate over the phone using their natural, visual-gestural language. These technologies have allowed deaf people to access and use telephone systems similarly to how non-deaf individuals use the phone. As a result, deaf people had fewer barriers to employment opportunities when the use of the telephone was required on the job.

Prior to the establishment of the telephone and the TTY, was the invention of Braille for individuals with visual impairments by a teenager from north central France. Blinded at age three, Louis Braille invented a universal system for reading and writing for people with visual impairments (Roth & Fee, 2011) by the time he was fifteen years old. He published the first Braille book, *Method of Writing Words, Music, and Plain Songs by Means of Dots, for Use by the Blind and Arranged for Them*, in 1829.

Braille is read by passing one's fingertips over characters made up of an arrangement of one to six embossed points. These points represent different alphanumeric characters and can be written with a Braillewriter or by using a pointed stylus to punch dots through paper using an instrument called a Braille slate. An essential tool used by people who are blind or visually impaired, it has been translated into almost every known language (Roth & Fee, 2011). Since the development of Braille almost 200 years ago, the point system has evolved and can be printed from a brailling device attached to a laptop or personal computer. In addition, screen readers have been developed to allow interface between the computer's operating system, its applications, and the user. The user sends commands by pressing different combinations of keys on the computer keyboard or braille display to activate the speech synthesizer, and to speak automatically when changes occur on the computer screen.

People with other disabilities have also experienced innovations historically impacting this population. For example, prior to the end of World War II, most individuals who experienced limb amputations used crude attachments made of steel, wood or aluminum. These devices often had limited range of motion, particularly joint movement. During the US Civil War, the number of amputations rose dramatically, prompting the development of improved prosthetics. James Hanger, an amputee and veteran of the Civil War, patented the "Hanger Limb", made from whittled barrel staves (Norton, 2007). This was considered a vast improvement over previous models. After World War II, a significant turning point occurred in the field of prosthetics. A group of military personnel, orthopedic surgeons, engineers and prosthetics formed the Committee on Prosthetics Research and Development (CPRD) to address a growing need for advances in limb prosthetics (Childress, 1985). A result from this research group was the Vaduz hand. A result from this research group was the Vaduz hand. Heralded as an advanced device ahead of its time, the Vaduz hand contained components found in today's electronically powered limbs. Due to the pioneering work of Hanger and CPRD, persons with amputations today select from an overwhelming array of near-perfect prosthetic models, which allow for near-natural movement.

In addition, with the rise of computers, laptops and smart phones, more applications, or apps, have been developed to assist people with disabilities. Individuals who have cerebral palsy, have had strokes, have spinal cord injuries, or other progressive diseases that impact speech (i.e., Amyotrophic Lateral Sclerosis or Lou Gehrig's disease) are able to use digital applications that allow for text-to-speech by using hands -free devices or systems with eye-tracking capabilities.

WEARABLE TECHNOLOGIES FOR PEOPLE WITH DISABILITIES

Today's most popular wearable technological devices, or "wearables" are most frequently tied to fitness and personal health trackers (Johnson, 2015). Devices range from fitness trackers, watches, glasses, and smart clothing to capture data and send it to an application on the wearer's mobile device or computer (Singer & Perry, 2015). The wearer then accesses the data, along with a summarized analysis.

Wearables have been endorsed for job safety and communication. Sixty-eight percent of the global technology and business leaders see wearables as a priority for their companies. Seventy-three percent of working adults worldwide see at least one benefit of wearables in the workplace, according to the Work Force Institute (Johnson, 2015). According to a report by PricewaterhouseCoopers (PWC), more than three-quarters (77%) of respondents said an important benefit of wearable technology is its potential to make them more efficient and more productive at work. Seventy percent expected their workplace to permit the use of wearable technology, and 46% thought their employer should fund wearable technology.

People with disabilities, however, have a different perspective on the importance of wearables. Personal health tracking is important, but when it comes to wearables, often the user with a disability is using a wearable to equalize their workplace environment in order to perform essential job duties, or preventive measures to avoid a life threatening health episode. Below are descriptions of wearables either on the market or in the process of being approved to assist PWD. People who do not have a disability may commonly use some of these devices.

DEVICES FOR PEOPLE WITH PHYSICAL DISABILITIES

WPAL

Wearable Power Assist Locomotor (WPAL) is a motorized device for patients with paraplegia (Tanabe, Hirano, and Saitoh, 2013). It is designed to assist those with paraplegia with standing and walking. The device consists of a wearable robotic orthosis. The wearable robotic orthosis is based on a medical system with wheelchair compatibility. In addition, motors located at hip, knee and ankle joints, reduce heart rate during walking. The purpose behind WPAL is the person with a disability will fluctuate their usage of the wearable robot and wheelchair. For the majority of the time, patients wear only the flexible inner thigh cuffs in the wheelchair. The other part of the robot and walker are placed at a convenient location suitable for walking (e.g. kitchen, standing work space, etc.). While this wearable may seem cumbersome and time consuming to put on and take off, the implications for improving health on the part of someone with paraplegia are important. Standing and occasional walking improve circulation, decreasing the likelihood of muscle atrophy, and reduce the occurrence of sores from sitting or lying for long periods of time. Long-term sitting in a wheelchair causes medical problems such as pressure sores or osteoporosis. Circulation improvement can be beneficial on the job, reducing the number of days the worker is hospitalized, interrupting schedules at work.

Google Glasses

One of the most recent inventions tested, as a prototype over the past few years has been Google Glass. It is also a device receiving a significant amount of media coverage (Metz, 2015). Users are able to use minimal head and hand movements, or use voice, to take photos and videos while viewing the world through a pair of rather large glasses. In addition, Google Glass allows people to read and answer texts and e-mails, look up information on Google, and use other functions similar to a computer or handheld devices.

Users of Google Glass, however, state the device is not totally hands-free. Lin (Stewart, 2015), who was excited about the release of the glasses, set to work on developing ways to modify the device for people with disabilities. Some of the concepts included hooking the glasses to alternate controls and switches to interface through a joystick on a power wheelchair, helping people who may not have fine motor skills or a discernible voice (such as people who have cerebral palsy) to control the functions of Google Glass.

Initially Google Glasses were a limited release with an estimated $1,500 retail value. Although the device started out strong, with a number of people interested in purchasing the device, Google Glasses stopped the consumer Glass 1.0 Explorer program using the initial prototype. They recently announced the release of the new Glass 2.0 Enterprise version for industry. There has been other version of glasses that were released, with similar functions, such as Huawei's Honor Lumus DK-40 Smartglasses, Meta Pro's AR Glasses, Vuzik Smart Glasses and Atheer Air, Gesture-based, 3D AR, glasses. These devices range from $549 to almost $3,000.

TDS

The Tongue Drive System (TDS) is a wireless and wearable AT for people with severe disabilities to assist them with accessing personal computers. Through a study (Kim et. al, 2014) conducted for the prototype, participants chose to have a magnetic tracer directly glued on their tongue, or used an embedded titanium tongue-barbell attached to the tongue through piercing. It has been determined even the most severe mobility disabled are able to move their tongue, which is why the tongue was sought for a site of control with the TDS. In addition, the prototype used in the study consisted of a headgear holding the magnetic sensors near the cheeks of the user on a pair of goosenecks, which facilitate their positioning.

Sensors were sampled at 50 Hz, and the raw data was transmitted wirelessly to a PC or smartphone (iPhone, Apple; Cupertino, California) at 2.4 GHz. A real-time sensor signal-processing algorithm, running on the PC/smartphone, cancels the external magnetic interference and detects tongue positions for up to six user-defined commands plus its neutral/resting position. Such a wearable increases accessibility to work computers and other motor functions necessary to complete on the job tasks.

The conclusion of the study, which consisted of questionnaires and qualitative information gathered in addition to testing the actual performance of the TDS system, was that overall, the participants were not concerned about the tongue piercing and would be willing to wear them throughout the day. A larger sample will be tested soon.

DEVICES FOR PEOPLE WHO ARE DEAF, DEAFBLIND OR HARD OF HEARING

Technology to improve hearing has progressed significantly since the first hearing aid was developed in the late 1800's. Digital hearing aids, personal FM Loop systems and the Cochlear Implant are several devices that have become available over the last few decades. While all may be considered wearable in their own right, most do not connect to smart technology, and in the case of Cochlear Implants, require invasive, albeit routine, surgery in order to use. However, in at least one case, hearing aids has been connected to smart technology in order to improve its function. Starkey's Halo hearing aids are controlled through an iPhone app and a Pebble smartwatch, which allowed the user the ability to adjust the volume through these smart devices (Miller, 2014). There are a number of prototypes that have emerged to improve the communication access for sign language users who seek the ability to convert a visual language into written means, so that individuals might interact with those who are not well versed in visual communication.

Gesture Vocalizer

Four Indian students of the Holy Grace Academy of Engineering, Mala, recently developed a device that can translate sign language into spoken words that benefit not only deaf individuals, but also individuals who are speech impaired (Desai, 2015). Intended to reduce the amount of time normally used to communicate with deaf or speech impaired individuals, the device has several components. A glove that is covered with sensors detects the location and movement of simple hand movements, including the bending of fingers. An additional accelerometer can pick up on gestures through the tilting of hand movement. An audio processor attached with a speaker and an LCD display displays the output as well as verbally translates the gestures. The processor is portable and can be powered by a 9V battery. No cost estimates for manufacturing the translating components have been given.

Enable Talk

A group of Ukrainian students won first place in the Software Design Competition of the 2012 Microsoft Imagine Cup. The prize netted $25,000 for students to continue development (Hill, 2012). The team came up with the idea after observing college students' who were deaf and struggling to communicate with other students. Initially, the team looked at commercially available gloves to modify to interpret a range of signs, but in the end, they opted to develop their own.

Their glove had a total of 15 flex sensors in the fingers to determine the motion of the glove through space. The sensor data was processed by a microcontroller on the glove sent via Bluetooth to a mobile device and translates the positions of the hand and fingers into text when the pattern is recognized. The team continues to work on expanding the vocabulary while gathering funds to release the device. At this time, it is estimated that each device can be manufactured for $75.

Jafari's Motion and Muscle Tracking Approach to ASL Translation

A yet-unnamed device has been developed to translate ASL into speech by Roozbeh Jafari, associate professor at Texas A & M University's Department of biomedical Engineering (O'Brien, 2015). Stating that some systems for translating ASL to speech uses visual cameras, Jafari's project applies an external motion sensor and wearable muscle-tracking sensor that measures electrical signals into numerical values that are interpreted by computers. Currently, Jafari and team are working on improving the sensitivity and accuracy of the devices, particularly since the system must be retrained to match the specific style of ASL vocabulary used by the wearer. In addition, while the system recognizes individual words, a pause is regarded between each word (or sign), which impacts the timeliness of outputting the translation. The immediate goals is to output whole phrases and sentences, as well as increase the number of vocabulary words of the device.

Finger-Braille

Finger-Braille (Matsuda and Isomura, 2012) is a method of tapping a Deaf-Blind's person fingers to provide tactile information to mimic Braille. This method is used in Japan to communicate with the country's more than 20,000 Deaf-Blind individuals. A similar communication method is used in the United States called Tactile Communication. In the US, however, Tactile Communication is provided in several ways—through close vision communication in which sign language is used near the person's personal space and does not move outside an agreed-upon square like area within the vision of the Deaf-Blind individual. Another form of Tactile Communication is through fingerspelling each word, with the Deaf-Blind individual using their hands to receive information from each finger spelled word.

Glove-style interfaces were used for this prototype because they covered the palm or the fingertips, which has the highest tactile sensitivity. Two types of ring-shaped devices were developed: 1) vibration motor type, and 2) solenoid type. The vibration motor type consists of six small-size, lightweight DC motors, used for vibration of cellular phones. Solenoid type consists of six tubular solenoids. Each device is fixed onto the setting of the ring. Thereby, the palm can be used freely and the device never prevents the user from obtaining original tactile sense information. The initial version of this prototype connects to a parallel port, and the improved version connects to a USB port where the characters or messages for each prototype use the optimum position of the interface device.

In the initial experiment, the devices were firmly set to the root of the fingers. In this location, phantom sensation was experienced and some participants could not recognize the message at all. This experiment revealed the best place to attach the device for tapping in daily Finger-Braille communications. In the next experiment, the vibration motor type device was more readable than the solenoid type. However, many Deaf-Blind people believed, if improved, the solenoid type would be better since it is similar to the original Finger-Braille method.

Glove Prototype for Deaf-Blind

A similar device to Finger-Braille was developed by Tom Bieling, a German researcher, and doctoral candidate in the Design Research Lab at Berlin's University of the Arts (Schelmetic, 2013). Bieling developed a prototype-computerized glove to translate text into the Lorm Alphabet to help people who are both deaf and blind to communicate. Blind and deaf Austrian scientist, Hieronymus Lorm, invented the Lorm alphabet in 1881.

The glove, made of Gore-Tex material, has wires and sensors and is paired with a blue-tooth transmitter, and interprets the alphabet through "touches" that vibrate an incoming text message. The participants, who are deaf-blind, reported they could understand incoming Lorm vibrations on the tops of their hands just as well as they could when someone tapped out letters in their palms.

The device works in the following ways: the blind-deaf person taps letters onto the glove's palm side, and these touches are converted into digital text and sent via Bluetooth to a smartphone app to send the text or e-mail. Bieling is currently working on a model made of lighter material.

Sony Entertainment Access Glasses

Prior to the release of Sony Entertainment Access Glasses to over 6,000 Regal Cinemas Theaters (Rood, 2013), deaf and hard of hearing moviegoers were restricted to limited releases of captioned movies on the big screen, often only once or twice per week. Another device, which fit into cup holders of theater seats, allowed viewing of film captions through a small window on the device that often was out of range of the action occurring in the film. Moviegoers with a hearing loss wanted more movie viewing options and non-deaf movie attendees complained of being distracted by on-screen captioning. Due to these limitations, only 38% of Americans who are deaf or hard of hearing went to a movie at least once in 2012. By contrast, 72% of Americans without a hearing loss attended a movie at least once in 2012 (Herbert, 2013). Regal Cinemas worked with Sony to develop glasses that showed text directly into the lenses of the device.

The Sony glasses are similar to 3-D glasses. Captions are projected onto the glasses and appear to float about 10 feet in front of the user. The glasses are attached to a receiver box, which allows the user to adjust the display distance (near, mid, far) and brightness (on a scale of 1 to 5). For individuals with enough residual hearing, the glasses provide volume control (Herbert, 2013). The glasses are not only for people who are deaf or hard of hearing. For people with visual impairments, there is an option for audio tracks that describe the action on the screen. The glasses cost Regal Cinemas approximately $1,750 per pair and can be checked out for free prior to film showings.

ESight

ESight eyewear is a wearable, electronic assistive technology designed specifically for people with low vision. The device incorporates a headset and hand controller to adapt a live video stream, and optimizes an LED display placed directly in front of the eyes of a user (Maynard, 2015).

Controls permit the operator to adjust contrast, brightness, and the image color. ESight eyewear uses a high-resolution camera to capture what the user is looking at, and projects in real-time onto two LED screens situated in the eyewear. Some of the benefits of eSight eyewear include being able to view items when shopping, to see items above or below eye level, or to recognize familiar faces from a distance.

OrCam

The main function of the OrCam device is to provide text-to-voice for visually impaired users. OrCam has a small camera clipped to the wearer's glasses using a magnet and connected by a thin cable to a portable computer designed to fit in the wearer's pocket. A bone-conduction speaker bypasses the ear canal, providing voice-to-text capability as it reads aloud the words or objects pointed to by the user.

Unlike other systems, which use cumbersome devices to translate text into speech in limited environments, the OrCam system is designed to both recognize and speak "text in the wild," (Markoff, 2013). This term is used to describe newspaper articles and environmental text such as bus numbers, billboard signs, and changing stoplights. The software also helps users to identify familiar faces in a crowd. An additional benefit of the OrCam is the ability to enhance the vision of people who are blind by allowing visual signals to bypass damaged retinas and carry information directly to the user's brain.

SOCIOCULTURAL AND VOCATIONAL IMPLICATIONS

Wearable's and AT devices have a considerable impact on social, economic, educational, vocational, and independent living opportunities for people with disabilities. The WPAL, TDS, and Google Glass all offer people with mobility and speech impairments opportunities for greater independence. Many individuals who have SCI (Spinal Cord Injuries) are paralyzed from the neck down, with restricted movement capability. This is also true of other disabilities, such as the gradual deterioration of body functions experienced by individuals who have ALS or Multiple Sclerosis, individuals who have had strokes, or individuals who were born with cerebral palsy. While TDS and Google Glass do not present a perfect solution for independence from personal care attendants, they provide some range of movement for PWDs on the job, so that employers can expand their tasks and duties. As Glass and other smart glasses continue to evolve, it is anticipated that technological advancements will continue to broaden opportunities for workers with disabilities.

Wearables for physical mobility enhancement may not increase the sociocultural experiences of people with mobility impairments during interactions with people who are not disabled. However, there may be a significant impact when it comes to access to the world of work. Employers are often hesitant to hire people with severe physical disabilities out of concern about the costs associated with accommodating them. Employers can be shown alternatives making to extensive structural changes in the workplace, and wearables further mitigate physical barriers by providing virtual access to information and materials. .

In addition, there is a fascination with wearables and AT among co-workers and employers. While such technologies may make a worker's disability more noticeable, transparency about wearables and the need for technological solutions to counteract some of the limitations faced by workers with disabilities may allow persons with disabilities to engage in education and self-advocacy in social and vocational settings.

Wearables increase communication options and access for people who are visually impaired, deaf-blind, or deaf and hard of hearing, and may actually have a significant positive influence on the sociocultural and vocational experiences of these individuals. For deaf workers, communications with non-deaf workers who do not use sign language is a barrier often difficult to overcome. Using wearables to increase voice-to-text capabilities has the potential to expand social opportunities and improve employability.

The use of wearables to interpret between two languages can increase communication options and reduce worker isolation. For deaf and hard of hearing workers, the avenue for communicating with hearing co-workers provides opportunities for educating and self-advocacy. While most deaf people are experienced with the use of commonly known or universal gestures, having technology available to increase communication s will have a positive social impact. Viewing a movie using Sony Entertainment Access Glasses provides deaf people with access to communication in a group-oriented social activity.

Participation in the workforce is critically important for persons with disabilities. At the state and federal levels, an array of vocational rehabilitation services has been available for persons with disabilities (PWD) since the 1920s. Vocational rehabilitation (VR) counselors assist nearly one million PWD with pre-employment skills, employment placement, and employment supports annually (citation?). Included in these services are various assistive technologies. These technologies may be designed specifically for PWD, but in many cases existing wearable technology can be applied to resolve an employment need. Mobile devices, worn on the hip or in one's pocket, provide a rich selection of applications (apps). Young adults, minorities, people with no college-level training, and people from lower household income levels use smartphones as their main source of Internet access. Creative use of smartphone apps and digital music players can allow PWD to participate more fully in the workforce. For example, smartphone alarms can be set to remind people to take medication, or for transitioning to a new work task. Free and inexpensive apps can be downloaded to assist with monitoring diet and exercise, arranging task lists for completion at home or on the job, or to augment communications as needed. As wearables improve in mobility and connection to smart phones, it is possible they will assist individuals with disabilities by increasing access to workforce opportunities and adapting to various workplace demands.

It is recognized that a majority of PWD would benefit from wearables and AT to improve their social environment as well as their vocational options. During a report on International Disability Day, WHO representative Dr. Michel Thieren emphasized the need to harness the power of technology to promote the advantages of using assistive technology and accessible information to ensure full and equal participation of persons with disabilities in society and in shaping the future of sustainable development for all (Majority of the disabled facing discrimination, poverty Yet, poverty and lack of funds remain barriers to PWD when it comes to purchasing wearables and AT. Many PWD cannot afford wearables to assist them in employment settings, or in social situations. While many of the prototypes discussed earlier have not yet become available for public use, it is expected the debut prices will be high. Preliminary costs for Google Glasses were approximately $1,500 per unit, out of the price range for many potential users, with or without disabilities. As stated previously, there is a high unemployment and underemployment rate for people with disabilities (citation). The majority of PWD receive SSI or SSDI, and even with federal disability payments, they live at or below the poverty level (citation). Without additional support from VR, Veterans Affairs, or other state-federal programs to fund technology for PWD, it is likely this population will be unable to acquire emerging technologies with the greatest potential for helping them.

WEARABLES IN THE WORKFORCE

Under the Americans with Disabilities Act (1990), employers are not required to provide personal devices such as handhelds or wearables to employees with disabilities. However, it is not unusual for corporations to provide essential employees with technology in order to increase their productivity and availability. While it is difficult to discern the ways that persons with disabilities are utilizing or adapting wearables in the professional workforce, workers with disabilities' use of personal devices and apps in the service industries has been documented (Heasley, 2012).

In a study of workplace accommodation, workers who had Autism Spectrum Disorder (ASD) were equipped with iPods to assist them on the job (Gentry, Kriner, Sima, McDonough, & Wehman, 2012). Picture books and task lists were developed to clarify and demonstrate job responsibilities to workers; and, auditory tones, instructional videos, and verbal commands were used to prompt task changes and breaks, and to remind workers of safety considerations. Workers who participated in the iPod intervention used significantly fewer job-coaching hours than is typical for new workers with disabilities (Gentry, et al., 2015). A 2009 case study of the iPod as a task-prompting mechanism for a worker with developmental disabilities indicated similar findings (Van Laarhoven, Johnson, Van Laarhoven-Myers, Grider, & Grider, 2009).

In order to address a 15% unemployment rate among persons with disabilities, the federal government sponsored a disability employment-focused app design competition in 2012. Key apps introduced as a result of this campaign were a job bank for persons with disabilities, and a disability accommodations resource for employers. Because the designers of such apps are typically unfamiliar with specific disability needs, most app users with disabilities in the workforce are individuals who have adapted apps made for the general public.

Mobile technology provides primary or sole access to Internet resources for a significant number of younger persons, persons of color, and persons of lower socioeconomic status (Zickuhr & Smith, 2012). The architectural trend towards universal design, or physical environments built for use by older persons and persons with or without disabilities, has moved into education as universal learning design and is now appearing in computer science as universal access. It is anticipated that emergent development of wearables will require consideration of a broad range of user abilities and needs in the workplace (Emilliani & Stephanidis, 2005).

CONCLUSION

As technology evolves and becomes more connected to the needs of the public, there will be a niche for wearables for PWD. From smart glasses allowing visually impaired individuals to maximize their residual vision to trendy devices used in tongue piercings to control access to personal computers, there are devices of benefit for a wide range of disabilities.

While there are exciting and unique wearable prototypes in various stages of development, financial barriers still exist preventing the majority of PWD from obtaining access to the technology. The substantial numbers of people with disabilities living at or below the poverty level cannot afford wearables to assist them with communication or employment. It is important that the information on wearables is disseminated to programs serving PWDs. Such programs include Veterans Affairs (VA), State Telecommunications Assistance Programs and other state or federal programs that provide assistive technology supports. The more aware PWDs are of programs that distribute applicable technologies, the more likely they are to receive the benefits wearable technologies offer. In closing, we are hopeful the information provided in this chapter will be helpful to workers with disabilities, employers of PWDs, and the innovators and engineers working to develop wearable technologies for universal use.

REFERENCES

Basas, C. G. (2013). Universally designing the public sector workplace: Technology as disability access. *Working USA*, *16*(1), 69–86. doi:10.1111/wusa.12024

Bertoni, D., & US Government Accountability, O. (2012). Employment for People with Disabilities: Little Is Known about the Effectiveness of Fragmented and Overlapping Programs. Report to Congressional Committees. GAO-12-677. *US Government Accountability Office*.

Brault, M. (2012). *Health and disability statistics branch*. United States Census Bureau of the Census.

Childress, D. S. (1985). Historical aspects of powered limb prostheses. *Clinical Prosthetics & Orthotics*, *9*(1), 2–13.

Desai, A. N. (2015, June 4). Indian students develop device that converts sign language into speech and text—lending voice to the mute. *INQUISITR*. Retrieved from http://www.inquisitr.com/2144746/indian-students-develop-device-that-converts-sign-language-into-speech-and-text-lending-voice-to-the-mute/

Emilliani, P. L., & Stephanidis, C. (2005). Universal access to ambient intelligence environments: Opportunities and challenges for people with disabilities. *IBM Systems Journal*, *44*(3), 605–619. doi:10.1147/sj.443.0605

Gentry, T., Kriner, R., Sima, A., McDonough, J., & Wehman, P. (2015). Reducing the need for personal supports among workers with autism using an iPod touch as assistive technology: Delayed randomized control trial. *Journal of Autism and Developmental Disorders*, *45*(3), 669–684. doi:10.1007/s10803-014-2221-8 PMID:25212414

Heasley, S. (2012). IPod may ease transition for those with autism. *Disability Scoop*. Retrieved from http://www.disabilityscoop.com/2012/09/06/ipod-transition-autism/16398/

Heasley, S. (2014, September 18). Nearly 1 in 3 with disabilities in poverty. *Disabilityscoop*. Retrieved from http://www.disabilityscoop.com/2014/09/18/report-disabilities-poverty/19683/

Herbert, G. (2013). Regal's new closed caption glasses finally make movies enjoyable for deaf and hard-of-hearing. *Syracuse.com*. Retrieved from http://www.syracuse.com/entertainment/index.ssf/2013/06/closed_caption_glasses_movies_regal_theaters_deaf.html

Hill, D. (2012). Smart gloves turn sign language gestures into vocalized speech. *Singularity Hub.* Retrieved from http://singularityhub.com/2012/09/16/smart-gloves-turn-sign-language-gestures-into-vocalized-speech/

Hindhede, A. L. (2012). Negotiating hearing disability and hearing disabled identities. *Health, 16*(2), 169–185. doi:10.1177/1363459311403946 PMID:21540252

Jeonghee, K., Hangue, P., Bruce, J., Rowles, D., Holbrook, J., Nardone, B., & Ghovanloo, M. et al. (2014). Qualitative assessment of Tongue Drive System by people with high-level spinal cord injury. *Journal of Rehabilitation Research and Development, 51*(3), 451–465. PMID:25019667

Johnson, D. (2015). The revolution will be wearable. *Industrial Safety & Hygiene News, 49*(5), 1–30.

Livneh, H. (2012). On the origins of negative attitudes toward people with disabilities. In I. Marini & M. Stebricki (Eds.), *The psychological and social impact of illness and disability* (6th ed., pp. 13–14). New York, NY: Springer Publishing Company.

Majority of the disabled facing discrimination, poverty. (2014, December 3). *Nation (Pakistan).*

Markoff, J. (2014). Device from Israeli start-up gives the visually impaired a way to read. *New York Times*, A5.

Matsuda, Y., & Isomura, T. (2012). Finger Braille recognition system. In X. Ding (Ed.), *Advances in Character Recognition* (pp. 193–210). Rijeka, Croatia: Intech Open.

Metz, R. (2015). Google Glass Is Dead; Long Live Smart Glasses. *MIT Technology Review*, 79.

Miller, A. M. (2014). Wearable tech for people with disabilities. *U.S. News Health.* Retrieved from http://health.usnews.com/health-news/health-wellness/articles/2014/11/20/wearable-tech-for-people-with-disabilities

Norton, K. M. (2007). A brief history of prosthetics. *inMotion, 17*(7). Retrieved from http://www.amputee-coalition.org/inmotion/nov_dec_07/history_prosthetics.html

O'Brien, J. (2015, November 16). Breaking a different kind of language barrier: Sign language becomes sensor-based. *Mashable.*

Padden, C., & Humphries, T. (1988). *Deaf in America: Voices from a culture*. Cambridge, MA: Harvard University Press.

Ripat, J., & Woodgate, R. (2011). The intersection of culture, disability and assistive technology. *Disability and Rehabilitation. Assistive Technology, 6*(2), 87–96. doi:10.3109/17483107.2010.507859 PMID:20698763

Rood, R. (2013). New closed-captioning glasses help deaf go out to the movies. *NPR.org.* Retrieved from http://www.npr.org/sections/alltechconsidered/2013/05/12/183218751/new-closed-captioning-glasses-help-deaf-go-out-to-the-movies

Roth, G. A., & Fee, E. (2011). The Invention of Braille. *American Journal of Public Health, 101*(3), 454. doi:10.2105/AJPH.2010.200865 PMID:21307377

Ryan, E. B., Bajorek, S., Beaman, A., & Anas, A. P. (2005). "I Just Want You to Know That 'Them' is Me": Intergroup Perspectives on Communication and Disability. In *Intergroup Communication* (pp. 117–137). New York, NY: Peter Lang Publishing, Inc.

Scheltic, T. (2013). Glove Prototype Provides Mobile Communications Capabilities for the Blind-Deaf. *Wearable Tech News.* Retrieved from http://www.wearabletechworld.com/topics/wearable-tech/articles/343796-glove-prototype-provides-mobile-communications-capabilities-the-blind.htm

Schur, L. A. (2003). Barriers or opportunities? The causes of contingent and part-time work among people with disabilities. *Industrial Relations*, *42*(4), 589–622. doi:10.1111/1468-232X.00308

Singer, R. W., & Perry, A. J. (2015). Wearables: The well-dressed privacy policy. *Intellectual Property & Technology Law Journal*, *17*(7), 24–27.

Stewart, M. (2015). Through the Google GLASS. *Pn, 69*(2), 34-37.

Tanabe, S., Hirano, S., & Saitoh, E. (2013). Wearable Power-Assist Locomotor (WPAL) for supporting upright walking in persons with paraplegia. *NeuroRehabilitation*, *33*(1), 99. doi:10.3233/NRE-130932 PMID:23949036

Van Laarhoven, T., Johnson, J. W., Van Laarhoven-Myers, T., Grider, K. L., & Grider, K. M. (2009). The effectiveness of using a video iPod as a prompting device in employment settings. *Journal of Behavioral Education*, *18*(2), 119–141. doi:10.1007/s10864-009-9077-6

Zickuhr, K., & Smith, A. (2012). Digital differences. *Pew Research Center.* Retrieved from http://www.pewinternet.org/2012/04/13/digital-differences/

KEY TERMS AND DEFINITIONS

Contingent Work: Contingent work is a casual work environment including part-time or temporary workers where the employment is often not permanent or on going.

Identity: A set of qualities or characteristics distinguishing one person from another person or group of people.

Sign Language Interpreter: A person trained in interpreting between a spoken and a signed language. Someone who hears the spoken language then interprets it into sign language and/or voices the sign language.

Sociocultural: When combining cultural and social elements, sociocultural is a set of beliefs, customs, practices and behaviors within a population.

Teletype: A form of telegraph where code is transmitted and translated into printed messages. This was used most often during war times of WWI and WWII.

Visually Impaired: When eyesight cannot be corrected to a normal level, and it is below a functional level of vision.

Chapter 10
Mobile Technologies for Student Centered Learning in a Distance Higher Education Program

Lisbeth Amhag
Malmo University, Sweden

ABSTRACT

The aim of this study is to analyze mobile technologies for student centered learning in a distance higher education program with a focus on mobile online webinars (web-based seminars or conferencing) using mobile applications such as laptops, smart phones, or tablets. As wearable technologies continue to grow it could very well extend to smart glasses, smart watches etc. These tools can provide face-to-face interactions, recording flipped classrooms and parallel chat communications. The data collection consists of observations of ten online face-to-face webinars with 22 students, six interviews, and two surveys. Theoretically, the study joins the research tradition of Computer-Supported Collaborative Learning with emphasis on collaboration, and Computer Self-Efficacy concerned with individuals' media and information literacy. Important conclusions from the study demonstrated mobile interactions increased student centered learning on theoretical concepts, assisted in the ability to review information critically, and provided experiences bridging professional teaching practices.

INTRODUCTION

The aim of this chapter is to analyze and describe wearable technologies for student centered learning in a distance higher education program. We will focus on mobile online webinars using mobile applications such as laptops, tablets or smart phones using teacher-recorded inverted / flipped classrooms. The interactions also include verbal face-to-face (F2F) and parallel textual chat communications with other students and teachers.

DOI: 10.4018/978-1-5225-0069-8.ch010

Lage et al. (2000) describes the easiest definition of the flipped or inverted classroom: "Inverting the classroom means that events that have traditionally taken place inside the classroom now take place outside the classroom and vice versa" (p.32). The core idea of flipped classrooms are to invert the common instructional approach, as homework before the lesson instead of after, by using teacher-recorded videos with various briefings and interactive instructions. These teacher-recorded instructions and videos can be accessed from different locations by using mobile technologies, such as laptops, tablets or smart phones, prior to the class learning activities. In this study, the flipped classroom consists of two parts: 1) teacher-recorded videos with individual instructions outside the classroom; and 2) interactive group learning activities F2F inside the mobile online webinars. The concept removes the teacher-recorded videos and instructions from the online webinars in order to allow better use of the in-class time during the online webinars F2F with discussions, tutoring and scaffolding student centered learning activities (Bishop & Verleger, 2013; Lage et al., 2000; Long et al., 2014). Scaffolding knowledge processes is a kind of "thinking types" for fostering expertise in writing and argumentation, designed to provide procedural facilitation (Scardamalia & Bereiter, 1983).

"Class" becomes the room / place of the mobile online webinars integrating F2F conferencing with a predefined aim of using real-time experiences with guidance from a teacher. The potential of mobile online webinars F2F is that everyone can see and hear each other's verbally through discussions, while at the same time communicating via textual chat and notes. This concept assists learners F2F directly in problem solving, discussing theoretical concepts, reviewing the literature, engaging in collaborative group work, as well evaluations. Furthermore, the synchronous online webinars can be recorded for later asynchronous viewing online within the Learning Management System (LMS) in order to provide students with the opportunity to take a step back, reflect, self-assess and compare various contributions. In summary, several student centered learning activities can be a part of the course structure by using laptops, tablets or smart phones with teacher-recorded videos and instructions before the mobile online webinars F2F and parallel chat communications in order to mediate meaning and learning on a critical and higher-order level.

Examples of learning activities are:

- Tutoring and scaffolding, as well as peer learning processes F2F.
- F2F group discussions regarding theoretical concepts, literature review, authentic experiences, problem solving, and course assignments.
- Receiving and providing F2F peer feedback, co- and self-assessment.
- Collaborate with online chat, common notes, documents and whiteboards F2F.
- Sharing screens and software with others F2F.
- Examinations of course assignments and critical peer review F2F.

It is important to research how and in what ways mobile technologies, as in this study, by using laptops, tablets or smart phones, can support student centered learning and collaboration in a more digitalized higher education, instead of merely focusing on how new media techniques can make education more effective or detrimental (e.g. Amhag, 20011; 2012; 2013; Lee & Salman, 2012; Rockinson-Szapkiw et al., 2013; Sana et al., 2013).

BACKGROUND

Research on the emerging pedagogy of teacher-recorded flipped classroom videos in higher education are mostly based on popular press articles and conference proceedings. Bishop and Verleger (2013) have done a comprehensive survey of flipped classrooms to avoid a considerable amount of buzz in academic circles at all levels. The online buzz the authors have documented from mainly newspaper articles, particularly academically oriented ones, and online blogs. Furthermore, in complete websites dedicated to promoting the flipped classroom ideology. Moreover, several organizations are marketing materials, as screencasts, to help teachers who want to implement the flipped model in their classroom. However, most of the research studies explore student perceptions and work in single-groups, or mixed groups. The results show students overall are generally positive (ibid. 2013). A literature review by Roehl et al. (2013) demonstrates recorded flipped classroom videos are easy-to-use with mobile accessible technologies in order to provide prepared pre-class activities for planned in-class time. This design allows for an expanded range of learner-centered activities during class time and provides opportunities for greater teacher-to-student tutoring, peer-to-peer collaboration and cross-disciplinary engagement.

Likewise, the trial of flipped classroom videos by Wagner et al. (2013) indicate the recorded materials can offer a unique, yet challenging opportunity for universities to collaborate on maximizing the effectiveness of higher education. For example, the effectiveness of lab reinforced materials in comparison to lecture delivered materials and strategies to improve teaching with flipped lectures as a tool to enhance the presentation and retention of lecture materials. Some observations by Wallace (2014) are made regarding evidence on how video, text and instructions are balanced both synchronously and asynchronously according to need and practicality. The roles of the teacher and the student, and how these relationships become more complex and complementary in mobile online spaces, are also changing. The essence of flipped classrooms, according to Long et al. (2014), show more opportunities can be provided to students during in-class time. The teachers can prepare for their course implementing, both the pre-class phase and the in-class phase, applying their own technical literacy, and other available resources and supports. Even faculty contact time can be reduced by two-thirds using an active learning classroom according to Baepler et al. (2014). In this study, the students achieving learning outcomes performed significantly better, in comparison with the traditional classroom. Even student perceptions of the learning environment were improved. These results suggest pedagogically dialogue exchanges and active learning classrooms have a positive influence.

The impact of online synchronous webinars as demonstrated by Nelson´s (2010) study, found the learning outcomes of an online synchronous webinar versus classroom instruction among 224 nursing students found no significance difference between the groups. An online synchronous webinar was just as effective as classroom instruction. Rich´s (2011) investigation, measuring the impact of online synchronous webinar instruction at a National Science Foundation Advanced Technology Center. The study demonstrated collaborative factors but a lacked additional evidence on actual participation outcomes. Buxton et al. (2012) research evaluated pharmacists' satisfaction with and reasons for enrolling in a series of continuing education webinars. The online F2F webinars included presentations, lectures, workshops and online discussions. The results showed both positive satisfaction for the quality of the program efforts and the method of delivery, but lacked continued participation in subsequent webinars. Using Skype for online communications was examined by Kiriakidis (2010), who found it increased self-efficacy to school and district administrators' and found a gain in students' performance in various centered learning activities. The results showed self-efficacy can be increased through opportunities for ongoing, systemic, and logic discussions with other participants. Moreover, institutions of higher education, professional development providers, administrators' associations, school districts, and school leaders may benefit from having an awareness of how mobile online communication software can increase students' self-efficacy in the field of distance teaching and learning. However, all studies above focused on organizing and administrating the education from an individual perspective.

MAIN FOCUS

In this study, the main focus is on how mobile technologies such as laptops, tablets or smart phones with recorded teacher-created flipped classroom videos and mobile online webinars F2F can support active learning and collaboration in a more digitalized education. The tools are used for students' learning interactions and parallel chat communications with other students and teachers as a rising phenomenon. The following question will be illuminated from the study:

- How and in what way do students collaborate and share knowledge and experiences F2F during mobile online webinars using different multimedia tools in teacher-recorded flipped classroom, textual chat, and note taking as a mediating tool for learning and critical review?

The study monitored 22 student teachers (women=16, men=6) participating in half-time study in the first three continuing web-based courses over a eighteen-month period in Vocational Teacher Education Program, (VTEP). The student teachers were working to become authorized instructors in any of the twelve upper secondary school vocational programs in Sweden. VTEP in Sweden comprises 90 credits. The courses include 60 credits in educational sciences and 30 credits in teacher training at a partner school or within the students own teaching position as unauthorized teacher. The qualitative research design involved an in-depth data collection process and analysis of open-ended interviews and observations of mobile online webinars F2F. As a result, the researcher had the opportunity to obtain a deeper inquiry per individual and collectively in order to provide a more complete understanding. Two surveys' were conducted after completing the second course in June, 2013 and January, 2014 on whether the training contributed to their learning, improved activity engagement, increased their use of rich media and information literacy. Response options were on a 6-point scale: 1 = Disagree – 6 = Strongly Agree, and two open-ended qualitative questions.

THEORETICAL APPROACH AND ANALYSIS

Theoretically, the study joins the research tradition of Computer Supported Collaborative Learning, (CSCL) based on the premise that it is not possible to understand learning solely from individual actions, and Computer Self-Efficacy (CSE) concerned with the individuals' media and information literacy. According to socio-cultural theory, learning always arises as a product of a social community of practice in which our understanding of language, communication, culture, and various aspects of the social context for student learning and development is central. Wenger (1998) considered sociocultural theory of learning in terms of *communities of practices* constituted by students´ negotiated engagement and joint enterprise, as well shared repertoire. There is also a question of collective appropriation of tools through language, and how students use language as a tool for learning. Wenger describes assumptions of learning and the nature of knowledge with four premises: 1) we are social beings which is a central aspect of learning, 2) knowledge is a matter of the competence to evaluate, 3) knowing is a matter of participating in the pursuit of such valuation, and 4) meaning is our ability to experience the world and engage with it meaningfully (Wenger 1998. p. 4).

Methodically, the concept of Wenger (1998, p. 5) is divided into four components: 1) *meaning is learning as experiences;* a way of talking about our abilities, individually and collectively, to experience our life and the world as meaningful, 2) *practice is learning as doing*; a way of talking about the shared historical and social resources, frameworks, and perspectives to sustain mutual engagement in action, 3) *community is learning as belonging*; a way of talking about social configurations in which our enterprises are defined as worth pursuing by way of suggesting, and participating as a recognizable competence, and 4) *identity is learning as becoming*; a way of talking how learning changes the unique person we are and creates personal histories of becoming and development in and between moments of teaching practices and as professionals, e.g. as hairdresser, seller, chef, or vocational teacher. According to this perspective, individual's dialogue, interactions, and interplay constitute a determining factor for the individual's learning and knowledge development in higher vocational education.

The theoretical approach of self-efficacy is related to Computer Self-Efficacy (CSE), which has been used in research concerned with individuals' intentions to use information technology (IT). As in this study, teacher-recorded flipped classroom and participate in mobile online webinars F2F, as well chat. This concept give the opportunities to challenge different efficacies and make assessments of their ability to apply knowledge, and manage and evaluate the peer learning activities, individually and collectively (Bandura, 2002). Bandura stresses "self-precepts of efficacy influence thought patterns, actions, and emotional arousal" (1982, p. 122). Efficacy beliefs shape the students' outcomes, and expectations of their individual performance. Bandura (1984) distinguishes between the component skills in practice, for example the difference between driving a car in freeway traffic and navigating twisting mountain roads. It is comparable with the individual and collective efficacy using the recorded flipped classroom and participating in mobile online webinars F2F, as well the use mobile applications such as laptops, tablets. or smart phones to develop CSE, which can be seen as a level of *media and information literacy*.

Methodically, CSE has been used through Compeau and Higgins (1995) three interrelated dimensions: *magnitude*, *strength*, and *generalizability*. The *magnitude* of CSE can be understood as a reflection of the student's cognitive knowledge processes, based on the ability to analyze, communicate, manage information, understand concepts, meta-cognitive skills such as problem solving, interpreting, reflecting and evaluating. The *strength* of CSE refers to the self-confidence, self-esteem, and students' ability to perform various tasks. *Generalizability* of CSE reflects the degree online studies are limited to a specific area of a learning activity, such as *media and information literacy*, using a variety of software for the flipped classroom and computer systems for mobile online webinars F2F with different mobile applications.

IMPLEMENTATION

The data collection consisted of ten recorded mobile online webinars F2F (10 sessions, 3 hours long), of which five were examinations and the other five were prepared with teacher-recorded flipped classrooms videos during the spring and fall semesters of 2013. The student teachers (N=22) could also have parallel chat communications during the webinars. The students worked during the three web-based courses over an eighteen-month period in the Vocational Teacher Education Program, (VTEP) both individually and in groups with problem-based course assignments with deadlines. They were divided into five groups of four to five students in each. All webinars were scheduled with a date and time for each group and with pre-defined goals and questions from the course literature. During the mobile online webinars F2F students discussed theoretical concepts from the course literature and experiences between teaching practices and professions, as well as tutoring and scaffolding in course assignments. All ten webinars in courses 2 and 3 were recorded and accessible in the learning management system (LMS) in order to provide students with the opportunity to take a step back, reflect, self-assess and compare various contributions. When students were asked if they wanted to participate in the study they signed an agreement. They also got a guideline for the e-meeting system using Adobe Connect and received an online review of the tools (webcam, microphone / headset and chat). A summary of the design during the three courses is given in Table 1.

The data was collected after the three courses were ended. The data was complemented by semi-structured research interviews with six students about their participations with other students and teachers. In addition, the researchers wanted feedback on students' use of the teacher-recorded flipped classroom, and the content meaning and dialogue exchanges during the mobile online webinars F2F. Likewise, on the textual chat communications by using the mobile applications such as laptops, tablets, or smart phones. Before the research interviews, all 22 students were asked by email if they wanted to contribute. Six

Table 1. Summary of the design for a flipped classroom using mobile online webinars F2F during three courses in the Vocational Teacher Education Program, (VTEP) (Amhag, 2014)

Fall semester 2012 (half-time study)		**Course 1 VTEP** - Learning outcome about social relations, conflict management and educational leadership. Three group wise online webinars as a follow-up of course start and tutoring of course assignments.
Academic year 2013		(half-time study)
Date	**Length**	**Course 2 VTEP** - Learning outcome about syllabus, vocational didactics and assessment
02/15/2013	00:27:46	Flipped classroom: Theory of Pedagogy & Didactic, follow-up of course start
02/25/2013	01:16:27	Webinar: Teaching & Education, five groups, drop-in
03/15/2013	00:19:13	Flipped classroom: Theory of Formative assessment
03/21/2013	00:19:51	Flipped classroom: Theory of Assessment & Grading
03/21/2013	01:14:01	Webinar: Assessment & Grading, three groups, 25 min / group
03/25/2013	01:00:59	Webinar: Assessment & Grading, two groups, 30 min / group
05/07/2013	00:55:06	Webinar: Tutoring of the two last course assignments, five groups, drop-in
Date	**Length**	**Course 3 VTEP** - Learning outcome about development, learning and special education
09/09/2013	00:28:05	Flipped classroom: Theory of Read & writing disabilities, follow-up course start
10/14/2013	00:47:49	Webinar: Literature of Read & writing disabilities, five groups, drop-in
10/29/2013	01:04:57	Webinar: Examination of Read & writing disabilities, two groups, 32 min / group
10/31/2013	00:46:22	Webinar: Examination of Read & writing disabilities, one group
10/31/2013	00:42:36	Webinar: Examination of Read & writing disabilities, one group
10/31/2013	00:56:22	Webinar: Examination of Read & writing disabilities, one group
12/10/2013	00:34:24	Flipped classroom: Theory of Teaching & Education of vocational students
12/19/2013	00:39:50	Webinar: Tutoring of the last course assignments, five groups, drop-in

students enrolled (including both women and men). The interviews of six students and the whole group of 22 students' dialogue exchanges during ten webinars were transcribed. In the findings we have chosen five excerpts as examples of the analysis by using Wengers (1998) four components; *meaning*, *practice*, *community* and *identity*, complemented by Compeau and Higgins (1995) three interrelated dimensions: *magnitude*, *strength*, and *generalizability*. These patterns are summarizing the result of the study.

FINDINGS AND ANALYSIS

The first excerpt addresses student teachers self-efficacy and collaborations with interactions and interplay in mobile online webinars F2F. We examined how it might constitute a determining factor for the students learning and knowledge exchanges at distance.

Student 1

Student 1, a student in the hairdressing program, describes her experiences of webinars:

Without webinars, I don´t think I would have continued to study. In particular, all documents [papers and reports] posted to the LMS and preparations [flipped classroom] before the follow-up webinars, provided a sense of reality. Desire more lectures or other types of tutoring and scaffolding. I also think it has been convenient to be able to participate a little quickly and easily at work [own hairdressing salon] and was able to go away for a while and be with or look through a webinar afterwards. I think it's been good and would have liked additional!

Student 1's statements reflect the importance of social arrangements in the *community of practice* for meaningful learning at distance (Wenger, 1998). In the web-based learning environment, students can download research articles and reports from the National Agency of Education. The preparations offered teacher-recorded flipped classroom video lectures, prior to the follow-up webinars. Student 1 had developed more of a sense of reality when she managed the information from the literature. Student 1 had low self-efficacy in the beginning of the education program and is reflecting over her meta-cognitive skills. The magnitude of CSE might be gauged in terms of support levels required to undertake tasks (Compeau & Higgins, 1995). Student 1 felt the webinars helped her ability to continue the VTEP and not to drop out. She desires even more lectures or other component skills in teaching practices. They can be accomplished easier when she goes to work, and reflects on the online webinars afterwards. Like Bandura (1984) distinguishes between the component skills in practice through driving a car, the teacher-recorded flipped classroom videos gave student 1 the ability to negotiate new meanings and further development. The teacher-recorded flipped classroom videos and mobile online webinars F2F involves the whole unique person in a dynamic interplay of participation and reification on learning. A kind of *identity in practice* because learning constitutes trajectories of participation through connections in and between student 1's past and future in a process of individual and collective *learning as becoming* (Wenger, 1998, Ch. 3 & 6). This can be possible when student 1 directly meets other fellow-students and teachers F2F, and discusses how they shall deal with the theoretical concepts from the course literature and experiences between teaching practices and as a working professional. Student 1 has also practiced their teaching experiences within the teacher-recorded flipped classroom as unauthorized teacher:

I also think the "flipped classroom" is practical to use in my own teaching, a small group of pupils are gathered in the same place. In front of the computer, it is very clear who can or cannot. It quickly becomes a receipt of their understanding or not. Last week I was away from school, and they [the pupils] would find out for their own how to do a task and record and put out what they had done in the LMS. It's a very easy way! It is also a way for those who have difficulty in writing for some reason. They can show what they've done, even if they have not written four pages. They have learned something and can explain with pictures and speech instead.

Student 1's own teaching experiences after three VTEP-courses, highlight the importance of social and practical experiences in higher education in creating new identity through personal histories about becoming in and between one's own teaching practices and profession (Wenger, 1998). The progression of student 1's *magnitude* of CSE reflects her cognitive knowledge processes to manage digitalized teaching and evaluate the pupils' outcome (Compeau & Higgins, 1995). The *strength* of CSE has also advanced. Student 1 has the self-confidence to create new teaching using teacher-recorded flipped classroom and in planning the pupils' tasks in the LMS when she shall be away from school. The pupils can participate in practice together and record their achievement. Student 1 can even score a *higher strength* of CSE for the content on the computer, and realize how the pupils understood the task of problem solving. Furthermore, student 1 is also talking about her own meaning and learning with teacher-recorded flipped classroom videos and mobile webinars. It was discovered: "It quickly becomes a receipt of their [and own] understanding or not", and how the process can sustain mutual engagement in action (Wenger, 1998). Here student 1 understands what the pupils have done, even if they had not written four page. Instead, they can explain with pictures and speech. The statement provides an echo of student 1´s media and information literacy as an achieved level of *generalizability* of CSE (Compeau & Higgins, 1995) when using the recorded flipped classroom videos as a teaching tool.

Student 2

Student 2, a student teacher in Restaurant Management and Food, emphasizes the value of the teacher-recorded flipped classroom videos, but also the importance of the peer assessment processes among pupils at school:

Flipped classroom, I have used most of; I've even let pupils work in groups and record a particular lecture sequence when they were practicing in the dining room. I put up the recorded video sequences for peer feedback. I called the moment "find five errors" and if they could find five errors together, there are often some belongings. I could see the peer feedback was quite different. Pupils must learn how to give peer feedback as well. It is nothing I learned myself automatically, it is also something I have learned and practiced in VTEP.

In this excerpt, Student 2 raises using the teacher-recorded flipped classroom vidoes for group work, as well the possibility of recording the learning activities for peer feedback and self-assessment. Like student 1, her *strength* of CSE and self-confidence to use the flipped classroom and record lecture sequences for peer feedback has developed (Compeau & Higgins, 1995). She was aware of the value of peer feedback for learning and what pupils shall focus on when they find five errors together and how creative they must be when they are formulating peer feedback. This formative process does not happen automatically or on its own, as student 2 has experienced. It must be practiced in *learning as doing* (Wenger, 1998). As Wenger states, learning is fundamentally experiential and social, where pupils assist each other. The *meaning* of meta-cognitive ability is related to a wide range of skills, such as problem solving, communicating, reflecting and evaluating which strengthen the pupils´ confidence to work constructively, critically and creatively. It is measured by the extent they use their own and others meaning in their collaboration and dialogue exchanges'.

Student 3

The mobile education and collaborative learning activities, has also been very important for student 3, a student teacher in the Hotel and Tourism program, as she parallel with VTEP is working halftime as a flight attendant:

The webinars and flipped classroom, I have had huge benefit of! For me it is a bit special because I work as a flight attendant, but also for those students who may want to be able to close the gap for a time. Even the chat communications worked very well in the groups and it was fun with peer feedback and we had some time to respond to the course assignments. I liked it and the peer feedbacks were a great support!

In this excerpt, student 3 highlights the teacher-recorded flipped classroom videos, posted one week before the follow-up online webinars, which were also recorded and accessible afterwards in the learning management system (LMS). These recordings gave good margins for her to take part in the content meaning and discussions afterwards. She also emphasized having a week to provide peer feedback on some classmates' course assignments and their corresponding work in the group's discussion forums in the LMS. The peer assessment processes shaped *modes of belonging* because learning is a matter of engagement (Wenger, 1998). The students had opportunities to contribute actively to the *practice of community* with the recorded flipped classroom videos and mobile online webinars, as well textual chat communications, and give and receive peer feedback in the LMS discussion forums. The different software and collaborative learning activities provided value to students.as a degree of *generalizability* of CSE and *media and information literacy* (Compeau & Higgins, 1995). The students also practiced their understanding of the academic way of reading and writing, and making creative use of their higher-order thinking and critical review.

The two last excerpts address how the student teachers experienced the mobile online examination F2F of one course assignment in course 3.

Student 4

Student 4, a student teacher in the Handicraft program specializing in makeup and skin therapy, is overall positive:

I thought it worked well. Everyone can see and hear each other and ask questions. It is also easier to participate when you do not physically need to participate at the university. I wish more teachers can work in this way, absolutely. It is also the right time to have mobile web-based lectures and presentations online at distance. We as teachers want that too with our pupils at school and give them the possibility to show presentations from home. Then they do not miss anything if they not are at school for some reason.

In this except, student 4 emphasized the mobile online examination F2F and the importance of using modern technology in a flexible and mobile way in distance higher education and teaching. It is easier to participate from home than at a distance and she wishes more teachers would follow. According to Wenger (1998) it is important that learning involve the interplay between the local, as in this study the students own presentation from home during the examination on the literature about reading and writing disabilities. It provides a kind of global context online when the students can see and hear each other's argumentations from their different *practices* at distance. Therefore, the locality is connected with creation of learning *communities*, and depends on a dynamic combination of students' engagement, imagination and alignment to make this interplay between the local and the global to an engine of new learning.

Student 5

However, some student teachers considered the examination needing to have a clearer structure and teacher leadership. Student 5 points out some dilemmas:

The first participant could speak for ten minutes and then there was not much time left. In addition, the task was already described in detail in text, and archived [in the LMS]. It does not need a presentation of it too. Perhaps it would have been better if we had another issue to discuss, now it was a little cake-on-cake. You cannot say anything new in this little time, there was more to clarify about what you had written. I think that was the idea?

In this excerpt, student 5 raises the issue examinations for learning requires enough structure and continuity to gather students' higher-order thinking and critically review about the literature on reading and writing disabilities (Wenger, 1998). Learning is always a matter of engagement and it depends on the opportunities to contribute actively to the *practice of community* and to make creative use of each students' respective repertoires. Several researchers stressed academic education should place more value and emphasis on the processes of argumentation engaging in higher-order thinking (e.g. Amhag, 2011; Amhag & Jakobsson, 2009; Erduran & Villamanan, 2009; Matusov, 2007; Meyer, 2003; Richardson & Ice, 2010; Schellens & Valcke, 2005). This can be seen as "wise thinking" and to some extent "creative thinking" (Craft, 2006), and other forms of reflective interactions supporting students' higher-order thinking and motivation to collaborate in effective ways. Additional studies (e.g. Amhag, 2012; 2013; De Wever, et al., 2009; Dochy, et al., 1999; Scheuer, et al., 2010; Topping, 2005; 2009), have highlighted the fact that students not only need to "learn to argue", they also need to learn good argumentation practices, through argumentations about specific topics. Moreover, they need to use peer learning processes, such as peer assessment, self-assessment and co-assessment.

Some initially problem, showed in the observations who could occur at the start of the education with webinars online, were that the technology of e-meeting arrangements is unreliable. If the mobile connection is too low, the participants are logging out. A feeling of low *generalizability* of CSE easily arise (Compeau & Higgins, 1995). It was shown through students constantly were repeating, "Do you hear me?" The access of the voice audio requires a stable mobile connection. Additionally, if the teachers initially are unfamiliar with the technique online and how to support the students. This technical problem pointed especially one of the students out in the interview.

Two Surveys

In the two surveys, summarized in Table 2, given to students (N=22) in June 2013 and January 2014, the student answered the statement: *Did the following learning activities contribute to my learning in regards to?* a) course start lectures at the university, b) recorded flipped classroom, c) webinars F2F, d) active in webinars, e) seen the flipped classroom, and f) media and information literacy. The result in survey 1 (June, 2013) shows the majority of students (15 answers, 68%) assumed the flipped classroom (87%) and webinars (73%) contributed to their learning, in comparison with campus lectures at the start of the course at (67%). However, most of the students had retrospectively reviewed and reflected on the meaning contents of the teacher-recorded flipped classroom videos (80%). Likewise, the majority of the students considered they had been active during the webinars (80%) by discussing different perspectives. All students feel they need to have media and information literacy (100% strongly agree).

In the second survey, after the third course (January 2014), the majority of the students (12 answers, 55%) assumed the lectures at the start of the course (92%) contributed to their learning, the webinars (75%), and a little fewer for the recorded flipped classroom (66%). However, all students had retrospectively reviewed and reflected on the meaning contents of the recorded flipped classroom videos (92%). Likewise, the majority of students' considered they were active during the webinars (83%) by discussing different perspectives. All students feel they need to have media and information literacy (100% strongly agree). In the open-ended questions, some students required even more knowledge exchanges and interactive collaborations for learning and development.

In summary, the two surveys shows that all students (N=22) assumed they had *generalizability* of CSE [13] limited to this situated area at distance to use different software such as recorded flipped classroom videos and mobile online webinars F2F with parallel chat communications which shapes their *media and information literacy*. The majority of the students' show also *strength* of CSE through their self-confidence and self-esteem to take advantage of the recorded flipped classroom videos and thus be able to be more active in the online webinars F2F and perform various tasks in VTEP on an higher academic level.

Table 2. Two surveys results from June 2013 and January 2014 on: Did the following learning activities contribute to my learning? (Amhag, 2014)

Learning Activities	Survey 1 (68%) 1-3 Disagree- Partly Agree	Survey 1 (68%) 4-6 Agree – Strongly Agree	Survey 2 (55%) 1-3 Disagree- Partly Agree	Survey 2 (55%) 4-6 Agree – Strongly Agree
University lectures at course start	33%	67%	8%	92%
Flipped classroom	13%	87%	34%	66%
Webinars F2F	27%	73%	25%	75%
Active in webinars	20%	80%	17%	83%
Seen flipped classroom	20%	80%	8%	92%
Media & information literacy	0%	100%	0%	100%

SOLUTIONS AND RECOMMENDATIONS

Important conclusions from the results of the learning activities with teacher-recorded flipped classroom videos using mobile online webinars F2F, and textual chat communications are the time and space for learning and development expands. The teacher-recorded flipped classroom gave students more learner-centered experiences when they had the ability to prepare and negotiate new meanings with support from the teacher-recorded flipped classroom videos before and during mobile online webinars F2F by using laptops, tablets or smart phone for further learning and collaboration. The same results are shown in the research review on recorded flipped classroom videos by Bishop and Verleger (2013). In another study by Long et al. (2014) demonstrates students are generally positive to the working methods because it offers more preparation for the course content to support them complete the assignments and feel more prepared for scientific reading, academic writing, presentations, and examinations. The same results also appear in a study by Beapler et al. (2014) demonstrating students feel more learner-centered education when they are teaching in recorded flipped classroom videos, rather than traditional forms of monological lectures in a university classroom.

Fulton (2012) argues for a new way of teaching and learning, and has listed seven benefits of teacher-recorded flipped classroom videos: (1) students can work at their own speed, (2) homework before scheduled lectures provides teachers with better insight into students' reading and writing abilities, as well different learning styles, (3) teachers can more easily assume the learning outcomes and give feedback, (4) teaching time is used more efficiently and creatively, (5) the students' interest and involvement increases, contributing to better learning outcomes and development, (6) learning theories support new methods, and (7) the use of wearable technologies provide increased flexibility.

Although, the study of Herreid and Schiller (2013) among 200 teachers suggest that teacher-recorded flipped classroom videos brings more: a) discussions with and between students about the authentic problems, b) research-based work, c) those who miss teaching occasions can take part of in recorded flipped classroom videos and webinars retrospectively, d) the method promotes the design of in-class time and out-of-class time; e) students are more active and take more responsibility for their learning, and f) the students appreciate the working methods. Another study by Beapler et al. (2014) show the benefits in student grades were significantly better if they had blended learning, as with online webinars and recorded flipped classroom videos, combined with traditional seminars and lectures at the university, compared with a group using traditional teaching and individual tests in the classroom. The research review by Roehl et al. (2013) shows the technology in recordings of the recorded flipped classroom videos were perceived as easy to use by teachers. Likewise, observations by Wallace (2013) concerning the manner in which videos and text between synchronous and asynchronous media are used in teaching, shows how teachers and students relationships changed. They became more complex and complementary when teaching online based on the participants' conditions, needs and media literacy.

CONCLUSION

The design of distance teaching in higher education for students learner centered teaching can be considered an updated version of the "knowledge-building environments for progressive discourse" stated by Scardamalia and Bereiter (1994, p. 37-38). The integration and implementation of teacher-recorded flipped classroom videos and mobile online webinars F2F by using mobile technologies could also extend to smart glasses, smart watches, or other new devices. Twenty years ago, Scardamalia and Bereiter (1993; 1994) argued various programs and courses must integrate and implement e-mail, bulletin boards and presentation material more thoroughly in the education. They also pointed to the importance for students to see themselves as active knowledge builders, not as passive recipients. Moreover, it is essential for students to be able to identify their own knowledge needs, to recognize what is unclear, confusing, doubtful, incoherent, and so on. The role of the teacher is to give feed up, feedback and feed forward (Amhag, 2011, 2012; 2013), and show commitment to increasing the students' learning and development. In the modern design of learner-centered teaching at distance in higher education program, the teachers should include the following collaborative learning activities:

- Recorded flipped classroom videos before follow-up webinars F2F.
- Webinars F2F with group discussions on course content, literature, theories, concepts, problem solving and exchanges of knowledge and experiences.
- Tutoring and scaffolding F2F in scientific reading and academic writing, as well media and information literacy.
- Peer feedback, co- and self-assessment and critical review of their own and other's work.
- Collaboration with shared whiteboard, notes, and documents F2F.
- Split screen and software used with others F2F.
- Examinations online F2F of some course assignments.

These actions can be compared with "into an ongoing stream of activity", involving explicit and implicit "stimulus means" (Wertsch, 2007, pp. 180-184). They are explicit through tutoring, scaffolding through teacher-recorded flipped classroom videos or sharing other students´ expertise and experiences directly during mobile online webinars F2F and textual chat communication. Opposed to implicit through inner reflection and uncovering thoughts which are tending to be hidden and non-transitory and therefore, more difficult to detect.

REFERENCES

Amhag, L. (2011). *Students´Argument Patterns in Asynchronous Dialogues for Learning.*

Amhag, L. (2012). High school students' argument patterns in online peer feedback. E. Favaron, P. M. Pumilia-Gnarini, E. Pacetti, J. Bishop & L. Guerra (Ed.), *Handbook of Research on Didactic Strategies and Technologies for Education: Incorporating Advancements* (Vol. 2, Ch. 62, pp. 711-723). Retrieved from http://www.igi-global.com/chapter/high-school-students-argument- patterns/72113

Amhag, L. (2013). Creativity in and between Collaborative Peer Assessment Processes in Higher Distance Education. In *Creative Education* (pp. 94–104). Special Issue on Higher Education; doi:10.4236/ce.2013.47A2011

Amhag, L., & Jakobsson, A. (2009). Collaborative Learning as a Collective Competence when Students Use the Potential of Meaning in Asynchronous Dialogues. *Computers & Education, 52*(3), 656–667. doi:10.1016/j.compedu.2008.11.012

Baepler, P., Walker, J. D., & Driessen, M. (2014). It's not about seat time: Blending, flipping, and efficiency in active learning classrooms. *Computers & Education*, *78*, 227–236. doi:10.1016/j.compedu.2014.06.006

Bandura, A. (1984). Recycling Misconceptions of Perceived Self-efficacy. *Cognitive Therapy and Research*, *8*(3), 231–255. doi:10.1007/BF01172995

Bandura, A. (2002). Growing primacy of human agency in adaption and change in the electronic era. *European Psychologist*, *7*(1), 2–16. doi:10.1027//1016-9040.7.1.2

Bishop, J. L., & Verleger, M. A. (2013). The Flipped Classroom: A Survey of the Research. *Paper presented at the 120th ASEE Conference & Exposition.*

Buxton, E. C., Burns, E. C., & De Muth, J. E. (2012). Instructional Design and Assessment. Professional Development Webinars for Pharmacists. *American Journal of Pharmaceutical Education*, *76*(8), 1–7. doi:10.5688/ajpe768155 PMID:23129854

Compeau, D. R., & Higgins, C. A. (1995). Computer self-efficacy: Development of measure and initial test. *Management Information Systems Quarterly*, *19*(2), 189–211. doi:10.2307/249688

Craft, A. (2006). Fostering creativity with wisdom. *Cambridge Journal of Education*, *36*(3), 337–350. doi:10.1080/03057640600865835

De Wever, B., Van Keer, H., Schellens, T., & Valcke, M. (2009). Structuring asynchronous discussion groups: The impact of role assignment and self-assessment on students' levels of knowledge construction through social negotiation. *Journal of Computer Assisted Learning*, *25*(2), 177–188. doi:10.1111/j.1365-2729.2008.00292.x

Dochy, F., Segers, M., & Sluijsmans, D. (1999). The use of self-, peer- and co-assessment in higher education: A review. *Studies in Higher Education*, *24*(3), 331–350. doi:10.1080/03075079912331379935

Erduran, S. & Villamanan, R. (2009). Cool Argument: Engineering Students´ Written Arguments about Thermodynamics in the Context of the Peltier Effect in Refrigeration. *Educacaión quimica* (pp. 119-125).

Fulton, K. (2012). Upside down and inside out: Flip your classroom to improve student learning. *Learning and Leading with Technology*, *39*(8), 12–17.

Herreid, C. F., & Schiller, N. A. (2013). Case studies and the flipped classroom. *Journal of College Science Teaching*, *42*(5), 62–66.

Kiriakidis, P. (2010). How Does Skype, as an Online Communication Software Tool, Contribute to K-12 Administrators' Level of Self-efficacy? *International Journal of Leadership in Educational Technology*, *1*(9), 1–29.

Lage, M. J., Platt, G. J., & Treglia, M. (2000). Inverting the classroom: A gateway to creating an inclusive learning environment. *The Journal of Economic Education*, *31*(1), 30–43. doi:10.1080/00220480009596759

Lee, K., & Salman, R. (2012). The Design and Development of Mobile Collaborative Learning Application Using Android. *Journal of Information Technology and Application in Education,* 1(1), 1-8.

Long, T., Logan, J., & Waugh, M. (2014). Students' Perceptions of Pre-class Instructional Video in the Flipped Classroom Model: A Survey Study. *Proceedings of Society for Information Technology & Teacher Education International Conference*. VA: AACE.

Matusov, E. (2007). Applying Bakhtin Scholarship on Discourse in Education: A Critical Review Essay. *Educational Theory*, *57*(2), 215–237. doi:10.1111/j.1741-5446.2007.00253.x

Meyer, K. (2003). Face-to-face versus threaded discussions: The role of time and higher-order thinking. *Journal of Asynchronous Learning Networks*, *7*(3), 55–65.

Research Highlights in Technology and Teacher Education. (2011). *Ed/ITLib Digital Library*. Retrieved from http://storefront.acculink.com/aace

Rich, R. L. (2011). *A framework for synchronous web-based professional development [Thesis]*. Stockton, California.

Richardson, J. C., & Ice, P. (2010). Investigating students´ level of critical thinking across instructional strategies in online discussions. *The Internet and Higher Education*, *13*(1-2), 52–59. doi:10.1016/j.iheduc.2009.10.009

Rockinson-Szapkiw, A. J., Courduff, J., Carter, K., & Bennett, D. (2013). Electronic versus traditional print textbooks: A comparison study on the influence of university students' learning. *Computers & Education*, *63*, 259–266. doi:10.1016/j.compedu.2012.11.022

Roehl, A., Reddy, S. L., & Shannon, G. J. (2013). The Flipped Classroom: An Opportunity to Engage Millennial Students through Active Learning Strategies. *Journal of Family and Consumer Sciences*, *105*(2), 44–49. doi:10.14307/JFCS105.2.12

Sana, F., Weston, T., & Cepeda, N. J. (2013). Laptop multitasking hinders classroom learning for both users and nearby peers. *Computers & Education*, *62*, 24–31. doi:10.1016/j.compedu.2012.10.003

Scardamalia, M., & Bereiter, C. (1993). Technologies for knowledge-building discourse. *Communications of the ACM*, *36*(5), 37–41. doi:10.1145/155049.155056

Scardamalia, M., & Bereiter, C. (1994). Computer support for knowledge-building communities. *Journal of the Learning Sciences*, *3*(3), 265–283. doi:10.1207/s15327809jls0303_3

Schellens, T., & Valcke, M. (2005). Collaborative learning in asynchronous discussion groups: What about the impact on cognitive processing? *Computers in Human Behavior*, *21*(6), 957–975. doi:10.1016/j.chb.2004.02.025

Scheuer, O., Loll, F., Pinkwart, N., & McLaren, B. M. (2010). Computer-supported argumentation: A review of the state of the art. *Computer-Supported Collaborative Learning*, *5*(5), 43–102. doi:10.1007/s11412-009-9080-x

Topping, K. (1998). Peer Assessment between Students in Colleges and Universities. *Review of Educational Research*, *68*(3), 249–276. doi:10.3102/00346543068003249

Topping, K. (2005). Trends in Peer Learning. *Educational Psychology*, *25*(6), 631–645. doi:10.1080/01443410500345172

Wagner, D., Laforge, P., & Cripps, D. (2013). Lecture Material Retention: a First Trial Report on Flipped Classroom Strategies in Electronic Systems Engineering at the University of Regina. *Paper presented at the Canadian Engineering Education Association Conference*, Montreal, Canada.

Wallace, A. (2014). Social Learning Platforms and the Flipped Classroom. *International Journal of Information and Education Technology*, *4*(4), 293–296. doi:10.7763/IJIET.2014.V4.416

Wenger, E. (1998). *Communities of Practice - Learning, Meaning, and Identity*. New York: Cambridge University Press. doi:10.1017/CBO9780511803932

Wertsch, J. V. (2007). Mediation. In H. Daniels, M. Cole, & J. V. Wertsch (Eds.), *The Cambridge Companion to Vygotsky* (pp. 178–192). Cambridge: Cambridge University Press. doi:10.1017/CCOL0521831040.008

KEY TERMS AND DEFINITIONS

Computer Self-Efficacy (CSE): Is concerned with individuals' intentions to use information technology (IT) and the opportunities to challenge different efficacy beliefs, e.g. shape the students' outcomes, and expectations of their individual performance.

Computer-Supported Collaborative Learning (CSCL): Is the field concerned with how information and communication technology (ICT) might support collaborative learning in groups.

Distance Education: A specific context of education at distance depending on whether the teaching takes place synchronously in the same space and time, e.g. web-based seminars or conferencing, or asynchronously in different time and space, e.g. in text-based discussion forums and chat communications.

Distance Learning: Involves different learning activities at distance, e.g. peer assessment, give and receive peer feedback, critically review literature and each other's work, as well self-assess their efforts, regardless of time and space.

Flipped Classroom: Is inverting the common instructional approach by using teacher-recorded videos with various briefings and interactive instructions accessed from different locations using mobile technologies prior to the class learning activities.

Mobile Applications: Is also called mobile apps, by using Internet applications that run on laptops, tablets and smartphones or other mobile devices.

Online Webinar: Short for online web-based seminar or conferencing where group of students and their teacher have presentations, lectures, workshops or seminars face-to-face that is transmitted over the Web.

Wearable Technology: Refers to computer-powered devices or equipment that can be worn by a user, such as laptop, smart phone, tablet, smart watches, smart glasses, or similar items that have functions for to take photos and sync with your mobile devices.

Chapter 11
Google AdSense as a Mobile Technology in Education

Azizul Hassan
Cardiff Metropolitan University, UK

Donatella S. Privitera
University of Catania, Italy

ABSTRACT

Google AdSense, a cost per click (CPC) advertising programme for those wishing to place ads on their website. Arguable viewed as promising in business, this literature review identifies key features of this promise when applied to education. This chapter connects business and education to identify Google AdSense as a valid tool and subject matter to be considered for use within the mobile and wearable technology curriculum. This chapter suggests considering positive outcomes generated using Google AdSense for incorporation into an education curriculum. Google AdSense should be applied to create more reasons to attract diverse user groups. This chapter also identifies the importance of Google needing a set of regulatory policies and procedures to avoid misuse of Google AdSense. In cases where monetary transactions occur, a possibility of violating the set terms and conditions exists. The chapter concludes with a clear call for inclusion of Google AdSense in an education curriculum.

INTRODUCTION

Information and communications technology (ICT) continues to expand the boundaries of education into an “anytime/anywhere” experience. Mobile Technology - comes at an opportune time not only in terms of Internet technology but also business. Mobiles have already made many headlines across the entire world playing key new roles in education, health, entertainment, etc (Yang, 2004). But, part of the mobile telecoms revolution is also about building sustainable business models through the provision of services. From the business stand point, creating disruptive business models that will break monopolies through innovative solutions forms the last mile access point for the users and this is where opportunities reside. Fueling this trend is the fact that Search Engines have become popular both as information-seeking vehicles and as an online advertising media. There are many methods for online advertisements.

DOI: 10.4018/978-1-5225-0069-8.ch011

Rich media advertising companies use Banner Advertisements, Wallpaper Advertisements, Pop-ups, Video Advertisements, streaming audio, streaming video and mobile advertisements. Search Engines have created an era of context-based advertising. Google AdSense, Yahoo Publisher Network and MSN Ad Center provide contextual advertising to attract many online ad-publishers (Kumar & Kohli, 2007). Google AdSense, obviously, is a program owned and operated by Google (www.google.com/AdSense/). Within the Google network's content sites, this programme provides publishers many ways to offer images, texts, interactive media advertisements and videos. These offerings are targeted to the audience and site contents (Li et al., 2010).

This chapter reviews how Google AdSense is gaining popularity beyond the business communities, and why it needs to invade education and appeal to younger learners. The use of AdSense can be argued to have helped to pay for the development of mobile technology and could go a long way to helping to do the same for wearable technology given someone can figure out how to adapt it to that venue. In this chapter, it is suggested, on the ground of both theories and practices, Google AdSense to be included as a specialized subject matter for mobile technologies for students. From an industry perspective, anyone with a basic to advance knowledge of Google AdSense can see it as a reliable platform for earning/learning, where a tactile version can or must be adapted for expansion into wearable devices such as smart glasses or watches. Therefore, it is important to understand what Google AdSense is really doing, to have a basic knowledge of how it is doing it and to encourage people to find a way to adapt it for use in various media venues. It is believed, a high level of understanding and use can and will expand markets for various products and services while increasing employment and educational opportunities for vast numbers of people worldwide.

BACKGROUND

The geography of Internet users and online advertiser is diverse. The Internet and its associated new technologies have become an integral part of our lives making it easier to communicate, research information and purchase any kind of products and services. Online transactions are becoming simpler and faster, definitely making our lives easier since only one click can bring us in a few days whatever we want. However, the current young generation brought up with or by the Internet is the most at risk of developing compulsive buying behaviours, considering they are more accustomed to the online world than any other generation (Bighiu et al., 2015). Particularly of interest, a US study (Donthu et al., 1999) has revealed that students spend almost 12% of their personal budgets online. The web has an important role as much for consumers as the business operators seeking out those customers. Previous research has shown that innovation, organization and external factors influence a firm's decision to adopt e-commerce as a marketing and selling strategy (Abernathy & Clark, 2007; Guimaraes, 2012; Wan, 2002).

There are 3.4 billion Internet users in the world (see table 1) with significant penetration ratios in countries making up Asia (45.7%), Europe (19.2%) and North America (10.2%) (Internetworldstats, 2014). Internet users in Europe, specifically in the UK, Russia and Germany have high rates of use: in 2014 their percentage of users was 12-15%. Internet penetration in Europe is growing and travel related e-commerce represents the largest point of use, surpassing more traditional uses such as Internet banking or researching information for health. Between the top 20 countries with the highest number of Internet users there are Japan and Germany (86.2%), China (45.8%), Brazil (54.2%).

Table 1. Global Internet usage and demography in millions (Internet World Stats, June 2014)

World Regions	Population (2014 Est.)	Internet Users -2000	Internet Users 2014 (Millions)	Penetration (% Population)	Growth 2000-2014	Users %
Africa	1,125,7	4,5	297,9	26.5	6,498.6	9.8
Asia	3,996,4	114,3	1,386,2	34.7	1,112.7	45.7
Europe	825,9	105,0	582,4	70.5	454.2	19.2
Middle East	231,6	3,3	111,9	48.3	3,303.8	3.7
North America	353,8	108,1	310,3	87.7	187.1	10.2
Latin America -Caribbean	612,2	18,1	320,4	52.3	1,672.7	10.5
Oceania-Australia	36,8	7,6	26,8	72.9	251.6	0.9
World Total	7,182,5	360,9	3,035,7	42.3	741.0	100.0

The Internet or World Wide Web, an electronic medium based on broadcasting and publishing, facilitates all sorts of two-way communication. It integrates many of the positive advertising elements of major media especially with its visual impact. In fact, the growth and popularity of online media is in direct contrast to the declining value of rational media, as reflected in the decreasing use of TV, newspapers, and other sources (Kumar & Kohli, 2007). The Internet has the additional benefits of higher user involvement and user capture at opportune moments. Since the web is interactive, users play a much more active role in the communication process than they perform in traditional mass media (McElheran, 2012). Traditional mass media has been characterized by information push. In contrast, communication processes on the Web are driven by information pull (Turban et al., 2008). According to Mohapatra (2009), as such, consumers control what information they access, which parts they explore, what routes they navigate, and when they view. The hyperlink function then allows users to readily move from one source of information to another. This further enhances user control in the communication process allowing users to pursue their specific interests. Therefore, it has become important to study websites as places of business, specifically the ways they can be seen online (Hax, 2010). Web-site performance can be measured in various ways and there are many studies related to the tourism industry (Buhalis et al., 2011; Privitera, 2015) can help one see how it can affect just one industry. Otherwise, a web site is a new kind of information technology with three phases of marketing (pre, on-line and after sales) where web site success is related to four principal factors: quality of information and service, system use, playfulness and system design quality (Liu & Arnett, 2000).

Even though there are nearly 3 billion people with access to the Internet globally, that leaves 4 billion people unconnected to the Web, 90% of those are in the developing world. At the same time, we are currently undergoing a rapid shift from feature phones to smartphones, even in the world's less economically developed regions. Nowadays, almost everyone, from children on playgrounds to seniors, have mobile phones or/and tablets connected to the Internet with that number growing every year. At the end of 2014, half of the world's population had at least one mobile subscription, totaling over 3.6 billion unique mobile subscribers. The global mobile subscriber base increased by just over 5% in 2014 (GSMA, 2015). Millions are also converting to smartphones (Itu, 2014). In practice, the growing number of smartphones and tablets means significant changes in the behaviour of users (Kumar et al., 2006). This will affect the ways in which they spend their leisure time, the frequency of browsing the web, while increasing their appetite for mobile apps (Rayport & Bernard, 2004). Mobile is confirmed as a strong youth medium with 48% of users being between 18-34 years old, compared to 40% for the fixed Internet and 29% for the TV audience. Mobile is also more tilted towards men, who represent 63% of total users (GSMA, 2015).

It has been acknowledged (Yoon & Kim, 2001) that Internet advertising should be designed to consider the suitability of the medium based on consumer characteristics (e.g. personality traits). Ward and Lee (2000) and Lee (2002) suggested that Internet users react more favourably to familiar brands since they prefer well-known brands. As Internet users are constantly exposed to more advertisements, they still react more spontaneously, with positive attitudes, to banner advertisements for familiar brands. This supports Shapiro, MacInnis and Heckler (1997), who found advertisements displayed next to 'to be read' text in a scrolling display affects awareness of an existing brand. Mobile Advertising has naturally become the next frontier on mobiles, attracting big companies to invest, where the various advertising services fight for the mobile ad money which makes up a large part of the annual billions of dollars spent associated with the Internet. Success factors for mobile advertising campaigns, and potentially for post-purchase push communications, were affected by message content, management of media issues as device development, the fluctuating quality of transmission processing, product fit and media costs, measurement, and global campaign launch strategies (Ström et al., 2014).

Overall understanding of mobile advertising across Europe is strong but there are some key barriers to adoption, such as measurement and tracking (IAB Europe, 2015). Encouragingly, mobile is unlocking new advertising budgets as it becomes more integrated into the general population which could result in an increase in cross-media campaigns. The top 3 European spenders on mobile advertising shows mobile advertising is becoming a reality in United Kingdom, where a quarter of online display ad spending is allocated to mobile. In France and Germany, it constituted 9.2% and 4.9% of total online display advertising spending, respectively (Table 2). The situation in countries of Central and Eastern Europe is very similar. The dynamics of growth in many countries, such as Romania (244.9%), Serbia (153.3%), Hungary (126.1%) and Poland (110.6%), is still very high, yet mobile ad spending constitutes a small part of total online budgets, Table 2 below demonstrates the European advertising spending.

There are several challenges the markets have to face in order to increase the popularity of mobile advertising and monetize the mobile audience. Despite the fact that mobile advertising is still far from becoming a mainstream marketing solution in the UK, this process has already started and will be continuing due to the increasing usage of mobile devices among Internet users. The ongoing development of new technologies means that a growing share of the online population connects to the Internet via their mobile devices. A dynamic growth of mobile traffic could be observed in the past two years. In 2012, mobile constituted only 3% of the Internet traffic in the UK excluding in-app traffic. This number has grown to 10% at present, mostly due to the rapid increase of smartphone page views (Płomińska et al.,

Table 2. Top 3 European countries by mobile advertising spending (IBA Europe and IHS, July 2014)

	Mobile Display Adv Spend 2012 (€m)	Adv spending 2013 (€m)	Total Online Display Adv Spend 2013 (€m)	Mobile Display As A Share of Total Online Display (%)
UK	184,9	549	2,192	25.1
France	47,8	92,9	1,015	9.2
Germany	43,0	65	1,319	4.9

2014). Users tend to surf the web on their tablets or smartphones more and more often and this trend sets a new standard for website building. Despite this increasing demand, for most markets the majority of websites have not yet been adapted to the new mobile reality. Augmented Reality (AR) is a technology that merges computer-generated imagery into real environments. Several researchers have outlined the prospects of technology such as an AR application in diverse human knowledge areas including visitor management, destination image formation, education, tourism marketing and so on (Hassan & Dadwal, in press; Hassan, in press; Hassan, & Ramkissoon, in press; Hassan & Sharma, in press; Hassan & Jung, in press; Hassan, 2015; Dadwal & Hassan, 2015; Hassan & Shabani, 2015). On the other side, marketing activities seem to be more engaged with updated technologies (Hassan & Rahman, 2015; Hassan, 2012a; Hassan, 2012b; Azim & Hassan, 2013a; Azim & Hassan, 2013b). One technology seen as the most innovative is Google AdSense. Thus, this chapter concentrates on Google AdSense to be observed as a valid mobile technological concept for education.

TECHNOLOGY AND EDUCATION: GLOBAL TRENDS

Technology has been gradually merging with education. From the remote past, technology has been playing critical roles in advancing education and literacy across the world, regardless of geographical distances. Countries across the world have been stressing technical educations for centuries.

Education can benefit through concentrating on the most updated technologies and applications. This is clearly evident that, both the existing and prospective flow of technology led education systems have been emerging as influential. This is mainly due to the unprecedented role of Internet that has navigated and cross-tied almost all areas of human livelihood. Thus, technology and education has reached at an inseparable stage creating more opportunities for both for the present and future generations. Academic institutions across the world are constantly stressing technological education building more specialized establishments. For example, universities are increasingly supporting the education, not just internally for use within the academics but also, in their appearances aiming to leave welcoming messages to technology savvy learners (Figure 1).

Technology has become so much more interactive with education, resulting in new forms of learning systems having been developed, for example; eLearning, Massive Open Online Course (MOOC), or Moodle Learning Management System. In developing worlds, the entire academic and research systems have been placed within a technology-formatted system making academic activities much easier. Technology, so far, has been mostly practice based, resulting in the attachment of relatively new forms of innovative technologies to the class curriculum.

Technology is gaining supremacy over conventional education systems and that has been presented by leading research agencies in the world. Technology and in particular information technology is reshaping the global education system. Gartner (2014) shows that in coming years, technology is going to leave sweeping changes in academic and learning systems in the world (Table 3).

Gartner (2014) predicts that education as an industry and also in its primary, secondary and post-secondary stages is looking forward to restructuring and revamping institutional capacities and arrangements, creating more spaces and opportunities for information technology education.

In addition, the World Economic Forum supported by the Boston Consulting Group (2015) also shows that students require 16 skills in the 21st century.

Figure 1. A popular monogram reflecting its basic purposes
(The Florida Agricultural and Mechanical University, 2015)

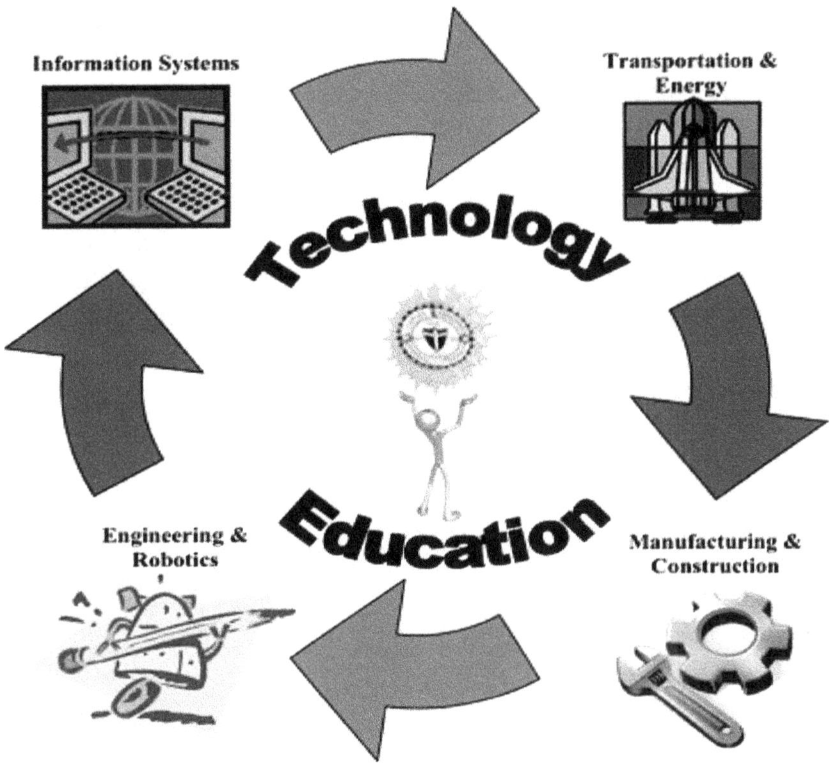

Table 3. Education trends in 2015 (Gartner, 2015)

Business Trends (What We See)	Strategic Technologies (What We Think You Should Do)
1. Student Success 2. Reinventing Credits 3. Global Competition for Students 4. (Re) Thinking Business Models 5. Retreating Political Responsibility 6. Outcome-Based Learning 7. Learning Analytics 8. Data-Driven Decisions 9. Consumerized Expectations 10. E-Research	1. Adaptable Learning Platforms 2. (Adaptive) E-Textbooks 3. CRM (Enroll., Ret., Alumni) 4. Big Data 5. Sourcing Strategies (Cloud) 6. Exostructure 7. Open Micro Credentials 8. Digital Assessment 9. Mobile 10. Social Learning

1. **Foundational Literacy:** Competencies Character Qualities How students apply core skills to everyday tasks: 1. Literacy; 2. Numeracy; 3. Scientific literacy; 4. ICT literacy; 5. Financial literacy; 6. Cultural and civic literacy;
2. **Competencies:** How students approach complex challenges: 7. Critical thinking/ problem solving; 8. Creativity; 9. Communication; 10. Collaboration;
3. **Character Qualities:** How students approach their changing environment: 11. Persistence; 12. Adaptability; 13. Curiosity; 14. Initiative; 15. Leadership; 16. Social and cultural awareness.

The question is how much developing countries are prepared to concentrate on technology education in its academic structure and with their limited capacities? Those countries with Gross National Income (GNI) or per capita income of US $11, 905 or below are defined as developing countries (World Bank, 2013). Following this categorization, a considerable number of countries in the world fall in this list including India, Bangladesh, Pakistan, Sri Lanka, and many more sub-Saharan and other countries of the world (Hassan & Forhad, 2013a; Hassan & Forhad, 2013b; Hassan & Burns, 2014).

Mobile technology adoption in developing countries is not uncommon. The activities of global mobile telephone companies have expanded rapidly contributing significantly to each respective country's socio-economic advancements. Developing countries across the world can be examples for the purpose. These countries have large populations with a significant number in the working and young age group(s). These people are mostly left uneducated with their potential capabilities never exploited at all. This leaves a considerable gap in terms of the demands and supplies of technically savvy people for the workforce. These countries can readily utilize its unemployed workforce by making available opportunities in technology centred areas. For that purpose, there is no option other than to transform their academic systems to become more technology centric.

Global literacy rates are clearly unequal, thus representing disparate differences between countries, in terms of their education systems, available resources and average literacy rates (Figure 2).

Figure 2. Literacy rates in developing regions
(European Parliamentary Research Services, 2015)

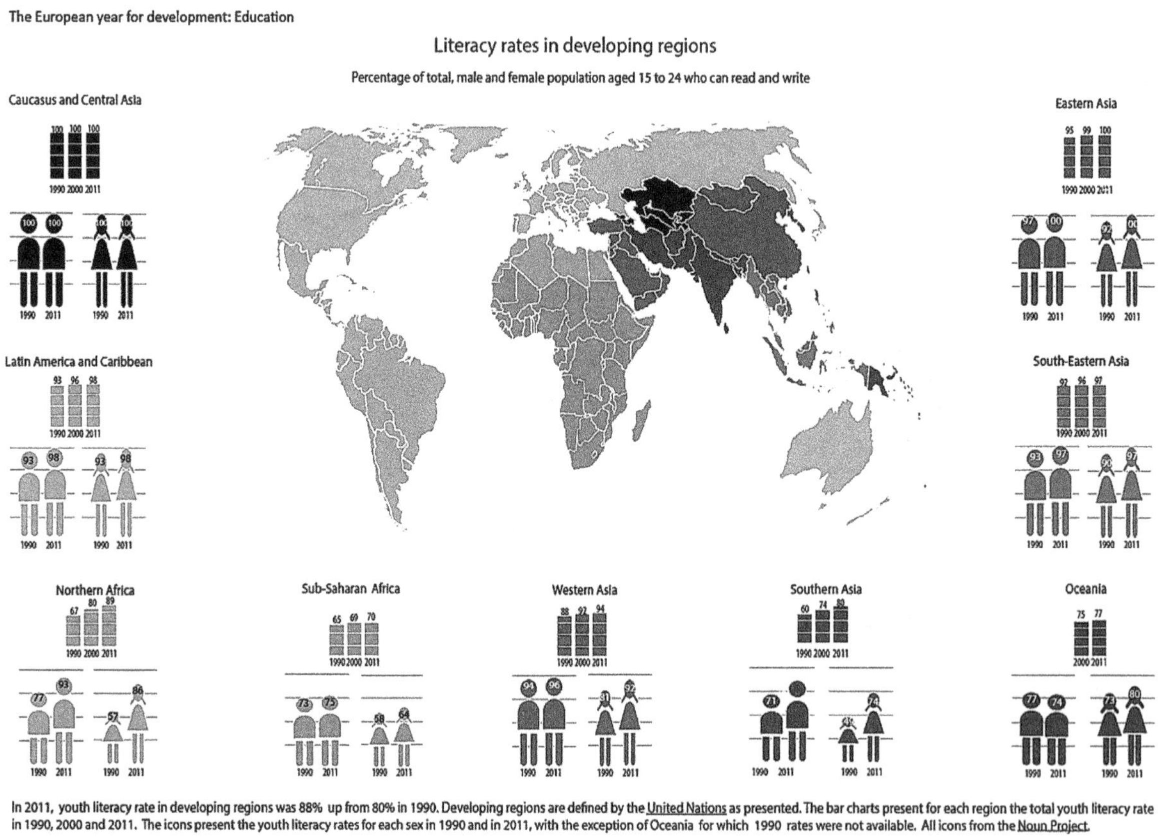

These countries can be classified as having emerging economies. They also have considerably large of young people as distinct population groups. Significantly, young age group populations are commonly left unemployed as well as most of them also do not have regular employment that leave them more vulnerable in significantly underdeveloped competitive business and market systems. In addition, there is a growing emphasis to bring them into the main economic platform. This is one of the reasons that necessitate the essentiality of technology use to serve this particular purpose.

Google is a well-known and well-accepted technological platform. Most of the world's young populations tend to have immense attraction towards using appropriate technology. The markets of many developing countries are also considerably large. In this particular case, it looks like Google is one of the leaders in finding ways to bring technology into definite use to attract young population groups to create new mainstream economic activities.

GOOGLE ADSENSE: IMPORTANCE AND FACTORS FOR ADOPTION

Research for this chapter clearly identified a severe lack of literature concerning Google AdSense. This chapter is particularly indebted to probably the only comprehensive book more focused on industry issues than covering theoretical aspects. Comm (2006) published a book in 2006, where relevant areas of Google AdSense in a more descriptive manner were discussed rather than having just a critical outlook. However, the book misses making a clear and concise definition of Google AdSense for both general and non-technical readers. Thus, the book mostly relies on Wikipedia (2015) to continue relevant discussions. According to Wikipedia (2015), Google AdSense is simply understood as a click and pay system. This programme is concerned with promoting advertising on the Internet by allowing marketers, business professionals or individual freelance users to get involved in rating/ selling/ marketing advertising and gaining income supports through the use of the click and pay concept. AdSense is set to link with a specific website of an identified marketer attracted to certain blogs to offer their customers more details about related products or services. In fact, a link exists between Google AdSense and the search engines. The Google search engine is useful collecting available information about any product, service, individual person or person as a customer, or any artefacts. For example, a customer searches on the website about a tourism product; advertising about that particular tourism product can then be shown along with specific providers of that particular product. Such advertising will then offer the chance to select a desired product or service. Google AdSense links selected products or services along with details of a number of business enterprises that can provide that service or product available to certain groups of customers allowing them to purchase, as they desire. This is the general pattern within which AdSense normally works. There are also other more expertise tailored applications for AdSense.

AdSense is first and foremost customer and user friendly. After signing up as a user of AdSense, users are allowed to design, prepare and to view the appearance of their proposed advertisement. There is a close relationship between website, advertisement and micro or blogging as used for AdSense. There is, however again, a serious lack of literature highlighting the definition, key aspects and features of Google AdSense. The only apparent acceptable academic definition of Google AdSense is offered by Gourishetty (2009). The author defines Google AdSense as allowing webmasters to dynamically serve content relevant advertisements on web pages. If the visitor clicks one of the Ad Sense ads served by the website, the website owner is credited for the referral. Google's AdSense program essentially allows approved websites to dynamically serve Google's pay-per-click AdWord results.

Website maintenance related to AdSense is very easy and requires very little effort. Webmasters need only to insert a Google generated java script "spider" into the webpage or website template. Google's spider parses the AdServer website then serves up ads relating to the website's content. Google uses a combination of keyword matching and context analysis to determine what ads should be served. The java script calls up the ad from Google then ensures ads are served each time a visitor goes to the web page. Early on, Google implemented a filtering system that allowed webmasters to prevent a specific domain's ads from being served on any websites in their account. Ad blocking meant that webmasters could dynamically prevent their websites from being used by competitors for competing ads. Google provides a wide variety of ad formats to match the most suitable options with each website. Webmasters can select from a handful of preformatted towers, inline rectangles, banners and buttons. The ad boxes can be modified by webmasters to resemble the website's colour scheme. AdSense has the capacities to appear and to be used as a reliable platform for online marketing (Table 3). The origin and characteristics of AdSense emphasize it is produced from Google AdWords advertisers and that Google is the main organization responsible for this programme. AdWords delivers AdSense to Google for displaying on sites it finds relevant for the purpose. AdSense has unique and incredible features for providing advertisements to targeted customers or marketers. AdSense appears in diverse forms through the use of text, graphics or even audio-visuals where, the typology of AdSense includes graphics, videos or audio advertisements. It is important to insert the advertising in a non-intrusive location on each viewers screen (Mei et al., 2007) proposed how to detect appropriate insertion positions in the video stream and subsequently within the image to reduce the intrusiveness. Also, AdSense advertisements for domains, mobile phones and feeds can be used to reach both marketers and their potential customers. AdSense is growing into a useful and ever more acceptable platform for online marketing (Table 4). This growth is not surprising as it has been commonly anticipated that technological innovation will expand the borders or reach of online marketing (MarketingWeek, 2015).

Using AdSense in tourism academia is probably one of the least researched subjects. However, its potential to become a research topic for marketing and tourism is immense and unquestionable. As argued by Buhalis and Wagner (2013), the least researched areas in tourism should be offered priorities in the academic domain to explore their potentials. According to Google (2015), AdSense as a concept can be interesting when applied to different areas of knowledge including tourism. AdSense advertisements are the result of Google AdWords advertisements.

AdSense as Google's online platform for advertisement is a relatively unfamiliar concept to most of the general Internet users. However, AdSense is not unknown among those who frequently use the Internet as their common market and reliable platform to generate income. AdSense can be a useful means for any Internet based advertising campaign for creating and maintaining marketing capable of reaching wider potential customer segments across the world. Also, AdSense is typically viewed as having the best potential to appear as a source of earning for different user bases like marketers, business professionals or freelancers. To be successful, marketers must be offered the best options to explore opportunities (Daim

Table 4. Summary of success factors in Google AdSense use (Hassan, 2015)

Online Marketing	Access to Advertisers	High Quality Advertising
A Mobile Strategy	Reporting Tools	Mobile Interface
To Earn Money	Easily Accessible Support	It Is Fast And Easy

et al., 2012; Miller, 2010). AdSense normally creates those opportunities by monitoring customers' direct purchases and related data. Also, AdSense creates more opportunities for a business enterprise's future businesses. The types of opportunities AdSense can create can be immense and can lead a business to become more able to be profitability.

This is useful for identifying the target customer bases by capturing and using their geographical locality, purchasing intentions and characteristic features. Google AdSense helps local customers to find and to gain access to any desired specific product or service providing business enterprise. Access also supports those businesses enterprises by generating responses from customers to cause changes in any aspect of the presentation or to modify that enterprise's website followed by details of the service or product offers. It thus becomes apparent there are various ways to transmit messages to audiences or customers. It is very well established that AdSense is capable of offering the maximum benefits from online platforms and offering some sort of freedom for them.

Google AdSense as an Opportunity

Since early times, marketers have been searching for a readily available and easily accessible means by which they can get their advertisements to reach huge customer networks. Through this is an utterly complex and technical process; AdSense has tamed the process to extend marketers advertising to ever-wider customer segments across the world, regardless of geographical or territorial boundaries.

The advertising and website design of any particular business organisation is expected to address each targeted customer base. The website and associated advertisements demand great content in terms of appearance and contents. Colour matching, control of resolution, contrasts and content are all important for an attractive website to create appeal to users while being capable of creating interests within diverse groups of concerned marketing clients. AdSense makes sense: an easy way to earn money from online content. It shows the right ads on website, videos, or even web-based games and permits users to make money. Witness Bart and Zach, Co-founders of SideReel, a firm helping users find, track and watch shows online, affirming: '...simply put, Google doubled our revenue. AdSense is a major revenue driver for us. It's a high-volume channel of quality ads' (2014).

The consequences of 'pay per click', commonly known as PPC, are all outcomes of the website's attractiveness with its contents and outlooks. Such content in a website needs to be attractive both for the advertisements and the users, at the same time. Any website has to be meaningful to be able to attract the desired potential traffic and then to persuade them to click simultaneously. Capacities of a website to attract potential customers are essential. Without genuine appeal, no user will like to click on a website that does not conform to those standards needed to completely persuade users to click. Rather, users might have a glimpse or brief viewing then leave if that particular website fails to create any interest or excitement in them. The demand for AdSense users or customer traffic needs to be carefully defined as Google AdSense relies on simple clicking.

AdSense relies on understanding the tastes, choices or preferences of identified groups of those who can be viewed as potential customers and then offers them the most personally suitable product and service options. This is one of the ways in which AdSense capitalizes on advertising as a process to support a particular business enterprise to manage its advertisement activities and to reach target customers. Through AdSense, marketers become able to pass information about their product or service offerings to mass customer bases by applying effective advertising strategies. For online marketers, AdSense becomes a solid useful platform for fulfilling their objectives to reach their potential target customers. AdSense enables online marketers to prepare answers for customer queries, providing detailed information about their services or products. This is one of the ways in which AdSense is turning into a reliable platform for reaching the target customer bases, thus allowing customers to get sufficient information about an offered product or service then persuade them to make the final purchase.

The establishment and engagement of a firm relationship with any customer base can hardly be seen as easy; while also having to cope with the individual customer traits and preferences that offer their clear dissimilarities. Altogether, the creation of a relationship with existing or new customer bases is rather a difficult type responsibility to perform. Still, in practice, AdSense can manage to create a relationship with its strong presence using Google's technological supports. Specifically, this is important for a small organization in an emerging market.

On the other side, there can be complexities in terms of technology use when selecting the most suitable technology to use in a volatile or challenging market structure. It is very often believed that Google AdSense allows users to earn money in relatively easy ways. However, the success of AdSense relies on numerous factors and its use can very often be hampered by diverse issues. A detailed and proper understanding of this system as an online platform is essential to ensuring desired outcomes and to serving specific purposes. Without having useful operational knowledge, AdSense can hardly bring any fruitful result or even to generate money for users. Over time, AdSense has become easier to use, and the new tools to control ads of several types have been great (Google, 2015). The operation and management of Google AdSense programme require expertise in relevant knowledge areas that can become a bit difficult for new users because of those customer bases previously described as being normally diverse with such different characteristics. In general, the success of a new technology partly relies on customers' close involvement. Google AdSense can hardly be expected to generate income for marketers in any cases where the desired outcome or responses from customers can appear as opposed to and not necessarily matching with marketers' pre-assumptions. Also, a considerable number of marketers cannot reap maximum outcomes from their use of Google AdSense as they find it difficult in practice, leaving it as an unreliable source of earning. The frequency and amount of earning generation also depends on appropriate use of AdSense. Thus, and this is generalized, Google AdSense is supportive of those marketing users' for income generation following relatively easy actions on the Internet.

AdSense enhances the reach of advertising and maximizes its potential by using effective advertisement based marketing strategies. The introduction of Google AdSense thus can be identified as an efficient platform to understand the capacity, necessity and application of effective advertising for generating business prospects. This certainly helps in expanding the scope and realizing the potential of a business enterprise. Meaning, the use and importance of AdSense can hardly be ignored in both the current and future marketing worlds.

It is important that Google AdSense be used within a set of clear operational and legal policies and procedures to restrict any possibility of misuse. Also, the safeguarding of both marketing user and customer rights is important in most cases. Marketing users should not violate set terms and conditions of use; particularly when financial transactions are concerned. Any negative adjustments or misuse needs to be addressed carefully to avoid any type complexities. Any incident of defamation or threat to personal safety should lead to legal penalties. A panel of experts monitoring all legal, safety and security aspects need to play key roles to prepare the rules and regulations to stop any possible misuse and to ensure the maximum benefits for all from Google AdSense. After the legal experts have done their jobs, the role of computer webmasters becomes important in terms of their expertise in relevant areas. Any negligence and carelessness on the part of webmasters when performing their roles can lead to discomfort and harm to the users, especially marketers and business professionals. In principal, Google needs to respect the webmasters and all of the users of AdSense with their rights and freedoms of use. Unexpected cases can be eliminated by taking positive measures against websites aiming to generate money by abusing AdSense.

The basis for creating benefits from Google AdSense is to attract diverse sets of interest groups that are theoretically related to the web/ Internet traffic to be generated. Internet traffic is particularly important because, the success of making money from Google AdSense largely relies on leading oncoming traffic to a user's website. In a simple manner, there have to be customer users to click on each website to make AdSense beneficial. Without these clicks, there cannot be any visible or effective result.

AdSense involves advertising that attracts customer responses and this is important for marketers to understand they must have the required capacities to understand these responses and follow them, accordingly. Customer responses need to be monitored and attended to carefully if marketers' capabilities are going to be developed sufficiently to adopt marketing strategies for diverse customer bases (Rust et al., 2004). Thus, the demand for understanding and addressing customer responses is a necessity. The marketing process provided by AdSense is seen as rewarding those who follow through the systematic procedures. Marketers are required to become aware that AdSense can become beneficial for the entire business through exploring and finding ways to work a relatively wide online marketing area. In order to become more aware about AdSense to ensure financial gains, marketers also need to take specific actions. Those actions should be realistic and should meet customer demands. Those who apply AdSense for promotional and marketing activities follows the most effective set of procedures are the ones expected to get the most out of it.

As we have stated before, AdSense follows a click and pay system that relies on customer responses. Advertisements should be most informative, offering the best information on key features of a product or service, while not being too general and widely descriptive. In an AdSense application, keyword selection for advertisements directs customers to find, make choices and purchase advertised products or services. AdSense advertising is thus required to be attached to keyword searching mechanisms to assist customers to get clear ideas about the offerings of a business enterprise. However, the misuse of Google AdSense when related to financial aspects needs strongly regularized. Google itself should have clear both legal and policy frameworks to reduce any sort of unexpected circumstance.

According to Mihart (2012), marketers need to generate enticing attractions to motivate customers to view their advertisements. Such advertisements should motivate and persuade customers to purchase a desired service or product. It is very natural that customers often times do not even see the best advertisements. Thus, it is also common for what is thought by the marketers to be advertisements to not always attract potential customers, as expected. Srisuwan and Barnes (2008) show that enjoyment is an important factor among the younger, socially aware and somewhat hedonistically-oriented consumers of the magazine. Firms should track consumer preferences from registration and then should try to create activities that give consumers enjoyment. There should be a fine balance of information, design and attractiveness as prerequisites for an effective advertisement. Thus, it becomes important for customers not to simply see selected advertisements but to have those advertisements revealed to them.

Those Blogs created solely for marketing purposes by advertisers are usually quite concise, short and clear to attract the potential customers (Thorpe et al., 2014). These blogs combine valid information from diverse sources. Marketers use blogs prepared by them to inform customers about their products, services or user experiences. Blogs are required to appear with information and detailed contents from sources that can be numerous and diverse. As long as blogs are relevant with AdSense, they should appear in various forms to offer customers the maximum possible opportunity to take in the information and to persuade them to want to obtain the offered products or services. AdSense is responsible for generating interests among potential customers, to then persuade them about how good the offered products or services are, then based on the blogs created by the marketers, to help them make important decisions to purchase those advertised products or services. AdSense is attached to payments; it is then paid for per click. This is one of the reasons why AdSense marketers are required to create a series of blogs providing constantly updated information. This is a concern to Google as it is cracking down on AdSense uses that result constantly in relation to mini blogs Google considers as spam. This makes the use of AdSense a bit difficult for marketers in relation to Google's restrictions. Thus comprehensively, there is growing attention being directed towards Google AdSense from numerous groups of beneficiaries. It demands instructors, teachers, and professors to include to include Google AdSense as a specialized component of academic curriculum in areas of business or marketing. More specifically, it is becoming essential for young learners to get maximum academic access to learn about AdSense and how it may lead or control them while they, the students learn to apply it themselves, in practice.

Evidence of AdSense Success

The two most popular uses of the Internet are; one- searching for and through data and information, and two- finding and purchasing products and services. As a result, of course, we see almost every company advertising their products and services online. As we stated earlier, these companies have or are learning they can quickly benefit from changing advertising scripts giving them greater capability to segment their markets while doing it at relatively low costs. Due to Internet advertising's proliferation, it is important to examine the factors that affect its effectiveness (Tavor, 2011). Ronald and Barbara (2002) found the effectiveness of online advertising is derived mostly from its interactive nature. They affirm there is a positive relation between consumers' attitudes toward each website and their relation towards their advertisements.

A series of success stories are made available on the Google Ads/AdSense website. There, one may read and hear inspiring stories from AdSense publishers pursuing their passions to write content online, thus it is a good place to demonstrate the utility of Google AdSense. As with any project, the process of adopting and using AdSense has to be a long-term commitment and sometimes the results are not immediate. Table 5 provides links to some websites viewed as successful with their use of AdSense.

A few additional websites are presented as relevant synopsis cases below:

- www.c4caravans.com in 2002: an online marketplace to help people advertise their caravans and find buyers. The firm has gone from two staff members to ten since then, and in their peak season they record around one million page views per month.
- www.HistoryOrb.com is an 'on this day in history' website. It started in 2000 with historical events, famous birthdays and deaths and has added more content over time. They now have 200,000 entries and the site is the biggest of its kind. The main goal is to provide better historical information on the net. In Figure 3 is presented a webpage for Google AdSense.
- www.genkijapan.net is an offshoot of a product-based English-teaching site. The site is for learning Japanese, the 'genki' way (which means in a *'fun' or 'lively' way*). AdSense revenue from YouTube allows me to make regular videos about learning Japanese, and I've created sites for other languages, like Korean, Thai and Chinese.

Table 5. Some principal success cases and location (Authors' elaboration from Google Ad Sense, 2015)

Firms	Kind of Activity	Advantages	Years	Location
www.c4caravans.com	Online Marketplace	Market Share	2002	United Kingdom
www.historyorb.com	Education	Growing Historical Information	2000	New Zealand
www.genkijapan.net	Education	Reaching More People	2000	United Kingdom
www.blurtit.com	Social Community	Improving Services	2006	United Kingdom
www.japan-guidc.com	Tourism	Creating Content	1996	Japan
www.psychcentral.com	Mental Health	Growing Business	2003	USA
www.babynames.com	Database of Names	Get Partner; Growing Business	1996	USA
www.sanfranciscochinatown.com	Tourism	Diversifying Into New Areas	2002	USA
www.thehomeschoolmom.com	Education	Growing Revenue	2000	USA
www.cairnsdining.com	Tourism	Creating Websites	2003	Australia
www.golisbon.com	Tourism	Creating Stable Revenue	2004	Portugal

- www.blurtit.com is a social question and answer site, where users can ask and answer questions. The aim is to help people help each other – to ask, to learn, to share and to grow. We started in 2006 with a few questions and answers, which we opened up to users coming through search traffic. We have seen steady growth since and have around 300,000 visits to the site per day.
- www.japan-guide.com is an online travel and sightseeing guide for Japan. The site has existed since 1996 but in 2001 the site turned into a full-time business. Now the site registers about one million unique visitors a month. Consequently, it's grown from just a few pages into a 1,000-page guide.
- www.psychcentral.com was introduced in 2003 as one of the first mental health websites to be run by a mental health professional. It has been fundamental to growing the business. The site had approximately 200,000 visitors a month back then – now it's nearly 4 million.

Google AdSense and Concerns about Traffic and General Success Factors

Increasing and consistent traffic flow is the key to Google AdSense success. This is truly challenging. However, Gourishetty (2009: 7) again suggested some useful tips to maintain a good balanced traffic flow:

If you want to earn profit from your site through AdSense, you have to get traffic to your site. From the following points you can get a good number of traffic without investing a great deal of money. Always keep in mind that the traffic cannot be generated overnight, and you have to invest your time to make your site payoff.

- *On page Optimization*
- *Make your site search engine friendly (don't over rule search engine rules)*
- *Submit site to search engines.*
- *Submit your site to major and important directories*
- *Link popularity – One way and reciprocal links – Avoid Link Farms*
- *Write Articles and submit your articles to article directory*

Figure 3. An example of using Google AdSense on a website
(History Orb, 2015)

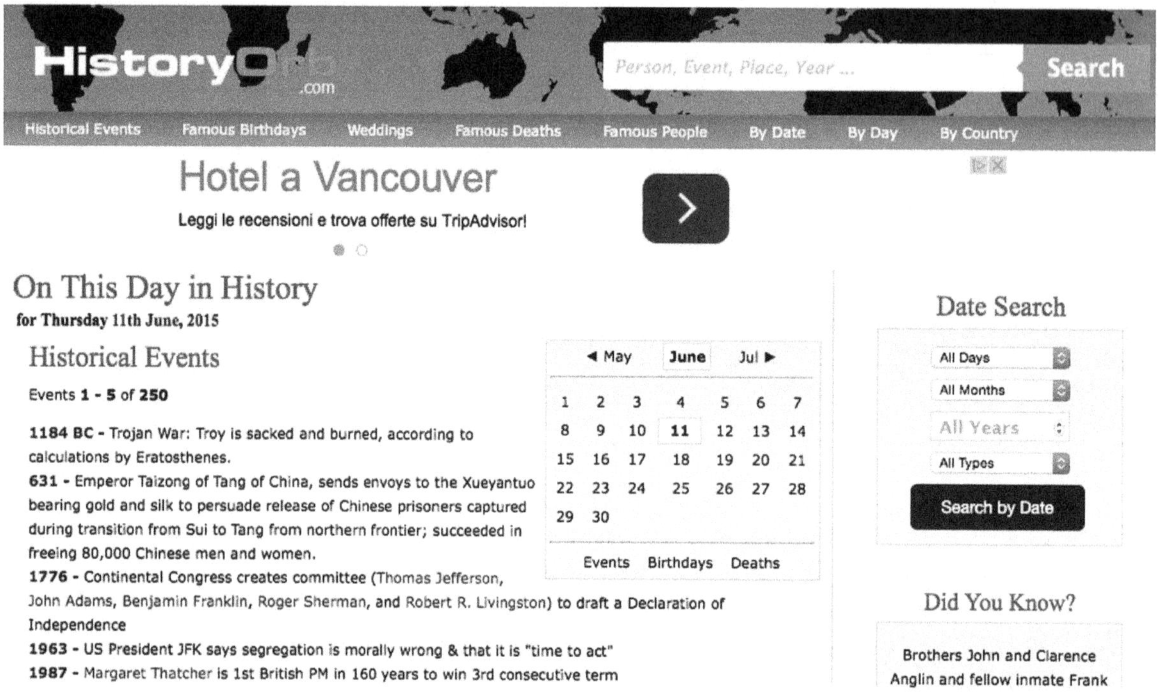

- *Email Marketing*
- *Auto responder marketing*
- *Blog Marketing*
- *RSS Feeds*
- *Create Forums and Massage Boards*
- *Social Media Optimization – Book Mark, Tell a friend, Digg News*
- *Give something to your customer (for example - eBook, software etc.)*
- *Get your customers reviews - Start poll on your site*
- *Get paid visitors – Arbitrage, PPC, AdWords Always refresh your contents.*

It is very unlikely that without having clear and concise operational and technical knowledge, a user can be able to maximize the benefits from the Google AdSense tools. An industry expert, according to Eriksson (2011, 39-40):

Here we are at the end of this report and you've learned a great deal. The new found wisdom will definitely make you ready to start cashing in with your own Google AdSense sites. Let's go over five vital points that I've covered...

- *You need more than one site, in most cases. One low traffic site is not going to yield you more than twenty bucks a month or so. If you choose high paying topics, you may net about a hundred bucks a month from one site. Ideally, you should have at least ten sites – but preferably, you will have a great deal more than that!*

Work out a schedule for yourself. Publish a certain number of new sites each month, and add a certain number of pages to existing sites each month. Keep adding until you've filled up the amount of time that you are willing to spend on your Google AdSense. Develop processes you can follow over and over again for success.

- *You must follow Google's Rules at all times! This is a huge pitfall for many. First, Google has regulations for your site, as to how functional it should be. Second, Google has rules pertaining to what your site may not include. Third, Google has rules about incentives for getting people to click on your ads, and they definitely have rules against clicking on your own ads.*
- *All of these rules are in place to protect the integrity of the program – not to prevent you from making money. Before you publish your first AdSense website, go and read the AdSense Program Policy, the Google Terms of Service, and the Webmaster Guidelines.*
- *You need traffic. You should be working on bringing traffic to your site each and every day. Write and submit articles, participate in the forums, get into social networking and social bookmarking. See if Google AdWords will work out for you. Do everything that you possibly can to bring traffic to your site. If you have no traffic, you have no Google AdSense revenue. It is as simple as that. Most AdSense publishers divide their time, spending about twenty percent of their time adding to existing sites, or building new ones, and spending the other 80% of their time promoting their existing sites.*
- *You must keep track of results, and compare results. You need to not only make changes to your pages, and track those results, but you must also track your advertising efforts as well. All About Residual Income With AdSense 39 of 41 Google AdSense Secrets Revealed.*

By testing and tracking results for your pages, you will find out what ad formats, ad placement, and ad colors work best for your site. By testing and tracking promotional efforts, you will be able to see what techniques are working best in driving traffic to your site.

When you know what works, you can focus your time and energy on those activities, and leave the techniques that don't work behind. This not only saves time, it increases AdSense revenue.

- *You can make a fulltime living from Google AdSense, but it takes time and effort. You apply for Google AdSense, get approved, and get the AdSense code on your site. The first day, you make ten bucks. You quickly add that up, and see that if you made ten bucks each day, you would make 300 per month from one site. You quickly add again, and see that if you had ten sites, all making 300 a month, you could quit your traditional job.*

Unfortunately, that may or may not be true. Before you quit your job, make sure that you have a years' worth of salary saved in the bank. This will prevent financial ruin in the event that Google makes program changes that affect your AdSense income.

Overall, if you are seriously considering ways of making a profit from running an online business, or are already running one and wish to expand, then it may well be worth your while taking a more in depth look at just what the AdSense program can offer you.

FUTURE RESEARCH DIRECTIONS

The lack of available literature on Google AdSense is acute, and that hampered this chapter in many ways. This chapter is primarily based on empirical evidence backed up by as much strong primary data as could be located and combined, it offers a comprehensive picture regarding Google AdSense in academia towards a defined destination. This can be defined as the key limitation of this chapter. Future researchers can replace with more definitive information to provide a better understanding of Google AdSense use in diverse areas of human knowledge. The traffic for Google AdSense is also a relevant area for concentrated attention to be studied and critically analysed, especially those cases where marketers violate and abuse the proper uses of Google AdSense. Such efforts can enrich literature in this defined area.

DISCUSSION AND CONCLUSION

Technology has been constantly influencing the conventional systems or curriculum of education. The number of competitors providing technology service or product is always increasing. Also, recently developed wearable devices such as smart glasses or smart watches are adding more technology users. Existing market competition becomes ever more intense with the involvement of increased numbers of technology or product service providers. Existing literature and evidences demonstrate advertising can be a great opportunity to grow online business. Search engines and the use of incentives can be useful but firms should also advertise their web site through offline channels or any media that can reach target consumers (Srisuwan & Barnes, 2008). It is evident; Google AdSense is gaining acceptability in different areas of business and marketing. Thus, Google AdSense can also become a solid platform for enhancing opportunities for employment. Google AdSense needs to be included in academic syllabuses to allow young learners to know about it and to capitalize from it. The case of developing countries obviously can become a pioneer example to view Google AdSense as rewarding. In terms of determining popularity and potential, this chapter suggests including Google AdSense and its Pay Per Click (PPC) approach in the academic curriculum. Such a course design needs to elaborate on all aspects of website designing, online marketing, email marketing as well as digital marketing to make learners capable of dealing with ongoing global employability and marketing challenges (Deighton & Kornfeld, 2007). Google AdSense can become a valid model for the purpose of both conceptual and practice based learning. Google AdSense can be included in the academic syllabus to both educate and train young learners to be better able to get the maximum benefits from such technology applications. It is true that the level of acceptance and availability of technological education is dissimilar across the world creating an imbalanced scenario in terms of literacy and skill acquisition. This situation worsens in cases where technological innovations are mostly resulting in those countries emphasizing technical educations.

This chapter concludes with how Google AdSense is gaining popularity beyond the business communities with education and younger learners. Google is a well-known and well-accepted technological platform in a world where most of the young populations tend to have immense attraction towards using appropriate technology. The potential markets in emerging countries (e.g. developing countries) are large and growing. In these developing countries, technology can be brought into definite use to attract young population groups with mainstream economic activities. It must be understood; the primary hardware for expanding this technological reach is the mobile cell phone. Thus, we have the mobile technology connection with this chapter.

The widening gap between developed and developing countries needs to be reduced significantly to bring greater balance between them. Thus, this chapter should leave one with the message that Google AdSense can become a solid platform and an effective tool to enhance opportunities for both socio-economic development and employment generation. Thus, Google AdSense needs to be included in academic curriculums to assist young learners to know how it is used and how they can get the most from it. This chapter suggests, on the grounds of both theories and practices, Google AdSense can be included as a specialized subject matter of mobile technologies for students. From an industry perspective, basic to advanced knowledge of Google AdSense can appear as a reliable platform for earning through tactile interventions that can possibly expand to wearable devices such as smart glasses or smart watches. This chapter mainly relies on literature review and technology industry analysis.

ACKNOWLEDGMENT

This chapter relates in general to a research project on ICT and Education of the University of Catania, Italy. This chapter is the result of the collaboration by the two authors. However, D. Privitera wrote sections 1, 2 and 4.2, whereas A. Hassan wrote sections 3, 4.1 and 4.3. The Conclusion is a joint contribution.

REFERENCES

Abernathy, W., & Clark, K. B. (2007). Innovation: Mapping the winds of creative destruction. *Research Policy*, *14*(1), 3–22. doi:10.1016/0048-7333(85)90021-6

Azim, R., & Hassan, A. (2013a, June 20-22). Understanding Recent Wireless and Mobile Technological Changes for Business Management Practices. Proceedings of the 6th International Conference on Business Market Management (BMM). The University of Bamberg, Bamberg, Germany. Retrieved from http://bit.ly/Ns5u1C

Azim, R., & Hassan, A. (2013b, June 20-22). Analysing the impact of mobile and wireless technology on Business Management Strategies. Proceedings of the he 6th International Conference on Business Market Management (BMM). The University of Bamberg, Bamberg, Germany. Retrieved from http://bit.ly/Ns5u1C

Bart, D., & Zach, M. (2014). *Make money from your site*. Retrieved from http://AdSensebrowse.blogspot.co.uk/

Bighiu, G., Manolic, A., & Roman, C. T. (2015). Compulsive buying behavior on the Internet. *Procedia. Economics and Finance*, *20*, 72–79.

Buhalis, D., Leung, D., & Law, R. (2011). E-Tourism: Critical Information and Communication Technologies for Tourism Destinations. In Y. Wang & A. Pizam (Eds.), *Destination Marketing and Management* (pp. 205–224). CAB International. doi:10.1079/9781845937621.0205

Comm, J. (2006). *The AdSense Code: What Google Never Told You about Making Money with AdSense*. Garden City, NY: Morgan James Publishing, LLC.

Dadwal, S., & Hassan, A. (2015). The augmented reality marketing: A merger of marketing and technology in tourism. In N. Ray (Ed.), *Emerging Innovative Marketing Strategies in the Tourism Industry* (pp. 78–96). Hershey, PA, USA: IGI Global. doi:10.4018/978-1-4666-8699-1.ch005

Daim, T. U., Letts, M., Krampits, M., Khamis, R., Dash, P., & Monalisa, M. (2012). IT infrastructure refresh planning for enterprises: A business process perspective. *Business Process Management Journal, 17*(3), 1–16.

Deighton, J. A., & Kornfeld, L. (2007). *Digital interactivity: Unanticipated consequences for markets, marketing and consumers*. Boston: Harvard Business School.

Dhir, A. (2004). *The digital consumer technology handbook*. Oxford: Elsevier.

Donthu, N., & Garcia, A. (1999). The Internet shopper. *Journal of Advertising Research, 39*, 52–58.

Eriksson, A. (2011). Google AdSense secrets revealed: All about residual income with AdSense. *Six-Bucks.com*. Retrieved from http://www.six-bucks.com/downloads/GoogleAdSenseSecretsRevealed.pdf

The European year for development: Education. (2015). *European Parliamentary Research Services*. Retrieved from http://epthinktank.eu/2015/02/27/the-european-year-for-development-education/

Gartner: 10 Business & Technology Trends for Education in 2015 [INFOGRAPHIC]. (2015). *Gartner*. Retrieved from https://mrmck.wordpress.com/2014/10/06/gartner-10-business-technology-trends-for-education-in-2015-infographic/

Goldsmith, R. E., & Lafferty, B. A. (2002). Consumer response to web sites and their influence on advertising effectiveness. *Internet Research, 12*(4), 318–328. doi:10.1108/10662240210438407

Success stories. (2015). *Google*. Retrieved from http://gsmamobileeconomy.covaibalem.

Gourishetty, N. (2009). *Google AdSense*. [Master in Business Consulting Thesis]. Hochschule Furtwangen University, Schwarzwald, Germany.

The mobile economy 2015. (2015). *GSMA*. Retrieved from http://gsmamobileeconomy.covaibalem

Guimaraes, T. (2012). Industry clockspeed's impact on business innovation success factors. *European Journal of Innovation Management, 14*(3), 322–344. doi:10.1108/14601061111148825

Hassan, A. (2012a). Key Components for an Effective Marketing Planning: A Conceptual Analysis. *International Journal of Management & Development Studies, 2*(1), 68–70.

Hassan, A. (2012b). Rationalization of Business Planning Through the Current Dynamics of Tourism. *International Journal of Management & Development Studies, 2*(1), 61–63.

Hassan, A. (2015). The customization of electronic word of mouth: An industry tailored application for tourism promotion. In S. Rathore & A. Panwar (Eds.), *Capturing, Analyzing and Managing Word-of-Mouth in the Digital Marketplace* (pp. 61–75). Hershey, PA, USA: IGI Global.

Hassan, A. (in press). Destination image formation: The function analysis of augmented reality application. In M. Khosrow-Pour (Ed.), *The Encyclopaedia of Information Science and Technology* (4th ed.). Hershey, PA, USA : IGI Global.

Hassan, A., & Ahmed, F. (2013b). The role of NGOs in preventing unemployment in Bangladesh. *Kuwait Chapter of Arabian Journal of Business and Management Review*, *2*(9), 1–12. doi:10.12816/0001227

Hassan, A., & Burns, P. (2014). Tourism policies of Bangladesh – A contextual analysis. *Tourism Planning & Development*, *11*(4), 463–466. doi:10.1080/21568316.2013.874366

Hassan, A., & Dadwal, S. (in press). Search engine marketing – An outlining of conceptualization and strategic application. In W. Ozuem & G. Bowen (Eds.), *Competitive Social Media Marketing Strategies*. Hershey, PA, USA : IGI Global.

Hassan, A., & Forhad, A. (2013a). The role of NGOs in the sustainable development of Bangladesh. *Present Environment & Sustainable Development*, *7*(2), 59–72.

Hassan, A., & Jung, T. (in press). Augmented reality as an emerging application in tourism education. In D. H. Choi, A. Dailey-Hebert, & J. S. Estes (Eds.), *Emerging Tools and Applications of Virtual Reality in Education*. Hershey, PA, USA: IGI Global. doi:10.4018/978-1-4666-9837-6.ch008

Hassan, A., & Rahman, M. (2015). Macromarketing perspective in promoting tourism: The case of the Buddhist Vihara at Paharpur. *Tourism Spectrum*, *1*(2), 13–19.

Hassan, A., & Ramkissoon, H. (in press). Augmented reality for visitor experiences. In J. N. Albrecht (Ed.), *Visitor Management*. Oxfordshire: CABI.

Hassan, A., & Shabani, N. (2015, May 26-29). eMarketing adoption in tourism and hospitality industry in London: Industry analysis and some narratives. *Proceedings of the 4th International Interdisciplinary Business-Economics Advancement Conference (IIBA)*, Las Vegas, USA. Retrieved from http://bit.ly/1BQqGnI

Hassan, A., & Sharma, A. (in press). Wildlife tourism: Technology adoption for marketing and conservation. In M. A. Khan & J. K. Fatima (Eds.), *Wilderness of Wildlife Tourism*. Waretown: Apple Academic Press, Inc.

Hax, A. C. (2010). *Reinventing your business strategy*. New York: Springer.

European agency snapshot study. (2015). *IAB*. Retrieved from http://www.iabeurope.eu/files/5014/2470/1844/IAB_Agency_Snapshot_Survey_European_Results_FINAL_2.pdf

Kumar, E., & Kohli, S. (2007). A strategic analysis of search engine advertising in web based-commerce. *Journal of Internet Banking and Commerce*, *12*(2), 1–13.

Kumar, V., Venkatesan, R., & Reinartz, W. J. (2006). Knowing what to sell when to whom. *Harvard Business Review*, May, 131–145. PMID:16515161

Lee, K., & Miller, K. E. (2003, December 1-3). A positive attitude-ad-brand relationship by customising banner advertisement design- An empirical study. *Proceedings of ANZMAC 2003 Conference* (pp. 25-33).

Lewis, W., Agarwal, R., & Sambamurthy, V. (2003). Sources of influence on beliefs about information technology use: An empirical study of knowledge workers. *Management Information Systems Quarterly*, *27*(4), 657–678.

Li, T., Mei, T., Niu, X., & Ngo, C. (2010, April 26–30). PageSense: Style-wise web page advertising. *Proceedings of the World Wide Web*, Raleigh, North Carolina (pp. 1273-1276).

Liu, C., & Arnett, K. P. (2000). Exploring the factors associated with web site success in the context of electronic commerce. *Information & Management*, *38*(1), 28–33. doi:10.1016/S0378-7206(00)00049-5

Online retail will drive innovation in interactive marketing. (2015). *Marketing Week*. Retrieved from http://bit.ly/1nbpjdM

McElheran, K. S. (2012). *Do market leaders lead in business process innovation? The case(s) of e-business adoption*. Boston: Harvard Business School.

Mei, T., Hua, X. S., Yang, L., & Li, S. (2007). VideoSense: towards effective online video advertising. *Proceedings of ACM MULTIMEDIA* (pp. 1075-1084).

Mihart, C. (2012). Impact of integrated marketing communication on consumer behaviour: Effects on consumer decision – Making process. *International Journal of Marketing Studies*, *4*(2), 121–129. doi:10.5539/ijms.v4n2p121

Mohapatra, S. (2009). *Business process automation*. New Delhi: PHI Private Limited.

On this day in history. (2015). Historyorb. Retrieved from http://www.historyorb.com/

Płomińska, N., Smagurauskaite, A., Platonova, E., & Gadzhuk, O. (2014). *Consumers go mobile in CEE- Mobile market overview.* Retrieved from http://www.iabeurope.eu/files/4414/1898/2815/Gemius_Consumers_go_mobile_in_CEE_2014_final.pdf

Privitera, D. (2015). Tourist valorisation of urban poverty: An empirical study on the web. *Urban Forum* (pp. 1-29). doi:10.1007/s12132-015-9259-3

Rayport, J. F., & Bernard, J. J. (2004). Best face forward. *Harvard Business Review*, *82*(12), 47–52. PMID:15605565

Rust, R. T., Lemon, K. N., & Zeithaml, V. A. (2004). Return on marketing: Using customer equity to focus marketing strategy. *Journal of Marketing*, *68*(1), 109–127. doi:10.1509/jmkg.68.1.109.24030

Shapiro, S., MacInnis, D. J., & Heckler, S. E. (1997). The effects of incidental AdExposure on the formation of consideration sets. *The Journal of Consumer Research*, *24*(1), 94–104. doi:10.1086/209496

Srisuwan, P., & Barnes, S. J. (2008). Predicting online channel use for an online and print magazine: A case study. *Internet Research*, *18*(3), 266–285. doi:10.1108/10662240810883317

Ström, R., Vendel, M., & Bredican, J. (2014). Mobile marketing: A literature review on its value for consumers and retailers. *Journal of retailing and consumer services*, 21(6), 1001-1012.

Tavor, T. (2011). Online advertising development and their economic effectiveness. *Australian Journal of Business and Management Research*, *1*(6), 121–133.

The Florida Agricultural and Mechanical University. (2015). *Welcome to the technology education program at Florida A&M University.* Retrieved from http://www.famu.edu/index.cfm?TechEd

The World Economic Forum & Boston Consultancy Group. (2015). *New vision for education unlocking the potential of technology*. Retrieved from http://www3.weforum.org/docs/WEFUSA_NewVisionforEducation_Report2015.pdf

Thorpe, M., Holm, J., & van den Boer, G. (2014). *Discovering the decisions within your business processes using IBM blueworks live*. New York: International Business Machines Corporation.

Turban, E., McLean, E. R., & Wetherbe, J. C. (2008). *Information technology for management*. New Jersey: John Wiley and sons, Inc.

Ward, M. R., & Lee, M. J. (2000). Internet shopping, consumer search and product branding. *Journal of Product and Brand Management*, *9*(1), 6–20. doi:10.1108/10610420010316302

AdSense. (2015, May 22). *Wikipedia*. Retrieved from http://en.wikipedia.org/wiki/AdSense.

Least developed countries: A UN classification. (2015). *World Bank*. Retrieved from http://data.worldbank.org/region/LDC

Yang, C. C. (2004). Exploring factors affecting the adoption of mobile commerce in Singapore. *Telematics and Informatics*, *22*(3), 257–277. doi:10.1016/j.tele.2004.11.003

Yoon, S. J., & Kim, J. H. (2001). The effects of perceived consumer characteristics on the choice and use of Internet ads. *Brand Management*, *8*(4/5), 346–364. doi:10.1057/palgrave.bm.2540033

ADDITIONAL READING

What is Technology Adoption? (2015). *Bridges-to-technology*. Retrieved from http://bit.ly/T5kFnb

Consumer buying behaviour. (2015). *Business Dictionary*. Retrieved from http://bit.ly/1kICBJX

Cantoni, L., & Zheng, X. (2013). *Information and communication technologies in tourism*. Berlin, Heidelberg: Springer.

Chang, M. K., & Cheung, W. (2001). Determinants of the intention to use Internet/ WWW at work: A confirmatory study. *Information & Management*, *39*(1), 1–14. doi:10.1016/S0378-7206(01)00075-1

Dasgupta, M., & Sahay, A. (2011). Barriers to diffusion of innovation: An empirical study in India. *International Journal of Indian Culture and Business Management*, *4*(3), 325–346. doi:10.1504/IJICBM.2011.040168

Duffy, J. A. M., & Ketchand, A. A. (1998). Examining the role of service quality in overall service satisfaction. *Journal of Managerial Issues*, *10*, 240–255.

Galletta, D. F., & Zhang, Y. (2003). *Human-computer interaction and management information systems: Applications*. Oxon: Routledge.

Gartner technology research. (2014). *Gartner Incorporated*. Retrieved from http://gtnr.it/1nvU5Bb

Herbst, I., Braun, A.-K., Mccall, R., & Broll, W. (2008). *TimeWarp: Interactive time travel with a mobile mixed reality game*.

Karch, M. (2015). *Apps*. Retrieved from http://abt.cm/1l48Ojv

Li, N., Buhalis, D., & Zhang, L. (2013). Interdisciplinary research on information science and tourism: A systematic review. *Tourism Tribune*, *28*(1), 114–128.

Michopoulou, E., & Buhalis, D. (2013). Information provision for challenging markets: The case of the accessibility requiring market in the context of tourism. *Information & Management*, *50*(5), 229–239. doi:10.1016/j.im.2013.04.001

Mitropoulos, P., & Tatum, C. B. (2008). Forces driving adoption of new information technologies. *Journal of Construction Engineering and Management*, September-October, 340–348.

Morabito, V. (2014). *Trends and challenges in digital business innovation*. Heidelberg: Springer. doi:10.1007/978-3-319-04307-4

Morrison, A. M. (2013). *Marketing and managing tourism destinations*. London: Routledge.

Okazaki, S. (2005). Mobile advertising adoption by multinationals: Senior executives' initial responses. *Internet Research*, *15*(2), 160–180. doi:10.1108/10662240510590342

Factors identified that influence willingness to use new information technology. (2013). *Penn State*. Retrieved from http://news.psu.edu/story/267639/2013/03/07/ science-and-technology/factors-identified-influence-willingness-use-new.

Rogers, M. E. (2003). *Diffusion of innovations* (5th ed.). New York: Free Press.

Digital marketing-what is it and why it matters. (2015). *SAS Institute Inc*. Retrieved from http://bit.ly/1cRj6SG

Shen, Y., Ong, S. K., & Nee, A. Y. C. (2011). Vision-based hand interaction in augmented reality environment. *International Journal of Human-Computer Interaction*, *27*(6), 523–544. doi:10.1080/10447318.2011.555297

So, S., & Sun, H. (2011). An extension of IDT in examining the relationship between electronic-enabled supply chain integration and the adoption of lean production. *International Journal of Production Research*, *49*(2), 447–466. doi:10.1080/00207540903433866

Spencer, A. J., Buhalis, D., & Moital, M. (2012). A hierarchical model of technology adoption for small owner-managed travel firms: An organizational decision-making and leadership perspective. *Tourism Management*, *33*(5), 1195–1208. doi:10.1016/j.tourman.2011.11.011

Suh, Y., Shin, C., Woo, W., Dow, S., & MacIntyre, B. (2011). Enhancing and evaluating users' social experience with a mobile phone guide applied to cultural heritage. *Personal and Ubiquitous Computing*, *15*(6), 649–665. doi:10.1007/s00779-010-0344-2

Walters, D. (2002). *Operations strategy*. New York: Palgrave.

Werthner, H., & Klein, S. (1999). *Information technology and tourism - A challenging relationship.* Vienna: Springer-Verlag. doi:10.1007/978-3-7091-6363-4

Wu, I. L., Li, J. Y., & Fu, C. Y. (2011). The adoption of mobile healthcare by hospital's professionals: An integrative perspective. *Decision Support Systems*, *51*(3), 587–596. doi:10.1016/j.dss.2011.03.003

Wu, J. H., & Wang, S. C. (2005). What drives mobile commerce? An empirical evaluation of the revised technology acceptance model. *Information & Management*, *42*(5), 719–729. doi:10.1016/j.im.2004.07.001

Wynne, C., Berthon, P., Pitt, L., Ewing, M., & Napoli, J. (2001). The impact of the Internet on the distribution value chain- The case of the South African tourism industry. *International Marketing Review*, *18*(4), 420–431. doi:10.1108/EUM0000000005934

Youngdahl, W., & Loomba, A. (2000). Service-driven global supply chains. *International Journal of Service Industry Management*, *11*(4), 329–347. doi:10.1108/09564230010355368

Yu, C.-S., & Tao, Y. H. (2009). Understanding business-level innovation technology adoption. *Technovation*, *29*(2), 92–109. doi:10.1016/j.technovation.2008.07.007

KEY TERMS AND DEFINITIONS

Google AdSense: Google AdSense is a cost per click (CPC) advertising programme for those wishing to place ads on their website. A Google product providing a way for website publishers of all sizes to earn money by displaying targeted Google ads on their sites. The publisher chooses where to place ads into their sites and then earns money when visitors click on (or view, in some cases) the ads.

Mobile Technology: The recent development of Smartphone has revolutionized the technological world. Mobile technologies have been expanded to both non-wearable and wearable technological gadgets like smart watch. Surely, may more technological innovations will be released in the followed years.

Pay Per Click (PPC): This is a model of Internet banking. Following this model, each time an advertiser gets a click on an ad, the advertiser has to pay. This is a process of purchasing visits to an advertiser's site rather than earning those visits organically. The most common type of pay per click is considered search engine advertising. When someone clicks on your advertising, the program counts that as a click. Clicks can help you understand how well your ad is appealing to people who see it. In your account statistics, you will see the click through rate (CTR), which tell you how many people have seen your ad and end up clicking on it. A good CTR is relative to what you're advertising and the subsequent networks.

Technology in Education: Education can benefit from new and emerging and applications meeting the learning needs, goals, and objectives.

Chapter 12
The ScavengAR Hunt:
An Augmented Reality Teacher Training Case Study Using Mobile Devices

Daniel Watanabe
Olympic College, USA

ABSTRACT

This chapter explains the design and execution of a pre-service teacher training case study using Augmented Reality (AR), Quick Response (QR) codes, and social media mobile applications installed on iPads. The ScavengAR Hunt activity centered on a story narrative inspired by works of fine art using trigger images from the AR mobile application ARART® and incorporated elements of the board game Clue®. Pre-service teachers in the study were divided into groups of 4-6 and assigned specific, individual roles related to a mobile application used in the ScavengAR Hunt, and completed specific tasks while exploring the campus of a midwestern university. The research monitored the groups in real-time through reports submitted on Twitter and responses from QR code scans. The ScavengAR Hunt served as a model for designing a mobile learning activity incorporating multiple mobile applications.

INTRODUCTION

This chapter focuses on the design process of an activity utilizing mobile devices and mobile applications in an advanced instructional technology course informed by research on Augmented Reality (AR), game design, mobile device usage in the classroom, and instructional design. The ScavengAR Hunt sought to incorporate multiple mobile devices and mobile applications into a collaborative group activity, where each group member contributed to group success by completing specific tasks at four different on-campus locations requiring the use of mobile devices and mobile applications.

Content for the ScavengAR Hunt revolved around target images for multiple AR mobile applications, a technology unfamiliar to the pre-service teachers. The ScavengAR Hunt was intended to serve as a model for pre-service teachers to potentially duplicate in the future, adjusting the specifications of the activity to meet the needs of the content and audience. For most pre-service teachers, this meant elementary school students. Five different sections of pre-service teachers in an advanced instructional

DOI: 10.4018/978-1-5225-0069-8.ch012

technology course participated in the ScavengAR Hunt over the course of one academic year. The ScavengAR Hunt was designed as an instructional game, focusing on the ability of pre-service teachers to use mobile devices effectively in a socially interactive context. Its design incorporates three basic elements: rules of a game, the use of mobile devices (iPads and smartphones) and mobile applications to complete the game, and a collaborative activity allowing the instructor and peers to provide feedback to each other during the activity.

Design

Schute, Rieber, & Van Eck (2012) acknowledge there is currently a lack of experimental studies examining the effectiveness of using games in learning environments. However, they identify a conducive learning environment as one that is active, provides ongoing feedback, challenges learners appropriately, and is able to maintain and sustain learner attention. In addition, they list a series of "must haves" for educational games: (a) conflict or challenge, (b) rules of engagement, (c) goals or outcomes (plus subgoals), (d) continuous feedback, (e) interaction with the environment, (f) compelling storyline. Table 1 below shows how "must haves" were incorporated into the design of the ScavengAR Hunt.

The ScavengAR Hunt follows the Evidence-Centered Design model, which originated from Messick (1994). Evidence-Centered game design seeks to elicit evidence from its participants, showing a degree of competency or skills. It also emphasizes the importance of establishing principled interpretations of the evidence so the competencies or skills are assessed appropriately.

The ScavengAR Hunt uses the ECD action model to obtain evidence from pre-service teachers using mobile devices and applications. The action model can be described as a sequence of actions and what constitutes success of those actions (Schute, Rieber, & Van Eck 2012 p. 328). Groups of four or five

Table 1. "Must have" educational game elements in ScavengAR Hunt (Shute, Rieber & Van Eck, 2012)

"Must Have"	ScavengAR Hunt
Conflict or challenge	Usage of mobile devices and applications is monitored and assessed through submission of Twitter posts and Google Form submissions while the groups are on campus completing the activity. Completion is determined when pre-service teachers submit the Case File, an artifact the pre-services create over the course of the activity, to the instructor. Pre-service teachers must complete the activity within 90 minutes.
Rules of engagement	Groups are provided written instructions and tasks to complete at four separate on-campus locations. The Case File confirms whether the group members followed instructions and completed their tasks.
Goals or outcomes	Each group member uses mobile devices and mobile applications to complete a series of specific tasks. All members must complete their tasks at each location for the group to succeed in completing the activity. Instructor can provide feedback to the group remotely, while group members are expected to collaborate and assist each other to complete group tasks.
Continuous feedback	Instructor can communicate with group via Twitter, providing feedback based on posts and Google Form submissions as the groups complete their assigned tasks. Instructor has access to solutions to provide feedback on task completion, potential technology issues, and deviating from the correct location.
Interaction with the environment	Augmented reality mobile applications allows for technology-based interaction. Pre-service teachers must complete tasks at four separate physical locations across campus.
Compelling storyline	Groups follow a narrative based on works of art used as triggers for the augmented reality mobile applications. The story also forces the group to solve a crime using the criteria of the board game Clue© by identifying a suspect, weapon, and location.

pre-service teachers are provided instructions they must complete using a mobile device. Each member is responsible for contributing evidence to the group, aligning with their role within the group. Evidence constitutes a Twitter post, a Google Form submission, images, and videos created with a mobile device.

Design of the ScavengAR necessitated the use of an instructional design model for the instructor to follow. The ADDIE model was chosen primarily because it outlines the phases designers often experience when creating learning activities. The ADDIE model allows for flexibility and provides structural guidance, since it is not a rigid model in itself, but encompasses multiple design models sharing a common structure (Branch & Merrill, 2012 p.9) By adopting the ADDIE model, the each element incorporated into the ScavengAR Hunt became clearer in form and function, allowing a smoother integration into the final version of the ScavengAR Hunt.

The ADDIE model suggests a thorough analysis of the audience and content, leading to the creation of clear learning objectives. The goals of the ScavengAR Hunt were divided to two basic categories: using mobile devices and integrating mobile applications to complete a task and allow pre-service teachers to work in collaborative groups to complete small tasks as a part of larger, group task, while creating an opportunity to solve problems creatively. Incorporating the ADDIE model with the method-not-media hypothesis proved a useful framework for the instructor during the design process. Below are the steps integrated into the ADDIE model used.

- **Analysis Phase:** Gather background information on the audience and identify knowledge, skills, attributes, and needs. Analysis of the content and determine appropriate level of depth for the audience. Create clear, measurable learning goals or objectives based on content and audience analysis.
- **Design Phase:** Determine what is needed for the activity to align and meet learning goals or objectives, and what content or materials need to be obtained, created or utilized to facilitate achievement of the learning goals or objectives including formative and summative assessments.
- **Development Phase:** Plan and create appropriate materials and components necessary for the activity to succeed in aligning with the learning goals and objectives.
- **Implementation Phase:** Install formative and summative assessments to verify achievement of learning goals and objectives. Provide feedback on activities with opportunities for revision or enhancement.
- **Evaluation:** Verify whether the learning goals or objectives were met and adjust activities as necessary.

BACKGROUND

The instructor wanted the ScavengAR Hunt to be a model pre-service teachers could adopt or duplicate once they had their own students and classrooms. The instructor did not expect the pre-service teachers to create an activity as complex as the ScavengAR Hunt, especially since the users (elementary or secondary students) possess drastically different needs and technological capabilities than the pre-service teachers themselves.

The instructor sought to incorporate the premise of Clark's method-not-media hypothesis, where "the most promising approach to learning is to assume it is caused by instructional methods embedded and presented by a variety of media," into the fundamental design of the ScavengAR Hunt (Clark, 2005 p.

99). The instructor understood that a classroom set of iPads were available and sought to utilize them in an engaging, interactive learning activity. Peluso (2012) explained making iPads available to educators does not automatically result in learning; an instructional design methodology must accompany the introduction of iPads and by extension, the mobile applications iPads use. Jung and Mayer (2013) also point out there is currently a lack of research regarding the adaptation and application of effective instructional methods to mobile technology and whether it is necessary to adapt.

Augmented reality (AR) served as the content for the ScavengAR Hunt. Use of other mobile applications and technologies was based on how the AR mobile applications functioned within the activity. As a technology, AR mixes virtual experiences with real world experiences (Klopfer and Sheldon, 2010). The ScavengAR Hunt, however, placed a stronger emphasis on AR as a concept that could enhance or facilitate learning. According to Wu, Lee, Chang, & Liang (2012) when AR is utilized in combination with other technologies, its educational potential is extended and increased. A cloud-based file sharing mobile application, Google Drive, and a social media mobile application, Twitter, served as the primary companion technologies to AR in the ScavengAR Hunt.

The ScavengAR Hunt was designed as a game-based activity centering on tasks. Prominent features identified by Squire & Jan (2007) appear in the activity's design, including roles, a contest, and embedded tools. The tasks were also designed to incorporate an element of fun and curiosity, another common element to AR games that emphasize tasks (Wu, Lee, Chang, & Liang, 2012). Specht, Ternier, and Greller (2011) conducted a case study on an AR based learning activity similar to the ScavengAR Hunt, called Locatory. In the activity, students traveled to locations across a university campus playing a game much like Memory® collecting matching virtual cards. Players were described as "very enthusiastic" during the activity. However, the researchers also noted how the players appeared to be totally absorbed in the AR technology, with some players wondering into parking lots and streets searching for virtual cards.

Institution and Audience Background

The Midwestern University's pre-service teacher programs are an area of specialty attracting pre-service teachers from across the state and surrounding states. Pre-service teachers in the elementary education program were primarily female, Caucasian, and resided in the university's home state, especially from the state's three largest metropolitan areas, and came from a middle class or rural background. Each course section contained between 20-25 pre-service teachers. Approximately 4-5 pre-service teachers were male, and most shared a similar background with their female counterparts. Exceptions to the predominant pre-service teacher demographic included 1-2 African-American or Hispanic/Latino pre-service teachers per section, and 1-2 international pre-service teachers from China or South Korea.

The ScavengAR Hunt introduced pre-service teachers to the capabilities of mobile devices through applications incorporating familiar devices into a learning activity. The class had access to a classroom set of iPads, a technology the pre-service teachers were familiar with. Mobile applications, another familiar technology, were also used in the activity. The instructor purposefully designed the activity to take advantage of student familiarity and interest in mobile devices and applications to create a comfortable environment for the pre-service teachers to apply higher-level skills.

Prior to the introduction of the ScavengAR Hunt, applications of technology for the pre-service teacher education students was limited. Although pre-service teachers were required to purchase their own iPads during the academic year, and most did, pre-service teachers in the program did not have a structured experience using of the technology. Usage of the classroom set of iPads was limited to recording video and taking photographs for projects requiring the use of laptop computers. Aside from limited iPad use, pre-service teachers also had limited experiences using mobile applications due to cost. Only free or institutionally licensed mobile applications were available for use since mobile applications were linked to an institutional account.

The instructor administered a pre-course survey in the fall of 2014 to assess pre-service teachers' ownership, usage, and attitudes towards the course technology ultimately used in the ScavengAR Hunt.

Of the 67 pre-service teachers surveyed, 62 (92.5%) indicated they owned at least one technology used in the course. Mobile devices were the most commonly owned – 43 (64.2%) owned an iPhone, 41 (61.2%) owned an iPad, 32 (47.8%) owned an iPod (Figure 1).

Sixty-two of 67 pre-service teachers reported owning a smartphone, 43 owned an iPhone. The survey revealed the top five smartphone activities (other than making telephone calls) to be: Checking email or using the Internet (88.1%), taking pictures (86.6%), social media – Facebook, Twitter, etc. (85.1%), listening to music (79.1%), watching video (70.1%) (Figure 2).

Of the 67 pre-service teachers surveyed, 24 reported using their mobile devices "Constantly" (35.8%) or "Regularly" (32.8%). Further, pre-service teachers reported similar results regarding attitude and comfort level with mobile devices (iPad, iPhone, and iPod), expressing feeling "comfortable" or "very comfortable" with their mobile devices (Figure 3).

Mobile applications were identified by the pre-service teachers as the topic they wanted to learn about the most, with 47 (70.1%) giving mobile applications a score of four or five on a five-point Likert-scale, 31 (46.3%) gave a score of five (Figure 4).

The purpose of the pre-course survey was to create a composite of the pre-service teachers' attitude, usage, and comfort level regarding technology with special emphasis on course technology, so the in-

Figure 1. Course technology ownership for pre-service teachers enrolled in fall 2014 advanced level course

Do you own any Apple products?

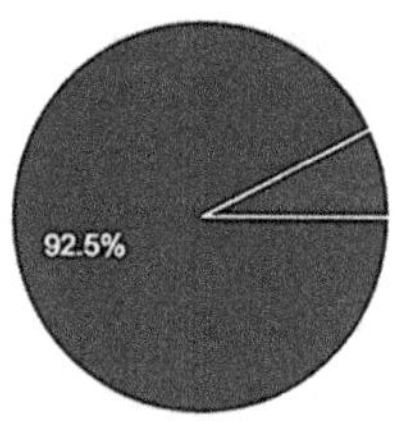

Yes	**62**	92.5%
No	**5**	7.5%

What Apple products do you own?

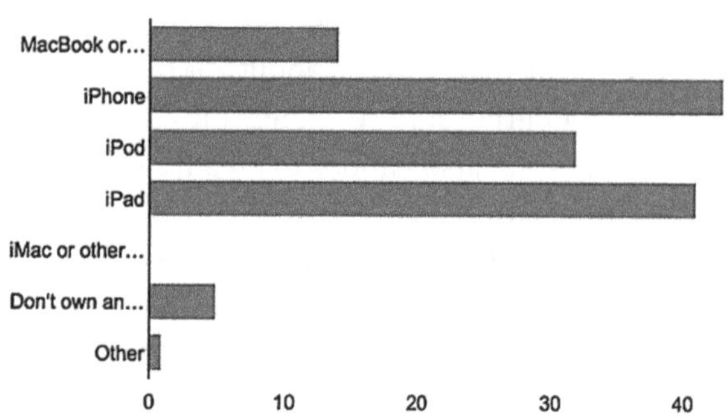

MacBook or other Apple laptop	**14**	20.9%
iPhone	**43**	64.2%
iPod	**32**	47.8%
iPad	**41**	61.2%
iMac or other Apple desktop computer	**0**	0%
Don't own any Apple products	**5**	7.5%
Other	**1**	1.5%

Figure 2. Course technology usage for pre-service teachers enrolled in fall 2014 advanced level course

How often do you use Apple products?

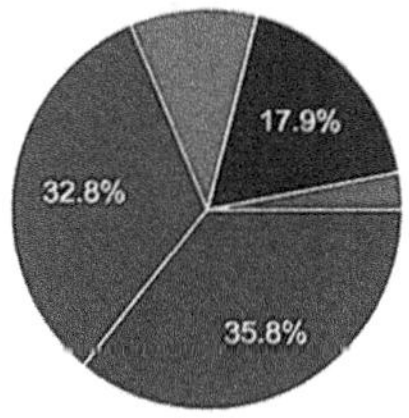

Constantly - You get separation anxiety from your iPhone (or other product)	**24**	35.8%
Regularly - You use Apple products just about every day for multiple purposes.	**22**	32.8%
Occasionally - Your usage of Apple products depends on your purpose, like listening to your iPod while working out.	**7**	10.4%
Rarely - Apple products are recognizable, but your usage and familiarity with them is limited.	**12**	17.9%
Never - Apple products don't factor into your existence much at all.	**2**	3%

Figure 3. Smartphone usage for pre-service teachers enrolled in fall 2014 advanced level course

What do you use your Smartphone for?

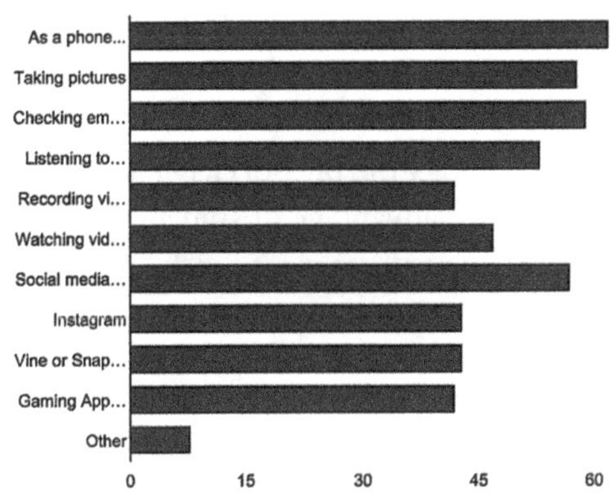

As a phone...	**62**	92.5%
Taking pictures	**58**	86.6%
Checking email or the internet	**59**	88.1%
Listening to music	**53**	79.1%
Recording video	**42**	62.7%
Watching videos	**47**	70.1%
Social media - Facebook, Twitter, etc.	**57**	85.1%
Instagram	**43**	64.2%
Vine or Snapcat	**43**	64.2%
Gaming Apps - Angry Birds, Candy Crush, Words with Friends, etc.	**42**	62.7%
Other	**8**	11.9%

Figure 4. Fall 2014 pre-service teachers rate interest in mobile applications for advanced course

How interested are you in Apps for mobile devices (Smartphones, tablets)?

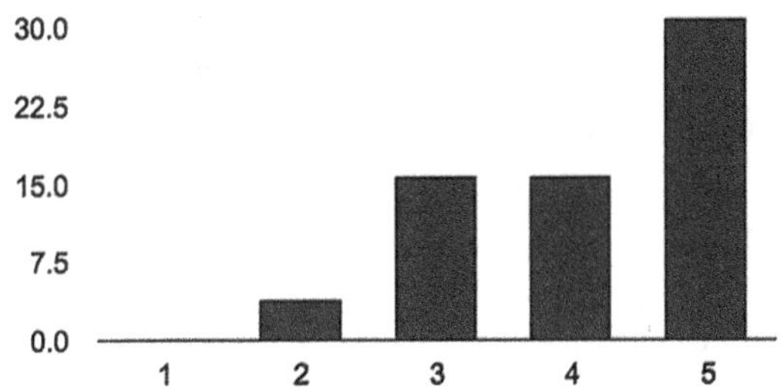

Not interested: 1	**0**	0%
2	**4**	6%
3	**16**	23.9%
4	**16**	23.9%
Very interested: 5	**31**	46.3%

structor could better understand the audience. By fall, the ScavengAR Hunt was already a planned and organized activity of the course. The survey data allowed the instructor to better anticipate and account for the pre-service teachers' accessibility to mobile devices, how pre-service teachers used mobile devices, and their self-efficacy towards mobile devices allowing the instructor to make necessary or useful adjustments to the activity.

CASE STUDY DESCRIPTION

The ScavengAR Hunt had two main goals: engaging pre-service teachers with technology and gain higher-level application skills and cooperating to complete a technology-based learning activity. Pre-service teachers applied their existing experience using mobile devices and applied it to a new technology, AR, working with other pre-service teachers to complete a series of unfamiliar tasks. The activity provided a social context where pre-service teachers could learn from each other as they completed the activity. The ScavengAR Hunt's design goals and sub-goals are listed below. Design considerations and decisions for each goal and sub-goal are described in depth.

Goals and Objectives

1. Pre-service teachers engage with technology and gain higher-level application skills.
   a. Pre-service teachers adapt familiar technology to unfamiliar task.

The advanced course had access to a classroom set of iPads, a technology the pre-service teachers were familiar with. The instructor purposefully designed the activity to take advantage of student familiarity and interest in mobile devices and applications to create a comfortable environment for the pre-service teachers to apply higher-level skills.

Usage of the classroom set of iPads was limited to recording video and taking photographs for projects requiring the use of laptop computers. Aside from limited iPad use, pre-service teachers also had limited exposure and experience using mobile applications due to cost. Only free or institutionally licensed mobile applications were available for use since mobile applications were linked to an institutional account. Pre-service teachers were required to purchase their own iPads during the academic year, and almost all did, pre-service teachers did not have structured trainings or experiences using mobile devices in learning activities.

Google Drive© was used extensively in the advanced course for file storage and project submission. Google Drive was chosen because every pre-service teacher possessed a Google account from the university and the applications allowed for large file storage, like video files produced during class projects. All major projects produced in the advanced course were submitted using Google Drive.

The Google Drive mobile application functioned differently than the web-based version, so it was necessary for pre-service teachers to experience and practice uploading files (images and videos) from the iPad Camera Roll to the Google Drive mobile application. During the ScavengAR Hunt activity, one pre-service teacher in each group, known as the Google Driver, was tasked with uploading image and video files from the iPad to the Google Drive mobile application.

QR codes were incorporated into the ScavengAR Hunt as a remnant from an earlier version of the activity. The previous version of the activity, created by another instructor, relied upon QR codes, which pre-service teachers would scan with an iPod or other mobile device. Pre-service teachers acknowledged awareness of QR codes, since they are prominent in the environment, frequently used to advertise products or causes. This awareness, in addition to building upon the previous version of the ScavengAR Hunt, led the instructor to include QR codes in the activity.

Adobe Voice© was incorporated into the ScavengAR Hunt in the fall trial of the ScavengAR Hunt. Adobe Voice is a mobile application allowing users to create short videos using a series of short, narrated slides. The instructor assigned a project where pre-service teachers narrated a short children's book, speech, or personal experience using Adobe Voice. The project was assigned during the early weeks of the fall course, several weeks before the ScavengAR Hunt. The instructor created the role of Videographer to incorporate Adobe Voice into the existing ScavengAR Hunt, providing the pre-service teachers an opportunity to utilize the mobile application during the activity. The Videographer was tasked with narrating and documenting their group's experience as they executed assigned tasks provided by the instructor.

b. Pre-service teachers demonstrate fluency with unfamiliar technology.

Pre-service teachers were expected to show their fluency with AR applications during the activity through interactions with the environment, usually by physically placing a trigger image on a wall or table and capturing the result in a photograph or video. These interactions usually fit within the context of the narrative established by the fine art trigger image used by ARART. For example, ARART used Hokusai's "Great Wave off Kanagawa" as a trigger image, which activated a fight between giant monsters. The String AR application used an image of a dragon as a trigger image, which would fly through a wall when scanned. These two trigger images were paired together, loosely unified with a "monster" theme.

When the instructor introduced the concept of AR to pre-service teachers, it was clear they had little or no awareness of AR. Identifying suitable AR mobile applications followed a basic criteria: AR applications and/or trigger images had to be available for free, AR applications needed multiple trigger images or multiple superimposing options, and the applications needed to function properly when tested. The search for AR applications yielded a multitude of content subsequently used to build a narrative for the activity. When writing the narrative for the ScavengAR Hunt, the instructor selected existing locations across campus and paired them with the content used by the AR applications. The ARART application used famous works of fine art as trigger images superimposed with animations. The famous work of art served as the foundation for the narrative each group of pre-service teachers followed during the activity and related directly to the animation activated when the trigger image was scanned with the application.

The instructor identified locations around campus where the pre-service teachers would visit and complete specific tasks using AR applications, and were built into the narrative. By doing this, the instructor allowed the pre-service teachers to utilize and show competency with unfamiliar AR technology at a familiar campus setting (Figure 5, 6).

2. Pre-service teachers cooperatively complete a technology-based learning activity.
   a. Pre-service teachers self-selected roles within an assigned group appropriately.

Five to six pre-service teachers were assigned to each group. Group selection was not random. The instructor selected each group based on observations and interactions with the pre-service teachers over the course of the semester. Group selection criteria included performance on earlier projects using

Figure 5. Pre-service teachers scan the ARART® "Mona Lisa" painting by Da Vinci, triggering the augmented reality animation

mobile devices and other course technology, assessed through creative and technical guidelines with an accompanied evaluation rubric; qualitative observations of the pre-service teachers personalities, like competitiveness, and group dynamics, identifying the pre-service teachers who appeared to work well with others or pre-service teachers likely needing additional peer support to complete the activity. Within each group, there were five unique roles. Each role specialized in the usage of one or two specific mobile applications. After the pre-service teachers were divided into groups by the instructor, they choose a role best suited to their abilities. In most situations, only one group member was permitted to serve in each role. Specific instructions accompanied each role at the four locations of the ScavengAR hunt. In order to complete the activity satisfactorily, each group member was required to complete his or her assigned task at each of the four locations.

After each pre-service teacher selected their role in the group, the group submitted the Case File to the instructor, listing each designated member's role to allow the instructor the ability to monitor each member's contributions. The instructor was able to use formative assessment throughout the activity to gauge how well students had self-selected their roles and offer direction or correction to enhance the student's experience in their self-selected role. Additionally, some group members were able to assume leadership or advisory roles since group selections had been carefully considered by the instructor. This

Figure 6. Pre-service teachers scan the ARART® "Sunflowers" painting by van Gogh, triggering the augmented reality animation

balance of structure or latitude given to students enabled them to grow as a result of the participation in the learning activity.

- **Twitter Raven:** The Twitter Raven followed the instructor's Twitter account using their own personal Twitter account for reporting on the group's progress during the ScavengAR Hunt in real time. In addition, Twitter Raven served as the communication contact with the instructor if the group encountered difficulties or had questions. Pre-service teachers with a Twitter account were made available for this role and were given permission to use their own mobile devices.
- **Google Driver:** Google Drive was used extensively in the course, and was the primary method of project submission. Each pre-service teacher had a personal Shared Folder created by the instructor used to submit assignments. The Google Driver's role was associated with the Google Drive mobile application, and the pre-service teachers uploaded the image and video files to their personal Shared Folder so the instructor can view and verify all groups' tasks were completed.
- **Thespian:** The Thespians were pre-service teachers comfortable being on camera and served as the group's actor, the member who was featured in the images or videos created during the ScavengAR Hunt as the group members completing their assigned tasks.
- **Courier:** the Courier was responsible for maintaining and transporting the group's Case File and materials, like the markers. The Case File contained all the initial target images for all of the applications used in the activity with clues to guide the group. As the group progressed through the activity, they were required to add content to the Case File. A complete Case File was then submit-

ted to the instructor at the conclusion of the activity and was necessary for successful completion. The Couriers role required minimal technical skill, aside from scanning the QR code to update the instructor on the group's progress.

- **Videographer:** the Videographer used the Adobe Voice mobile application to narrate and document the group's experiences during the ScavengAR Hunt, taking photos and using the application to add narration. In the original version of the ScavengAR Hunt, the Videographer role did not exist, since the Adobe Voice mobile application was unavailable. The role was added during the fall semester for the ScavengAR Hunt after the instructor included an Adobe Voice assignment into the course.

b. Pre-service teachers assumed leadership roles appropriately.

While designing the ScavengAR Hunt, the instructor also wanted to promote opportunities for pre-service teachers to assume leadership roles, allowing them to provide feedback to other group members should the mobile devices or applications not work properly or to provide help to those who struggle using mobile devices. Several criteria were used in group selection to accomplish this objective. Academic performance, behavioral tendencies, and personality traits were all used in determining the composition of the groups. Academic performance was assessed using class projects and assignments prior to the ScavengAR Hunt activity. By the time the ScavengAR Hunt activity arrived on the course schedule, pre-service teachers had already completed at least two projects using mobile devices. The projects were unrelated to the ScavengAR Hunt, but used components from the ScavengAR Hunt such as Google Drive and the Adobe Voice mobile application. Projects were assessed with a quality rubric and examples including both technical and creative criteria. Technical criteria typically included basic requirements and expectations specific to the application or software and demonstrated technical proficiency. Creative criteria was more abstract, like applying a consistent layout or theme throughout the course of the project. Students received grades on their projects with both criteria weighted equally. Behavioral tendencies and personality traits referred to the pre-service teacher's general behavior during class and interactions with the instructor. The instructor observed and considered typical classroom behavior such as punctuality, attendance, the ability to stay on task, asking questions and receiving feedback, handling adversity appropriately, and disruptive or distracting behavior tendencies. The instructor also noticed groups of pre-service teachers who sat near each other and those who appeared to have positive and negative relationships with one another. Using the following three criteria, the instructor placed the pre-service teachers into four categories:

- **Potential Leaders:** The instructor identified potential leaders as those who showed a strong proficiency with mobile devices and consistently produced excellent projects. Potential leaders also possessed qualities like reliability (punctual, communicates well with peers) and a respectful attitude towards other members of the class. Further, competitive tendencies were also considered, since they can be advantageous in an activity like the ScavengAR Hunt. For example, one class contained two student-athletes who consistently performed well in the class and were seen as potential group leaders. The instructor made it a point to place these two pre-service teachers in separate groups.
- **Capable Members:** Capable members were those who performed consistently and solidly in all three of the instructor's criteria. They produced good but not necessarily outstanding projects and

proved to be capable of understanding the technologies used in the projects. They tended to show up to class, perform, and contribute consistently but the instructor could not clearly identify an area of exceptional strength according to the three criteria.

- **Positive Influencers:** Positive influencers were pre-service teachers who were not the most technically proficient, but possessed potentially beneficial personality traits or interests to their group during the ScavengAR Hunt activity. Positive influences were often were creative, had an accommodating disposition, unique interests like drama or art, and typically shared their interests substantially in their projects. These members also had the potential to boost group morale if the activity did not go as planned.
- **Conditional Members:** Conditional members were the pre-service teachers who would likely require assistance from their peers while completing the ScavengAR Hunt. Relationships with other members played a key role when assigning these pre-service teachers into their respective groups. These members often found it difficult or challenging to grasp the technology in the course or it presented a barrier to be accounted for during group selection. For example, two separate course sections contained international pre-service teachers, and were placed in groups where they had a familiar or positive relationship with at least one other peer. Frequently, these members chose the Courier or Thespian role, which required minimal use of technology.

Groups were assigned during the class period prior to the activity itself, and the groups were given the opportunity to meet each other and choose their roles within the group. The instructor provided a description of each role and each group was expected to submit a paper form with each of the roles (Twitter Raven, Courier, Google Driver, Thespian, Videographer) and the group member serving each role. After assigning roles, the groups were then assigned two iPads by the instructor and told to install the mobile applications needed to complete the activity. After installation, the groups had the opportunity to practice using the iPads and all the mobile applications used in the activity to ensure they functioned properly. The Google Driver person also had the opportunity to log in to his/her account in preparation for the activity. Besides the two iPads, the groups could make use of their personal mobile devices, usually smartphones, to complete the activity. All pre-service teachers serving as Twitter Raven used their smartphone. During this process, the instructor observed certain group members would guide the rest of the group through the process of deciding on group roles. The instructor noticed some predictability in the role selection, like those identified as Conditional members choosing roles requiring minimal use of the mobile devices. Pre-service teachers identified as Potential leaders did not necessarily assume leadership roles within the group, nor did there seem to be any strong tendency with the roles Potential leaders selected within the group.

c. Pre-service teachers learned from each other appropriately.

Once the ScavengAR Hunt began, the pre-service teachers were physically separated from the instructor until the activity was completed. Although communication with the instructor was still possible through the Twitter Raven, the pre-service teachers were left to follow the instructor's written directions to complete specific tasks for each role.

The pre-service teachers spent time during the previous class period installing and using all the mobile applications used in the ScavengAR Hunt with iPads. This allowed the pre-service teachers to experience the functionality of the mobile applications, so they would know what to expect once the activity began. The instructor provided basic feedback on each mobile application's functionality, so the pre-service teachers could identify potential technology issues and trouble shoot appropriately, like physically moving an iPad higher, so the iPad's camera could better capture the trigger image for AR applications.

Directions given to each group were designed to allow the groups to have a certain degree of freedom and the ability to choose where and how each specific task was completed. At times, the directions forced the pre-service teachers to engage in creative problem solving using additional mobile applications. In this instance, the instructor required the Twitter Raven to record a brief video update at the final location, stating "We've solved the crime." However, the instructor did not realize that Twitter did not allow video messaging. To address this problem, multiple pre-service teachers completed the task by utilizing the Vine© mobile application, which proved to be ideal, since Vine records six-second videos – plenty of time to say, "We've solved the crime." As a result, the instructor incorporated Vine into all subsequent versions of the ScavengAR Hunt.

Assessment

Summative Assessment

Summative assessment of the ScavengAR Hunt consisted of each group submitting a completed Case File to the instructor after visiting the four on-campus locations and returning to the classroom where the activity began. Completion of the Case File was dependent upon each individual group member completing the tasks assigned to their role within the group using mobile devices.

The pre-service teachers chose individual roles in their groups. Each of the five members (Courier, Google Driver, Thespian, Twitter Raven, and Videographer) was expected to complete a specific task at each of the four locations of their ScavengAR Hunt route according to their chosen role. Each group received a series of four marked envelopes containing written directions and materials needed to complete the tasks at each assigned location. These directions and materials were housed within the Case File. Each role was expected to complete the same tasks, however they did not necessarily complete these tasks at the same point during the activity. The instructor created eight unique sets of written directions. No two sets of written directions were exactly the same, nor were the solutions explained through the narrative. At the conclusion of the activity, it was the responsibility of the Courier to submit the group's Case File. The Case File contained all the physical materials: target images for AR mobile applications, envelopes, suspect page, locations page, and weapons page, and QR code the instructor needed to verify completion according to the written directions. Members used colored markers to chart progress in the Case File and complete tasks in the directions.

Pre-service teachers needed to indicate individual progress, so each role within the group was assigned a color with a corresponding marker. The most important factor was the colors remained common and consistent for each role so pre-service teachers and the instructor understood who completed the task. Color-coding each role also created the opportunity for members to form an identity within their group, one instrumental to completing the ScavengAR Hunt. Submission of the physical Case File, in addition to responses on the form linked to the QR code and Twitter feed formed all of the elements necessary for completion of the activity.

In the context of the course, each pre-service teacher was awarded 100 points (there were approximately 2000 possible points in the course) for participation in the ScavengAR Hunt. Bonus points and incentives were awarded to groups based on the order they completed the ScavengAR Hunt. Each member in the first place group was awarded 20 bonus points and a gift card valued at $5. Each member of the second group to finish received 15 bonus points, while the members of third and fourth place groups each received 10 bonus points for completion.

The instructor felt the pre-service teachers should not be at risk of losing points during the ScavengAR Hunt, which might adversely affect the students' performance in the course. Rather, the instructor felt awarding bonus points was a desirable alternative, since every pre-service teachers would receive a bonus as long as they completed the activity. The size of the bonus depended upon how quickly the group completed its assigned tasks.

The designed elements in the ScavengAR Hunt created a multitude of variables, and thus potential hindrances to group performance. For example, pre-service teachers could encounter technology problems like failure to connect to the university's Wi-Fi network, which would adversely affect the group's ability to complete their tasks since all mobile applications used in the activity required an Internet connection to work properly. While measures were taken to anticipate and reduce the likelihood of such issues, the pre-service teachers practiced scanning AR targets and installed all applications on the group's iPads before the activity, with the possibility for technological failure the instructor felt the pre-service teachers should not be penalized for circumstances beyond their control.

Formative Assessment

The instructor provided real-time feedback to the group based on posts on Twitter and Google Form responses. Groups were expected to communicate with the instructor at each location stop during the ScavengAR Hunt using Twitter and through Google Form responses linked to a QR code they scanned with their mobile devices. The instructor would then redistribute the responses to the groups through Twitter, so groups could monitor each other's progress, especially given the substantial incentive of completing the ScavengAR Hunt first.

The tables below were part of the "answer key" used by the instructor to provide relevant feedback, to the groups as they roamed across campus like ensuring groups visited the correct location and answered questions correctly. Table 2 lists what task each group was expected to accomplish at each of its four on campus locations. Table 3 lists the solutions to the "crime" that needed to be verified in the Case File.

The Twitter Raven served as the group's primary communication liaison to the instructor. The Twitter Raven would communicate with the instructor through posts on group progress, allowing the instructor to monitor the group in real time. The Twitter Raven was instructed at the briefing session before the activity to post any questions or issues encountered during the group's experience on the ScavengAR Hunt. It was impossible for the instructor to anticipate every potential problem, so an open channel of communication was a necessary component of the ScavengAR Hunt's design.

The instructor was able to identify issues and communicate with the groups, allowing the groups to make adjustments and complete the activity. For example, a group misread the written directions and went to the wrong location. The instructor identified the issue after the group submitted a response to the Google Form, meaning the group scanned the QR Code with their mobile device. Upon verifying the group was at the wrong location after referencing the "answer key," the instructor communicated with the group's Twitter Raven, instructing the group member to call the room where the instructor was monitoring the activity. The instructor mentioned the group was at the wrong location, and suggested the group re-read the written directions. The group identified its mistake, and continued to the next location.

Table 2. Location stops and tasks for ScavengAR Hunt groups (Watanabe, 2015)

Group	Personnel	Stop 1	Stop 2	Stop 3	Stop 4
1	Courier Google Driver Thespian Twitter Raven Videographer	Library ARART Sunflowers	Cafe Group Photo	Ballroom STRING Dragon	PE Building colAR Fireworks
2	Courier Google Driver Thespian Twitter Raven Videographer	Resource Center ARART Mona Lisa	PE Building colAR Fireworks	Cafe Group Photo	Library STRING Boffswana
3	Courier Google Driver Thespian Twitter Raven Videographer	Café Group Photo	Ballroom ARART Girl w/Earring	PE Building colAR- Carousel	Resource Center STRING Shoe
4	Courier Google Driver Thespian Twitter Raven Videographer	PE Building colAR Bird	Library ARART Great Wave	Cafe Group Photo	Dining Hall STRING Shoe
5	Courier Google Driver Thespian Twitter Raven Videographer	PE Building colAR Carousel	Resource Center ARART Mona Lisa	Cafe Group Photo Paint U	Music Hall STRING Shoe
6	Courier Google Driver Thespian Twitter Raven Videographer	Resource Center ARART Girl w/earring	PE Building colAR Fireworks	Library Aurasma Dalek AR Dinopark	Cafe Group Photo
7	Courier Google Driver Thespian Twitter Raven Videographer	Café Group Photo	Library ARART Sunflowers	PE Building colAR Carousel	Lounge Aurasma Gorilla Paint U
8	Courier Google Driver Thespian Twitter Raven Videographer	Library ARART Great Wave	Cafe Group Photo	Plumb STRING Dragon	PE Building colAR Bird

The instructor also identified issues through lack of reporting. During the fall trial, multiple groups failed to consistently report their findings at each location by scanning the QR code, resulting in the instructor sending tweets to the groups, reminding them to scan the QR code at each location. When Twitter Ravens failed to report, the instructor responded through Twitter, asking if the group was having any issues. In one instance, a group was held up by an AR application, colAR, failing to work properly - resulting in a large loss of time. Since the ScavengAR Hunt took place during a single, 90 minute class period, the instructor advised the group to move on to their next location.

Real-time monitoring through Twitter and Google Forms in conjunction with the written "answer key" proved invaluable in the formative assessment of the ScavengAR Hunt. Since the group members were

Table 3. Solutions to each group's ScavengAR Hunt Case File (Watanabe, 2015)

Suspect	Location	Weapon	Artwork (Trigger Image)	Group Number
White	Lounge	Rope	Sunflowers	1
Green	Library	Knife	Mona Lisa	2
Red	Kitchen	Candlestick	Girl w/Earring	3
Yellow	Dining Hall	Lead Pipe	Great Wave	4
Blue	Music Hall	Revolver	Mona Lisa	5
Red	Library	Wrench	Girl w/Earring	6
Purple	Ballroom	Bare hands	Sunflowers	7
Purple	Lounge	Wrench	Great Wave	8

expected to complete specific tasks according to their chosen role, accurate reporting through Twitter and Google Form the responses meant the pre-service teachers were actively participating and using the technologies chosen for the ScavengAR Hunt. While additional communication and assistance from the instructor was available, the majority of groups did not find it necessary. Monitoring allowed the instructor to identify potential issues when it became clear the groups were either unaware of a problem or unable to solve the issue on their own.

Learning Activity

Group Assignment

For the ScavengAR Hunt, analysis of the audience and the mobile applications used in the activity needed to be in working order and firmly grounded in the context of the course. The instructor accounted for previous activities and projects completed in the course, which included use of Google Drive, Adobe Voice, and a general search and evaluation of mobile applications. Pre-service teachers explored and evaluated mobile applications designed for their future target audience, elementary or secondary students. Feedback from the mobile application search and evaluation revealed the pre-service teachers found most mobile applications to be supplementary in nature, useful for practicing skills or competencies, but not something they would regularly build a lesson or activity around. Pre-service teachers noticed these mobile applications lacked assessments, like the ability to check progress or offer feedback to the user.

The instructor carefully analyzed the pre-service teacher feedback and incorporated it into the design of the ScavengAR Hunt in the form of real-time remote monitoring through Twitter and forms created with Google Drive. Twitter and Google Drive allowed the instructor to receive and offer immediate feedback. Both Twitter and Google Drive were appropriate for the pre-service teacher audience, but Twitter would not necessarily be appropriate for elementary or secondary students. Forms created with Google Drive, however, are easily adaptable for any audience, function well within the Google Drive mobile application or mobile web browser. This feature alone justified the Google Drive's inclusion in the ScavengAR Hunt.

Creating a narrative, offering special incentives, and gamifying the ScavengAR Hunt were key components in the activity's design. These components were intended to enhance the experience and add a layer of competition and enjoyment to the activity. Incorporating these elements requires a solid understanding of the target audience and their interests, and is independent of the pre-service teacher's technical proficiency.

Rewarding all participants with full points for completion reduces the stakes of the activity. Pre-service teachers would not be punished for technology issues or circumstances beyond their control. Instead, the activity became a competition between groups to see who received the largest incentive in the form of bonus points or a Starbucks coffee gift card – a special reward. Appropriate incentives aligned with the needs or interests of the target audience.

The narrative element of the ScavengAR was based on the target images used by AR mobile applications, which were famous works of fine art or images from popular culture. The instructor used these images as a prompt for writing a story, much like creative writing exercises. The narrative included elements from the board game Clue, where participants solve a mystery involving a suspect, location, and weapon. On-campus locations replaced locations from the Clue board game, and suspects were renamed. These combined elements created a narrative involving familiar locations around campus and familiar works of fine art or images from popular culture while introducing pre-service teachers to the unfamiliar AR technology within Clue's mystery-solving game context. Further, the narrative provided context for the trigger image content, which was more desirable than simply having pre-service scan the trigger image with the AR mobile application.

Creating a level playing field for a competitive activity like the ScavengAR Hunt was a necessary design consideration. The instructor provided each group with an equal opportunity to not only complete the activity first, and earn extra incentives, but also create a relatively stable group dynamic. Creating such groups is difficult to accomplish in a relatively short amount of time, especially given the intangible and highly variable nature of personality traits such as "leadership qualities" or "getting along well with others."

To increase participation in the ScavengAR Hunt, a group activity, the instructor created five separate roles within the group: Courier, Google Driver, Thespian, Twitter Raven, and Videographer. Each group would have one member serving in each role, and roles could be duplicated if groups had more than five members. Each role also had an assigned color, used to verify the group member completed the assigned task in the directions. For example, the Twitter Raven used purple. The group members would use colored markers to complete the Case File, a binder used to house all the materials necessary to complete the activity, including AR target images and a listing of all the potential locations, suspects and weapons used in the Clue game.

Gamifying the activity created a competitive element to the ScavengAR Hunt enhanced by incentives. Clue was chosen since it had simple rules which were familiar to most pre-service teachers. Since Clue relies on deduction, it allowed the instructor to decide on a solution for each group, which was unique for each group, and to control the amount of information revealed to each group (Table 2). At each of the four locations, groups received more information for its Case File. By the fourth and final location, groups possessed all the information necessary to "solve the crime."

Throughout the ScavengAR Hunt, the groups reported their findings by scanning a QR code with a mobile application linked to a Google Form, tracked by the instructor. Each group received directions in envelopes, including the instructions to the tasks groups needed to complete at each location and the story narrative. The instructor provided hints about the location, suspect, and weapon in the narrative, and revealed the information in full, beneath the narrative paragraph. The instructor recognized the locations, suspects, and weapons were meant to enhance and provide context to the ScavengAR Hunt, however "solving the crime" was not the primary goal of the activity – group collaboration and utilizing mobile applications remained the primary goal.

Developing all of the materials for the ScavengAR Hunt was a significant undertaking for the instructor, who created and prepared the majority of the materials without assistance. Since the instructor taught two sections of the advanced course, eight separate Case Files were prepared, all unique. Each group followed a unique narrative, and with it, a unique solution to their Clue mystery reported through a common QR code. All groups visited four locations, and followed a unique order and path around campus. Two locations were common to every group, where they would obtain envelopes containing directions from graduate teaching assistants or student employees to continue the activity. In creating the Case File, the instructor obtained eight report binders and eight sets of colored markers. The instructor took photographs of seven separate locations around campus and created avatars representing suspects and weapons, creating three separate printable pages for inclusion in each group's Case File. Each group received three AR target images, using three different AR applications, all eight combinations were unique. The ScavengAR Hunt necessitated the creation of a course Twitter account.

The instructor recognized creating eight different Case Files was unnecessary; the activity would function effectively with four. The decision to create eight separate Case Files was designed to not give any one group of pre-service teachers an advantage, to create a unique experience, and to allow for flexibility when duplicating the activity in the future. The instructor did not expect pre-service teachers to create an activity with the same level of depth or number of components, but instead create an activity representing what was possible through careful consideration and thought using the ADDIE instructional design model.

Remote Monitoring

The instructor was separated from the pre-service teachers for the duration of the ScavengAR Hunt as the pre-service teachers explored campus and completed their assigned tasks. While completing their tasks, the instructor received updates on group progress as the pre-service teachers in real time through Twitter and the Google Drive form responses installed on the group member's iPads or smartphones. This design consideration allowed the instructor to monitor all the groups from the classroom, a centralized location where the instructor could give feedback or guidance if necessary. The instructor used Twitter to update all of the groups on each other's progress, which was important given the incentives built into the activity.

Technology Components

The ScavengAR Hunt's utilized mobile devices and mobile applications, focusing on use of the iPad and supplemented by the use of smartphones owned by the pre-service teachers. Approximately 65% of pre-service teachers taking the course in fall 2014 reported owning an iPhone (Figure 1). Initially, the iPad was chosen as the primary mobile device for the ScavengAR Hunt because the instructor had access to a classroom set, which could be distributed to groups of pre-service teachers. The number of iPads was adequate to supply each group of pre-service teachers with two iPads. The decision to give each group a pair of iPads was driven by the necessity of having a replacement iPad available should one not function properly and by the fact the instructor needed a mobile device available to record the group's participation in the ScavengAR Hunt. The first trial of the ScavengAR Hunt, in spring 2014, occurred at a time when the classroom set of iPads were the only readily available source of iPads. In the following semester, iPads were purchased by every pre-service teacher in the elementary education program as a part of a university technology initiative. This development did not influence the design of the ScavengAR Hunt, as far as the allotment of iPads for each group. The limit of two iPads per group remained.

The instructor allowed the use of the pre-service teacher's smartphones in the activity because it was a simple way to increase the number of mobile devices available to each group since the number of iPads available for distribution was limited during the initial trial in spring 2014. In briefing sessions prior to the ScavengAR Hunt, the instructor afforded pre-service teachers the opportunity to install any of the mobile applications used in the ScavengAR Hunt onto their smartphone. Smartphone usage proved essential to successful completion of the ScavengAR Hunt since pre-service teachers performing the role of Twitter Raven. Every pre-service teacher in that role had the Twitter mobile application installed on his or her smartphone for personal use. During a briefing session before the activity, the instructor recommended pre-service teachers use their personal Twitter account on their smartphone instead of using the class iPads. The instructor recognized potential issues could arise if a pre-service teacher used Twitter on a class iPad, notably the possibility other people could access the account. Twitter offered real-time text and image updates easily tracked and redistributed, along with a non-invasive way of communicating with pre-service teachers, desirable features for an activity like the ScavengAR Hunt. Social media and social media applications were also an emphasized topic in the advanced course's syllabus, further reinforcing Twitter's inclusion.

The narrative for the ScavengAR Hunt was based on the trigger images used by AR mobile applications. The instructor investigated and evaluated multiple AR mobile applications before settling on a list of suitable applications. All mobile applications used for the ScavengAR Hunt needed to be available for free, and the AR applications were no exception, so all participants could have access to them. In addition, trigger images for the AR applications needed to be available for free. Once found, paper copies of the trigger images were printed for use in the activity.

ARART was an AR application using famous works of fine art as trigger images. The fine artwork was incorporated into a story, the pre-service teachers were expected to solve. When scanned by the application, the image would become animated as long as the trigger points were identified by the iPad camera. For example, Da Vinci's "Mona Lisa" would move its fingers and crumble when scanned, revealing a portrait of Da Vinci underneath (Figure 8).

Aurasma was an AR application used by numerous corporations, so trigger images available for the application featured prominent figures from popular culture. The instructor chose a trigger image featuring a gorilla from a recent Rolling Stones album and a Dalek from the British Broadcasting Corporation (BBC) program "Dr. Who." When scanned, the gorilla would emerge from the album cover, while the Dalek would activate a 3-D model moving and making sounds when touched on the iPad.

The colAr application used coloring pages as trigger images to be downloaded for free from its companion website. When scanned, the trigger image would transform into a 3-D, interactive model. Since the trigger image could be colored, the 3-D model included all of the coloring, meaning if pre-service teachers colored the carousel trigger image entirely red, the 3-D model of the carousel would also appear entirely red when scanned.

The String application functioned in a similar fashion to Aurasma and was chosen because it offered different trigger images. When scanned, String's trigger images produced interactive 3-D models or animations, including an animated dragon that emerged from the trigger image and would fly away.

Paint U and AR Dinopark were AR applications not requiring a trigger image; rather they superimposed an image (Paint U) and 3-D animation (AR Dinopark) in the environment. Paint U was designed to identify points on a person's face, and superimpose a mask over the person's face. AR Dinopark allowed users to superimpose an animated dinosaur in the environment which would respond to commands when the user touched the iPad.

The instructor chose to use the Google Drive mobile application as the submission method for all the media created demonstrating the participation and completion of tasks in the ScavengAR Hunt, since Google Drive was the primary method of submission in the course. The Adobe Voice mobile application, used to create narrated slideshows, was included in the fall 2014 trial of the ScavengAR Hunt after being used in an earlier course assignment.

Technology Concerns

Access to the university's Wi-Fi network was a considerable concern. Quite simply, if the iPads or smartphones could not connect to the Wi-Fi network, none of the mobile applications would work. A reliable Internet connection was fundamental to the success of the activity. Without an Internet connection it was still possible to complete some of the tasks since the groups still possessed a physical Case File, washable markers, and access to the iPad camera. However, integration of the mobile applications and their proper usage was the objective of the activity.

The spring 2014 trial did not encounter any issues with Internet connectivity. In fall 2014, an influx of iPads created caused problems for the university's Wi-Fi network. The number of IP addresses was limited and at peak class times, pre-service teachers were unable to obtain consistent Internet connectivity. In this instance, the university's Information Technology department addressed the issue shortly after fall classes began. However, the event made Internet connectivity the primary concern for the successful completion of the ScavengAR Hunt.

Augmented reality applications rely on trigger images to function properly. Each trigger image contains specific points the camera must recognize to activate the super-imposed media. Typically a quality scan requires a specific set of circumstances to activate the media properly, including lighting conditions, and placing the iPad at the appropriate height above the trigger image. Pre-service teachers spent time scanning trigger images with augmented reality applications before the activity in order to identify potential concerns during the class period before the ScavengAR Hunt, as well as install the applications. ARART required considerable time to install due to the animated, super imposed content, which could prove detrimental to a group if it occurred during the activity.

The instructor understood from experience in teaching the advanced course that mobile device or application failure was always a distinct possibility, and needed to be accounted for despite the fact that it was impossible to predict what technical component would fail and why. This understanding was a primary reason for assigning each group two iPads with all applications installed and why two systems of real-time tracking, Twitter and Google Drive were employed.

SOLUTIONS AND RECOMMENDATIONS

The trail version of the ScavengAR Hunt completed during spring and fall 2014 featured a multitude of components: a story narrative, gamification, social media, real-time data tracking, mobile devices, and augmented reality mobile applications in a single group activity. Preparing the materials, individual roles, groups, activity rules, and narrative based on the AR content, proved to be an overwhelming task for the instructor, especially during the initial trial. In a sense, the ScavengAR Hunt elevated the previous, QR code-based activity to the opposite extreme. The ScavengAR Hunt contained so many components it became difficult for a single individual to monitor and account for every possible variable, making a relatively simple activity much more complicated. It would be difficult for others to design an exact duplicate the ScavengAR Hunt in its current form without thorough due diligence following the ADDIE model of instructional design. The ScavengAR Hunt was a highly sophisticated activity in its design, preparation, and execution perhaps reaching the limit of what would be realistic developing a single mobile learning based activity.

There were many mobile applications used in the ScavengAR Hunt serving the same purpose or function. Real-time tracking and feedback was achieved using Twitter and Google Forms, however the ScavengAR would be still be effective if only one was used. In future trials, deciding which mobile application to use would depend on the target audience. Aurasma and String performed the same function in the ScavengAR Hunt, as both AR applications used trigger images overlaid with multimedia. The decision to use both was a result of the fact that Aurasma and String offered variety in the target images but ultimately underlined the functionality of AR applications. Future trials could eliminate these redundant elements by choosing one AR application over the other, depending on how the trigger images fit in the context of the activity.

The ScavengAR Hunt could be streamlined further if an instructor desired to eliminate the gamification element or reduce the number of possible solutions by not creating eight separate Case Files. Eliminating incentives would not be recommended. Participants should feel free to explore using mobile devices and applications especially since AR applications offer a unique experience and perspective on existing content and have the ability to alter the environment not to mention the inherent unpredictability of factors, like Wi-Fi connections, allowing mobile devices to work properly.

Future trials could incorporate AR content created by the instructor or the audience. Currently, Aurasma offers a web-based program called Aurasma Studio. Users can upload their own trigger images and multimedia overlays and set parameters such as expiration date and set viewing permissions to public or private, potentially allowing the audience to participate in the activity's design and development.

Monitoring future developments in AR and wearable technology would result in significant alterations in the design of the ScavengAR Hunt to match the capabilities of the technologies. Aurasma already counts multiple consumer corporations among its client base, and it seems likely AR will develop into a prominent technology altering the way people perceive and interact with the environment. Wearable technology is certainly an emerging technology becoming more prevalent, although most wearable technology is restricted to devices like watches or wristbands. Adapting the ScavengAR Hunt for wearable technology is possible as long as the activity is grounded with a sound instructional framework, in the same way activities designed for iPads require sound instructional framework (Peluso 2012).

The ScavengAR Hunt has the potential to be customized for any audience as long as the goals of the activity properly align with the needs and capabilities of the audience and provide the instructor with manageable conditions monitored accurately. The ScavengAR Hunt is meant to be completed in groups and introduce students to mobile devices and applications, specifically AR applications. The trial version of the ScavengAR Hunt was meant to be a model of what is possible although the activity does not have to be replicated precisely to accomplish the same goals. In addition, creating an activity pre-service teachers enjoy was a considerable design consideration. The ScavengAR Hunt was designed to be memorable, and a chance for pre-service teachers to understand what could be done by employing instructional design in classroom activities.

REFERENCES

Branch, R., & Merrill, M. D. (2012). Characteristics of instructional design models. In R. Reiser & J. Dempsey (Eds.), *Trends and issues in instructional design and technology* (3rd ed.). Boston, MA: Pearson Education.

Clark, R. E. (2005). Five common but questionable principles in multimedia learning. In R. E. Mayer (Ed.), *The Cambridge handbook of multimedia learning*. New York: Cambridge University Press. doi:10.1017/CBO9780511816819.007

Klopfer, E., & Sheldon, J. (2010). Augmenting your own reality: Student authoring of science-based augmented reality games. *New Directions for Youth Development*, *2010*(128), 85–94. doi:10.1002/yd.378 PMID:21240956

Messick, S. (1994). The interplay of evidence and consequences in the validation of performance assessments. *Education Researcher*, *23*(2), 13–23. doi:10.3102/0013189X023002013

Peluso, D., (2012). The fast-paced iPad revolution: Can educators stay up to date and relevant about these ubiquitous devices? *British Journal of Educational Technology, 43(4)*, 125.-127.

Shute, V., Rieber, L., & Van Eck (2012). Games…and…learning. In R. Reiser, & J. Dempsey (Eds.), *Trends and issues in instructional design and technology (3rd ed.)* Boston, MA: Pearson Education.

Specht, M., Ternier, S., & Greller, W. (2011). Mobile augmented reality for learning: A case study. *Journal for Research Center for Educational Technology*, *7*(1), 117–127.

Squire, K., & Jan, M. (2007). Mad city mystery: Developing scientific argumentation skills with a place-based augmented reality game on handheld computers. *Journal of Science Education and Technology*, *16*(1), 5–29. doi:10.1007/s10956-006-9037-z

Sung, E., & Mayer, R. E. (2013). Online multimedia learning with mobile devices and desktop computers: An experimental test of Clark's methods-not-media hypothesis. *Computers in Human Behavior*, *29*(3), 639–647. doi:10.1016/j.chb.2012.10.022

Wu, H., Lee, S. W., Chang, H., & Liang, J. (2013). Current status, opportunities, and challenges of augmented reality in education. *Computers & Education*, *62*, 41–49. doi:10.1016/j.compedu.2012.10.024

KEY TERMS AND DEFINITIONS

ADDIE Model: A design process including the five basic phases of: analysis, design, development, implementation, and evaluation.

Augmented Reality: Technology that superimposes multimedia or interactive objects over a target image, allowing for altered perception of the environment.

Instructional Design: The process employed by instructors or trainers to create methods or activities to best suit the needs, skill set and knowledge of a specific learner audience.

Mobile Learning: Using and incorporating mobile devices and applications in learning activities or environments.

Remote Monitoring: Use of social media or cloud-based mobile applications to monitor participants during a learning activity from a centralized location.

ScavengAR Hunt: A group-oriented learning activity focusing on augmented reality (AR) mobile applications, where groups compete against each other to complete a series of specific tasks using multiple mobile applications and mobile devices.

Section 4

Mobile and Wearable Technology in Applied Sciences

Chapter 13
Trends in Wearable Technologies for Earth Science

Thomas A. Woolman
On Target Technologies, USA

ABSTRACT

In this chapter, we discuss the need for wearable technologies for earth science field research, a discipline that can be both extremely physically rigorous and even potentially dangerous at times. The field is also technically challenging because of environmental conditions as well as often being conducted in remote offline regions. Current trends in available wearable technology devices suitable for earth science applications are explored, in addition to describing possible modifications to existing wearable technologies in order to make them deployable, practical offline-capable geoscience systems. Offline Health and safety monitoring systems designed to operate on a low-powered personal body-area network system as well as offline (non-internet dependent) wearable communication technologies are particularly addressed. These types of wearable technologies represent some of the most complex aspects in this field as well being some of the most customizable systems with highly configurable components for remote back country fieldwork functionality.

INTRODUCTION

The key focus of this chapter includes sections related to enhancing the performance of field research or operational activities. Wearable technologies of benefit to field exploration and extractive operations include but are not limited to communications devices, safety and navigation systems, and interface systems allowing the researcher or manager to electronically communicate with larger less-portable systems such as laser range finding survey equipment, seismographs, satellite communications equipment, data recording technologies, field analysis and remote sensing equipment. In addition, wearable integrated power storage systems are explored along with safety enhancement devices. Flexible, light weight, small configuration wearable computer systems are also examined (U.S. Patent US5285398 A).

The section on wearable technology communications devices include specialized smart watch devices running operating systems designed to both enhance digital radio communications capabilities and run

DOI: 10.4018/978-1-5225-0069-8.ch013

sophisticated applications. In addition to the looming ubiquitous smart watch devices, wearable communications systems can have even more powerful capabilities related to field operations in remote and rugged terrains. This might include wearable polarized antennas including small portable integrated frequency tuners designed to allow for communication operations with orbiting satellites. This would be especially important for operations in locations where there are no operating cellular telephone networks, and normal UHF and VHF line-of-sight radio communications is made impractical due to terrain features or distances involved (Kaivanto, et al., 2011).

Wearable navigation and safety equipment would be related to devices capable of utilizing the Global Positioning System (GPS) to calculate reference point data using (UTM) to coordinate data and integrating the coordinates into an on-board application including a mapping database. As per Serra, et al. in 2010, the typical GPS-enabled smartphone navigation systems in 2015 require access to a cellular telephone network (3G or 4G connectivity speeds) in order to access a centralized navigation database. While on-board integrated navigation databases are available for some smartphone platforms (Pielot, et al., 2010) they are both uncommon and expensive on a smartphone platform. Fully integrated GPS databases on smartphones are also generally limited in scope and may not include sufficient reference data for off-road applications in remote terrains. Smartphone GPS antennas may also be susceptible to interference or lacking in sufficient sensitivity to receive weak GPS signals through dense foliage canopies or in rugged terrain blocking some weak navigation satellite signals utilize a digital UHF frequency band. More advanced and specialized devices within the wearable category for this case will be explored.

Safety systems in the class of wearable technologies for the earth sciences include Wireless Body Area Networks (WBANs) as per Muhammad, et al. in 2014, wearable fabric patch sensors for wearable health status, geolocation, monitoring and other uses as well. Such technologies enable a research manager or plant operator to monitor field conditions, personnel safety, health status, work environment conditions and operations in progress.

Remote sensing and cataloging wearable technologies includes both headset-based optical instruments like as well as more uniquely specialized systems. These systems allow direct wireless interfacing and control of larger, generally less-portable systems with common field applications including geophysics hardware such as radar systems, drilling equipment and induced polarization systems to name but a few possibilities. Support for wireless digital systems and non-networked electronic wearable systems will be analyzed. A review of the implications and potential applications for earth science education will also be discussed.

BACKGROUND

Recognizing that field research and commercial operations in earth sciences involves a unique degree of physical labor and potential risk related to working in difficult terrain (Higgitt & Bullard, 1999) combined with lack of infrastructure access in many cases, wearable technologies have the potential to reduce weight encumbrances. Because geoscience investigations often involve working on steep slopes, heavily forested terrain in remote locations, weight reduction associated with wearable technologies can be a tremendous asset. Risk of injury or even death can be a feature of field exploration geology and especially in related mining operations (Komljenovic et al., 2008), thus the ability to lighten the encumbrance of field staff without reducing the qualitative capabilities of their equipment is a welcome development.

Since off road vehicles cannot always be utilized to take the scientist directly to the locations being investigated, reaching sites on foot after walking considerable uphill distances is not unusual. The desire to utilize the best technologies available to aid in recording, cataloging and analyzing the various aspects of earth science data, whether it be meteorological, geological, hydrological, or biogeochemical surveys, the need for analytical survey technology is present. Additionally, the earth science-related extractive mineral industries such as mining are beginning to take notice of the potential cost benefits of utilizing wearable technologies for these remote location field operations (Mining.com, 2014). Some aspects of fieldwork employed by earth science practitioners involve unique health and safety risks potentially well suited to wearable technology solutions. All of the four areas of application analyzed below address some of those needs.

Four Areas of Application

1. Wearable Offline Communication Technologies for Earth Science

Digital communications devices, safety and navigation systems and interface systems are all classes of wearable devices aiding the geoscience researcher in their field endeavors. Wearable communications devices include technologies enabling the user to wirelessly communicate with colleagues without being burdened with traditional, bulky handheld transceivers (HTs) or relying on digital smartphones relying on cellular network infrastructures to allow for communications. Earth science research frequently takes place in remote terrains without cellular network service or even access to an electrical grid connection. Therefore, such wearable communications devices employed in such field work must be capable of independent point-to-point communication not relying upon a cellular network backbone (Gochfeld, et al., 2006). The technology should also have sufficient radio frequency (RF) power output (measured in either watts or milliwatts) and reception capability sufficient in the field, and lastly it should have sufficient power storage, either on the unit or by interfacing with a wearable power generation or battery technology.

It is important to understand the inherent limitations and capabilities associated with radio communications, particularly in the ultra high frequency (UHF) and to a lesser extent, the very high frequency (VHF) bands. Very high frequency (VHF) and ultra high frequency (UHF) both operate within the MHz (megahertz) frequency range, defined as millions of cycles per second. Common examples of VHF systems utilizing these frequency ranges include commercial FM radio and digital television broadcasts. MHz is defines as multiples of millions of cycles or 1,000,000 cycles per second, for example 91.3 MHz would be 91,300,000 cycles per second. Hertz (Hz) is defined as a cycle per second. VHF systems operate on the 1 meter band (the wavelength of one frequency oscillation cycle in meters and utilize an electromagnetic frequency of between 30 MHz and 300 MHz. UHF systems operate between the 1 meter band and the 10 centimeter band (e.g., length of one complete oscillation wave within the electromagnetic spectrum), and utilize an electromagnetic frequency of 300 MHz up to 3 GHz (gigahertz, or billions of cycles per second). Examples of typical UHF band applications include cellular phone systems, Wi-Fi networks, satellite television and computer communication networks.

The distinction between UHF and VHF systems is important for field operations as it applies to wearable technologies because UHF systems are slightly more amenable to wearables, as the smaller wavelength of the UHF band compared to the VHF band (1 meter for VHF and as low as 10 centimeters for UHF) allows for a smaller, more efficient (up to one full wavelength-long) antenna system for the

UHF HT. However one distinct disadvantage of UHF systems is backscatter and attenuation in both urban and tree canopied environments.

UHF band frequencies are often subject to a phenomenon of foliage attenuation, also referred to as multipath induced fading. GPS reception devices are similarly affected under tree canopies, particularly deciduous tree needles. This is due to the fact the needles are often close to ¼ wavelength in size and shape, making them an ideal natural absorber (attenuator) for radio frequency signals in that portion of the spectrum. Golden (1978) referred to factors increasing in number complexity when considering the performance of transmitting energy through foliage regions, in particular the role of multipath. Frequencies above 300 MHz are suggested to have the maximum RF signal attenuation with typical ground foliage conditions, with a loss of approximately 20 db.

Mitigation suggestions are possible in situations where foliage attenuation impairs UHF signal operations include greater antenna height of the repeater station above foliage canopies, operating at or below frequencies of 100 MHz (VHF "high" band) or removing tree clutter from approximately 100 meters around the perimeter of the repeater antenna when practical.

One of the unique capabilities commonly available to both the UHF and VHF radio bands is the employment of a frequency pairing mode of operation. The employment of frequency pairing necessitates the utilization of a radio repeater system, which will both enhance the capability of the user's radio but also make the technology more wearable. A radio repeater system is a mid-size unit typically either vehicle mounted or rated for outdoor use without being sheltered, and is capable of greater transmission output power ranges (typically in the 50-watt to 150-watt range, as compared to typical handheld transceiver units usually limited to 5-watt output). Equally important is the radio repeater is also designed to work with a large external antenna, in which the efficiency is maximized by the employment of a portable antenna mast. The portable antenna mast will characteristically be extendable anywhere from 2 meters in height up to 25 meters, depending on mounting options and wind, weather, hill slope and other field conditions.

Because of the noted propensity for UHF-band radio frequencies to attenuate and backscatter around ground object clutter such as trees, large boulders or mountain terrain, particularly under leaf canopy conditions (Golden, 1978). Emplacement of a repeater's antenna atop a tall portable mast mitigates much of this problem and substantially increases range and clarity of the entire radio network. It essentially allows a user with a small radio to transmit to other users as it he was personally carrying an antenna several meters tall and transmitting at 50 or even 150-watts instead of carrying an antenna one to two meters off of the ground and only transmitting at perhaps 5-watts.

The disadvantage of the use of a repeater system is the need for an additional piece of equipment to be brought out into the field, along with its associated portable antenna mast, tuned antenna and external power supply. The power supply on these repeater systems is typically either a solar generator (panels of photovoltaic cells coupled with either a lithium-ion battery or lead-acid battery for storing power to enable operations during darkness), or simply a large capacity Li-ion or lead-acid battery.

The radio repeater makes use of the VHF or UHF transceivers frequency pairing mode by flipping the signal pairs. That is, the transmitting handheld transceiver's transmit frequency becomes the radio repeater's receive frequency, and vice versa. The repeater will be the only unit on the network operating in this manner, and when it receives a signal on the receive frequency (the transmit frequency for the users HT), it re-broadcasts the same message on the repeater transmit frequency (the receive frequency on the users HT) in near real-time. The re-transmitted message is ten or more times greater in signal strength output, and at a much greater antenna height, and from a much larger and more efficient antenna unit.

Response transmissions from other HTs in the field will be handled in the same manner by the repeater, to process replies to the original user. Repeaters themselves can also be linked to other networks, such as telephone lines or satellite networks, allowing communications with personnel who may be many hundreds or even thousands of miles away, but still available on the local radio network during field operations. A typical repeater radio network topography diagram is shown in below in Figure 1.

A duplexer may also be integrated into this portable radio repeater system. The purpose of a duplexer unit is to allow the repeater to monitor one frequency and transmit on a separate frequency using the same antenna. This makes deployment of the radio repeater system in the field much easier, since only one antenna, antenna mast and antenna base anchoring system needs to be used. It also can make the radio repeater system more resilient and reliable for field operations, as only one antenna lead needs to be connected from the repeater to the antenna, greatly reducing the number of physical connections in the arrangement and the weight and bulk of the overall radio repeater system. Duplexer units are typically installed within the repeater unit as a module on board the digital motherboard circuitry. Installations are typically intended to be permanent and would not normally be done under field conditions. Radio repeater systems may also be trunked (connected) into other communications gateways, such as landline telephone systems, cellular telephone networks or digital satellite communication systems. This would allow communications with users in other parts of the world as if they were a part of the same local region wireless radio network, such as with corporate headquarters executives, or specialized research support staff working in laboratories.

Figure 1 represents a typical repeater system with an integrated duplexer for single antenna operations allowing a user to transmit information to another user on the other side of a terrain feature. Large terrain obstructions such as mountains can block radio frequency transmissions utilizing LOS (line of sight) frequencies such as with the UHF and VHF bands. An optional trunked satellite link is shown which allows the repeater to be integrated into a larger global network.

The result is the user can carry a smaller radio HT and a correspondingly smaller battery and antenna to meet the rigorous demands of remote field communication operations. While the employment of such a repeater system is not required for the utilization of wearable field communication technologies, it does maximize the capabilities of a wearable technology framework. Besides allowing for a much smaller HT radio due to reduced transmit power needs, it would also facilitate the use of a less efficient HT antenna. Typical UHF or VHF-band HT inflexible, extendible antennas are anywhere from 10 cm up to 40 cm.

Figure 1. Typical repeater system
(Public domain vector image by Mouagip, 2011 and modified by Thomas A. Woolman, 2015)

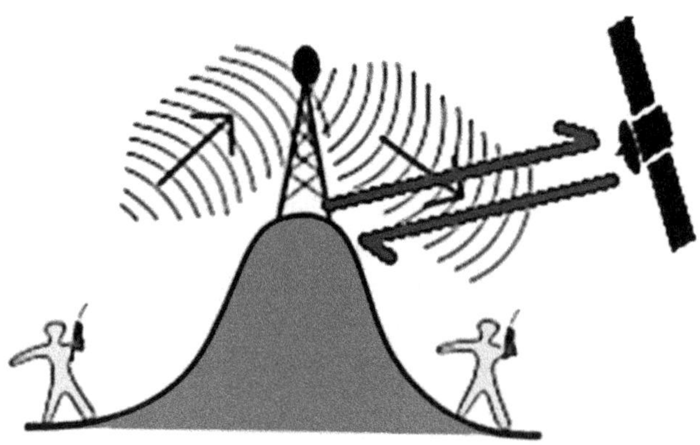

Of particular note with respect to wearable technologies is how ultra short UHF and even VHF antennas operating in the 68-to-470 MHz range may retain sufficient usefulness for field operations, because the influence of the human body on the performance of even short antennas at quarter-wave sizes and less. Andersen and Hansen (1977) described the results of antenna impedance, gain and radiation pattern calculations taking into account the close presence of an average-sized human body increasing antenna reception considerably, thereby making a very short and compact personal HT antenna practical especially when operated as part of a portable radio repeater network. Point-to-point HT radio communications are also enhanced even without the presence of a radio repeater network.

Flexible type antennas however are commonplace (but less efficient than longer traditional extendible antennas), and offer a reduced overall antenna length because thin wire inside the flexible material is coiled along the antenna core, giving the same overall antenna effective length. However even with a flexible UHF antenna in the ~8 cm length, it becomes a potential hindrance for field use and would not necessarily be considered wearable technology.

A wearable external HT antenna solution would be a product such as the body-worn VHF and UHF "octane" model antennas from Pharad, designed to be worn in a vest pocket, backpack pocket or under a field jacket. The antennas connect to the HT by the use of a TNC connector and a thin cable, and operate on (both transmit and receive) radio frequencies from 220 MHz up to 6 GHz. They are composed of flexible material, are completely waterproof, thin and lightweight. The wearable antenna weighs just 280 grams, and has dimensions of just 36 cm by 27 cm, while being less than 2 cm thick, allowing the antenna unit to fit under or inside clothing and remain operationally efficient (see Figure 2). The Pharad "octane" wearable VHF/UHF antenna shows all dimensions in inches. This wearable antenna provides near omnidirectional capabilities without significantly sacrificing capability (gain, frequency ranges or pattern).

The antenna unit also retains an omnidirectional signal pattern while maintaining a common vertical antenna polarization, for interoperability enhancement. The position of the user and the orientation of the wearable antenna within clothing thus does not hinder operations. Importantly for the earth science field practitioner, the ability to safely and (from a radio operation engineering perspective) efficiently employ an external antenna under or within field clothing removes the change of an antenna being damaged or destroyed by catching on tree branches or other field obstacles.

This also allows the small HT radio to be carried in more convenient locations, since the geometry of the HT is now changed with the large bulky antenna removed from the unit. As the Pharad octane antenna has a connecting cable length of approximately 94 cm, this would potentially allow the HT to be carried in a pocket on a backpack, on a belt or even inside a shirt pocket. Completely hands-free operation of the HT can even be obtained with the use of a Bluetooth-enabled in-ear microphone/speaker unit. Now wireless wearable handset units not requiring the use of battery power sources are now available to minimize bulk and weight. This is especially beneficial for field studies in challenging terrain, where weight reduction is essential. Such systems are based upon electromagnetic induction power supplies, and one such example is the "Magnetic Wireless Security Headset", available from S&L Works, LLC of the United States. This device utilizes a micro sized earpiece sitting inside the wearer's ear canal, or two such devices in each ear (providing biaural sound support). It also makes use of a small, sensitive microphone device allowing two-way communications through the HT unit.

A twisted copper coil induction loop is then worn on the body, typically around the neck. This copper coil induction loop provides a magnetic field powered by a worn core module. The earpieces and microphone are all Bluetooth-enabled to allow short-range digital communications with the HT device. Such

Figure 2. Wearable VHF/UHF antenna
(Pharad LLC, 2015)

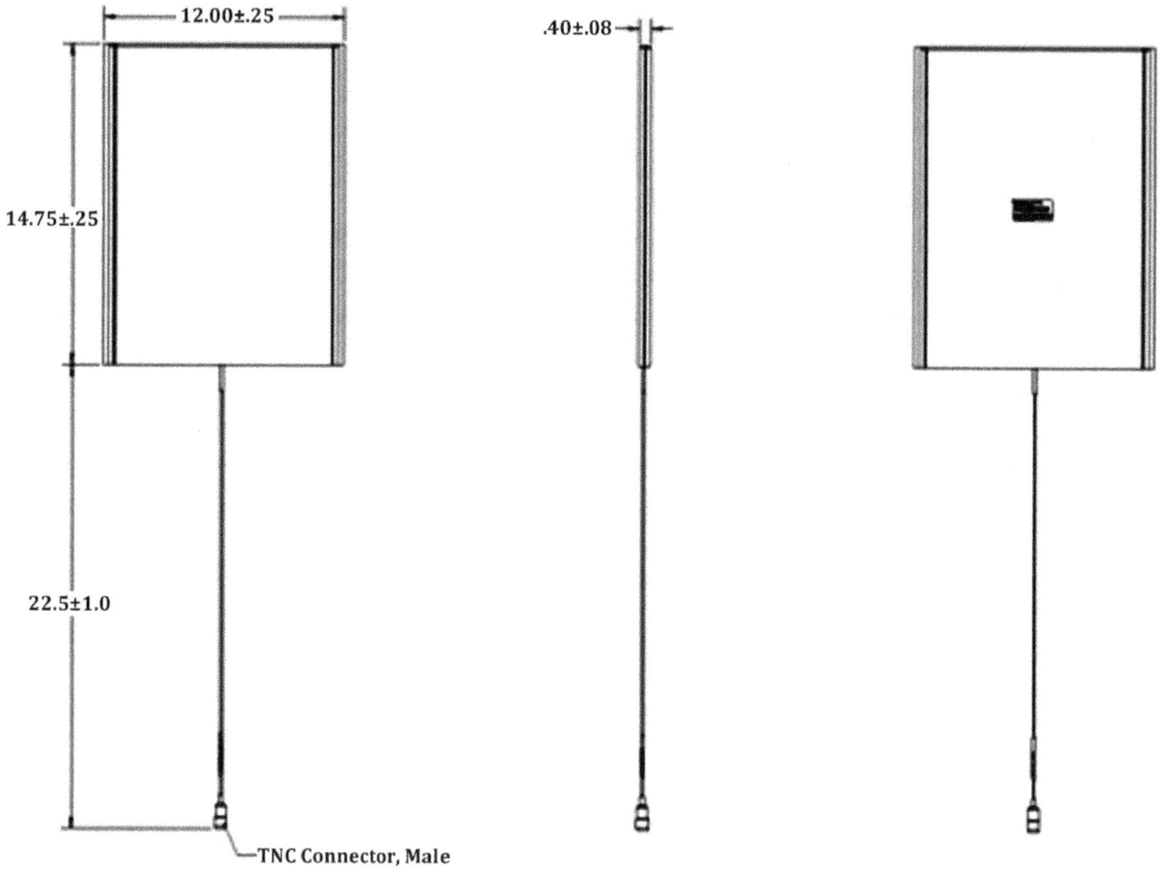

a system is ideal for field operations personnel desiring to minimize communication system bulk and to eliminate protruding cables and antennas potentially being damaged or snagged on obstructions during expeditions in difficult terrain, or even break when embedded into clothing such as shirts or external jackets. The ability to minimize embedded circuitry within flexible clothing material and simultaneously reduce the number of breakable physical electronic connections interfacing can potentially increase the reliability of the wearable electronic systems (Rosine, 1964).

One of the key benefits of the use of electromagnetic induction power transmission for wearable technologies is the ability to transfer power to an electronic component of a wearable device system on a separated, disconnected part of the wearer's body. This is known as "wireless power transfer" or WPT. WPT is therefore the ability to transfer electricity without the need for a wired connection. All current WPT technologies utilize the principle of a time-varying electromagnetic field. This is a particularly useful technology in situations where connecting wires may be hazardous or potentially subject to damage.

In the case of the above described in-ear system, the WPT power transfer method specifically used is known as a near-field WPT technique, making use of an electromagnetic field generated through an inductive coupling with coils of wire (Valtchev, et al., 2012). This technology involves the employment of Faraday's law of induction, predicting how a magnetic field will interact with an electric circuit,

Figure 3. Magnetic wireless security headset system
(S&L Works, LLC, 2015)

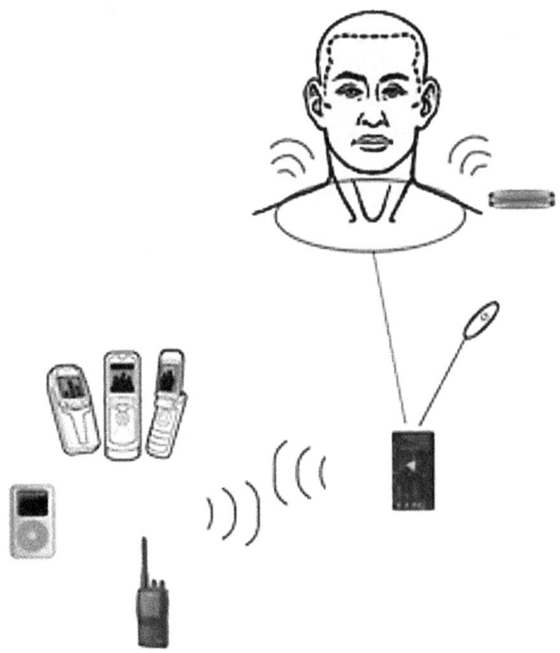

creating an electromotive force (Sadiku, 2007). Faraday's law states an induced electromagnetic force employed within a closed circuit is equal to the negative of the time rate of change of the magnetic flux enclosed by the path [of the circuit] (Jordan, et al., 1968). This is separate and distinct from the principle of radiated, or far-field WPT which is also referred to sometimes as "power beaming".

The principle of power beaming WPT typically utilizes larger scale energy transfers in the industrial power category, currently in the maximum range of tens of kilowatts over a distance of approximately one kilometer (Brown, 1984). The WPT technologies employed for power beaming are typically either lasers, microwave transmissions or powerful electromagnetic radiation, and must be aimed specifically at the receiving station to transmit power. While these systems can support substantial power levels over relatively great distances (Valtchev, et al., 2012) for purposes such as solar power satellites or wirelessly powered aircraft, the high power levels with the associated electromagnetic fields are potentially harmful to human health including the risk of serious injury or death, and are therefore unsuitable for wearable technology employments (Valtchev, et al., 2012).

The near-field WPT technologies employed for this wearable technology system have an approximate power transfer range of one wavelength (λ), utilizing magnetic fields employing inductive coupling between coils of wire using Faraday's law. Since these fields are discrete and non-radiative, the energies produced remain within a short range from the associated transmitter. Likewise, if no appropriate receiving device within the range of the magnetic field, no actual power will leave the transmitting device.

The range of this power transmission capability is governed by the Inverse Square Law, as dipole magnetic fields have a characteristic decrease of strength associated with distance r from the source, expressed as $1/r^2$ (Grant and Phillips 2008), where the quantity of energy transmitted is inversely proportional to the square of the distance from the source of the energy transmission. As a simple approximation of this law, doubling the distance from the transmitting induction coils in the case of the above described earphone system reduces the power to the earphones by one quarter. The power transmitted thereby decreased exponentially with distance, so if the distance between the induction coil and the ear pieces are greater than one wavelength of the magnetic field, very little if any power will be received. Thus, the disadvantage of near-field WPT processes is they are severely limited in power transmission range due to the relatively short distances of the electromagnetic fields produced. But, the same limitation may make near-field WPT an ideal candidate tool in some circumstances for wearable technologies. This ideal nature is due to the lower power electromagnetic transmission levels not associated with health and safety risk (World Health Organization, 2015), and the short distances associated with the human body.

It should also be noted the above HT and Bluetooth-enabled, WPT-powered in-ear speaker and wearable external antenna system can also be employed without the use of the radio repeater and its associated portable antenna mast and antenna system. However it would limit operational capabilities substantially under typical field conditions, perhaps to as little as 2 km using 5-watt power output, depending on various environmental factor variables.

Such a combination of a portable radio repeater network system with a portable antenna mast and optimized frequency-tuned antennas often allow for radio communication coverages of over dozens square kilometers or more, depending on factors such as repeater sensitivity and signal output, antenna height, antenna configurations and handheld transceiver variables. In combination with a wearable antenna and microphone/speaker systems provides both an extremely lightweight and greatly more effective communications capability for the earth science field practitioner and mine or well operators.

Local digital wireless networks, either indoor or outdoor, supporting either the 5 GHz 802.11a Wi-Fi standard or the 2.4 GHz band used by the 802.11b/g/n standard may also employ wearable shorter-range wireless communication devices such as the Vocera B3000n Communication Badge (see Figure 4). This device is a lightweight (less than 54 grams), voice-controlled with advanced voice recognition, wearable communication device enables instant two-way or one-to-many conversations using intuitive and simple commands.

The Vocera B3000n Badge is widely used by mobile workers including field operators who need wearable devices providing the convenience and expedience of being able to respond to calls without pressing a button. This can include mining or drilling equipment operators, field security personnel or exploration and technical staff who might otherwise be burdened with equipment preventing the use of normal radio communications gear. The restrictions and limitations associated with such a device are limited only by the Wi-Fi wide area network coverage, which is easily enhanced by the use of wireless network repeaters and routers. Such routers can even be mounted in off-road vehicles and connected to the network via the use of satellite uplinks.

Aspects of this wearable communications badge would be particularly appealing to earth science research and exploration including its ability to act as a hands free communications device. The unit contains an illuminated "halo" around the call button, providing a visual indication of the unit call status. It glows green when on an active call status, and pulses amber when on "Do Not Disturb" or on hold. The illuminated halo can also be used to indicate when text messages are waiting to be read, thereby potentially alleviating the need to carry an addition device for reading text messages over the network. The badge features an LCD display screen comprised of an 80 x 82 OLED bit-mapped display. It is capable of displaying up to 4 lines of text, with 15 characters per line allowing up to 150 characters per text message (depending on font selected). The control buttons on the unit include a call button, a hold or "do not disturb" button, and a combination volume/menu selection buttons.

Figure 4. The Vocera B3000n wearable communication badge
(Vocera Communications, Inc., 2015)

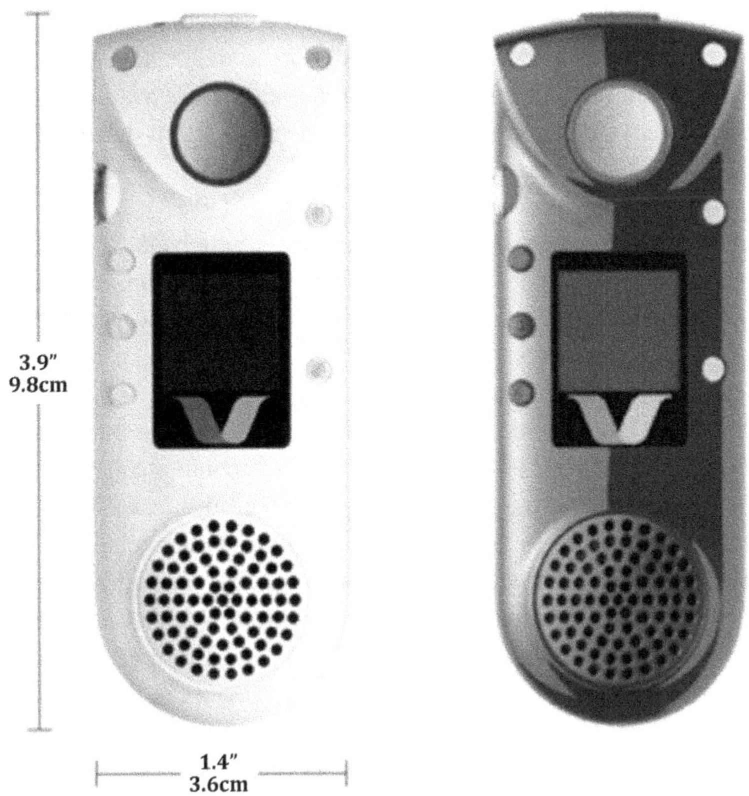

The unit also employs a robust, front facing speaker providing hands-free audio in order to clearly hear the conversation without having to hold the unit in the hand or place it close to one's ear. This is especially important when operating in an environment with frequent loud ambient noises such as mineral extraction or underground operations.

The devices also contain four separate microphones, designed to perform an integrated Acoustic Noise Reduction (ANR) to reduce background conversations and noise interferences to improve speech recognition in noisy environments including rainy season operations or at high altitudes where the sound of constant wind can be a hindrance to typical radio communication microphone systems. Peak speaker output is rated at 85 decibels at a distance of 10 cm. They include microphones within the unit and have a frequency response range of 350 Hz up to 3.75 kHz.

Additionally, a multi-user operation mode permits the sharing of a device. This permits a Vocera badge unit to be passed to another user at the end of a shift, allowing the new user to log into the Vocera System from the Vocera Badge and immediately begin using it. This reduces the amount of storage required in camp vehicles or to be kept in inventory at the facility, and further reduces average user weight encumbrance as every user will not be required to wear the communications badge at all times. The badge also features a button that puts it into a "do not disturb" mode to prevent unwanted interruptions. Private conversations are also possible with this device by the inclusion of an integrated 2.5 mm headset jack to allow the temporary installation of a microphone and earphone headset unit.

This communications badge is an example of a wearable network technology capable of both meeting the rigorous demands of harsh geoscience field applications as well as allowing staff to remain connected while keeping their hands free to perform regular job duties. It also utilizes a structurally rugged design including an independent support spine and a reinforced badge housing, to provide the durability required for extended field use. An optional special lanyard is also available featuring a breakaway safety feature, in case the lanyard becomes entangled by an obstruction during operational use. Power is supplied to the unit through a rechargeable Lithium ion polymer battery integrated into the unit, and an optional extended battery unit is also available for longer field use.

This standard battery provides approximately 3 hours of continuous talk time or 30 hours of standby time, while the extended battery provides approximately 5 hours of continuous talk time or 60 hours of standby time. Functionality in environments from 5% up to 95% humidity is provided, including operating in temperatures from 0 degrees Celsius up to 40 degrees Celsius. Encryption using 64-bit or 128-bit WEP, TKIP and AES-CCMP standards is supported.

Multiple communication groups within the network can be separately managed by creating separate operating channels. Using the 2.4 GHz Wi-Fi standard, up to 11 channels can be used with the U.S. FCC-governed network configuration, or up to 13 channels using the European ETSI network configuration standard. When using the 5 GHz Wi-Fi standard, up to 20 privacy channels can be employed using the U.S. FCC network setup, or up to 56 privacy channels with the European ETSI arrangement. A separate AC bulk battery allows for simultaneous recharging of multiple Vocera communication badges. This enables batteries to be swapped out of the unit after the end of a shift or at the end of a day, depending on the amount of transmitting being done. Communications badges can thus be online and operational virtually 24x7 without the need to shut down for recharging.

As a wearable communications device, this concept succeeds by utilizing an enterprise-class Wi-Fi network. It is able to deliver hands free, real-time voice communications whenever needed and potentially wherever they are located provided they have network Wi-Fi access including portable Wi-Fi routers connected to satellite uplink networks.

Wearable Navigation and Support Technologies for Offline Back Country Geoscience

One of the consistent needs and wearable-sized technology challenges for geoscience practitioners has been the need for accurate satellite navigation, in conjunction with topographical map data. Traditional smartphone GPS navigation is typically online and is connected to a navigation application host database via a 3G or 4G wireless data connection. While offline navigation databases are available for smartphones (Chen et al., 2013) the applications are both much more expensive and limited or even completely lacking in backcountry navigation features such as topographic raster data including elevation and geographical features. Furthermore, smartphone devices are typically a larger than ideal size for wearable technology.

While opportunities exist for customizable, programmable headset devices, the current generation of navigational applications supporting the platform are primarily limited to tools intended to support pedestrians or vehicle drivers within an urban environment (Sheehy, 2014). Opportunities exist for commercial off-the-shelf wearable optical interface systems aimed specifically at users who will be navigating offline due to lack of wireless Internet access but who will be carrying tools for stored topographical map database and receive GPS navigation signals. Being able to integrate such platforms into such a system would enable users to assess travel speed and distances as well as calculate slope angles and possibly even visualize terrain virtually that might otherwise be obscured due to snow storms, fog, smoke or sand storms.

Field operators could then note terrain changes recently been occurring via the wearable headset interface / offline navigation device combination system, and once back online with Internet, access the database update suggestions and shared on a centralized "Wiki-style" updatable navigation database. Such a system could be extremely helpful and even potentially lifesaving, as it would alert other operators to potential navigation hazards including flooded terrain and rockslide perils recently been discovered.

What currently exist for offline navigation wearable requirements are offline navigation smart watches such as the Garmin Epix (see Figure 5), combining a smart watch with accurate GPS reception capability and 8 GB of internal storage for topographical and satellite map data. The smart watch also allows the wearer to mark coordinates and plot routes to specific regions on the digital map. The ability to switch from a topographical map to a satellite view map is also extremely helpful for scientists who wish to examine areas they are exploring in different manners such as analyzing environmental conditions in portions of their region they cannot yet see first-hand due to distance or elevation, darkness or ecological conditions. Operators can then upload route data and marked coordinates to a laptop or other centralized navigation data repository when they return to the camp or office environment and have Wi-Fi access, or download new satellite or topographical map data for their next mission. The Garmin Epix Smart watch, featuring offline GPS navigation with both satellite and topographical map databases and the ability to store UTM point coordinates and plot routes is shown in Figure 5.

Of additional potential interest to the earth science field practitioner is the Belty device, which is a wearable appliance replacing a hiking belt (see Figure 6). The Belty (Newman, 2015) is a "smart belt" using a host of tension sensors to automatically loosen and tighten a wearer's belt as they sit, stand and move around in difficult terrain. While this may seem simple enough it could be extremely helpful for an operator who is walking up and down steeply forested terrain and carrying a heavy pack and multiple devices on or around their belt, and storing rock samples in either pockets or in backpacks. Shifting

Figure 5. Garmin Epix Smart Watch
(Garmin, 2015)

Figure 6. The Belty, a smart belt device
(Newman, 2015)

weight around while carrying a heavy load and then later returning to the base camp to sit and upload images and notes to a laptop all can contribute to the need to manage a belt and this device can be a welcome addition.

Additional wearable technologies aiding in earth science field research include both the "Sensoria", and the "Digitsoles" devices. The Sensoria is a pair of "smart socks" which record footprint information and analyzes which parts of the feet are getting the most pressure, potentially allowing the hiker to make changes in either footwear selections or in adding insole devices to relieve pressure problems before symptoms such as blisters or worse can begin to materialize. Keeping the researcher's feet in prime condition can greatly aid in the amount of field samples and data collected, thus helping to safeguard the success of the research mission. "Because blisters often limit individual mobility, their impact in a remote setting is potentially more serious. In addition, they provide a portal for infection. This risk is always higher in an aquatic environment" (Polliack, Scheinberg, 2006).

The Digitsoles device is a pair of heated insoles controlled by a Bluetooth-enabled smartphone or wearable computing device. The ability to manage foot temperature in extreme weather conditions such as rain or cold temperatures can be especially important. Feet can be among the first parts of the body to become susceptible to numbness and frostbite, especially if feet manage to get wet sometime during the trek. In a great many situations, a researcher needs to proceed into backcountry terrain on foot and leave their vehicle behind either because of steep conditions or lack of roads or impassable roads. Being able to maintain one's foot health is critical for safety and survival to return to their vehicle safely after they have gathered their data.

2. Wireless Body Area Networks for Health and Safety Monitoring in Earth Science

The new field of wireless body area networks (WBANs) is especially relevant because it is intended primarily for health and medical condition monitoring, an important and often overlooked aspect of field operations within earth science. The U.S. Federal Communications Commission (FCC) has allocated 40 MHz of radio frequency spectrum specifically for medical-related WBAN operations (Mearian, 2012).

This frequency spectrum allocation regulation dictates the use of low-power, wide-area radio links within the 2360 MHz to 2400 MHz range. One benefit of using this frequency range is not competing with or overlapping already congested Wi-Fi spectrums commonly employed around the globe (typically either 5 GHz or 2.4 GHz).

The use of WBANs allows for multiple related such devices (up to a maximum of 100 devices as per the IEEE-governed international standard definition of a WBAN) to be interconnected and communicating with each other for the benefit of clinical monitoring various aspects of a user's health. The WBAN protocol is a separate and distinct protocol from Bluetooth, which is typically a 1:1 device pair interconnection, for example a smartphone to a wireless ear piece, and of course uses a separate frequency allocation. WBANs intended for health and safety monitoring of geoscience expedition personnel having the additional stipulation of being able to function, at least temporarily, in an offline (non-Internet connected) environment. Such a specialized WBAN system will have to work while still being able to perform basic health and safety diagnostics using on board, wearable processing and data storage components.

A Wireless Body Area Network (WBAN) is formally defined by the IEEE in their 802.15.6 Wireless Personal Area Network (WPAN) working group as, "a communication standard optimized for low power devices and operation on, in or around the human body (but not limited to humans) to serve a variety of applications including medical, consumer electronics / personal entertainment and other" (IEEE, 2015). This standards definition allows different manufacturers to develop devices and associated software applications able to communicate with each other to fulfill specific tasks. WBANs are also sometimes referred to as WBSNs, or Wireless Body Sensor Networks. A portion of the current technical standard includes a specification allowing a maximum data communications rate of nearly 1 gigabit per second with the maximum signal output strength of 100 milliwatts (Karulf, 2008).

The type of sensors typically included in a WBAN network include heart (ECG) and respiratory (SpO_2) monitors and other devices capable of continuously measuring vital signs, such as EEG and blood pressure monitors (Darwish, Hassanien, 2011). Another possible application would be in continuous monitoring of a person's glucose level if the individual was classified as a diabetic. Devices of this nature could even be designed and programmed to auto regulate insulin levels with the use of an integrated pump system. Additional embedded physiological sensors could be developed for WBAN applications include temperature sensors on various portions of body extremities, accelerometers or another type of motion and impact sensors to monitor for sudden falls resulting in injury.

The entire personal WBAN system needed to deploy for an offline geoscience expedition would have to consist of a lightweight portable power supply, extremely efficient in both size and power consumption, reconfigurable from user to user due to different individual health profiles. This class of WBAN system would also need to have all sensors and components non-embedded inside the users body as the use of the system would be intended for the short-term expedition duration only.

In order for WBANs to successfully perform the intended task, the various individual health and safety monitoring sensor components must be wirelessly integrated with a processing unit, such as a wearable flexible computer, mobile phone or PDA device (the WBAN processor). The WBAN processor contains sophisticated software capable of not only monitoring the input from all of the users WBAN integrated sensors, it would be able to discriminate and ignore as needed the sensor telemetry from other nearby WBAN networks from other user operators. It will also need to diagnose patterns within the integrated sensor data stream indicative of an impending health problem before it actually starts (such as a heart attack or a stroke), and signaling the wearer to seek help and either slow down or stop their physical activities immediately.

The need for offline medical support WBANs for geoscience field operations should not be limited to older operators or even just those with known medical conditions. Field expeditions for the geosciences by their nature are often in some of the most remote and difficult terrains on Earth, and expedition team members are frequently separated by great distances during the data gathering aspects of the assignment. As such, any practitioner regardless of age and known health conditions is subject to external injury such as falls, abrasions, frostbite or even heart related problems when undertaking climbs across rugged terrain at altitudes unaccustomed to.

Warnings and user health diagnostic information can be digitally integrated via a Bluetooth connection to the personal HT field communications radio worn by the scientist. In the event the scientist cannot request emergency help due to injury or other medical emergency, this integration between WBAN and HT would allow the automated transmission of emergency medical information and GPS coordinates of the operator back to colleagues at the base camp or support vehicle. In extreme situations, the WBAN can be used as the basis for emergency telemedicine if the wearer becomes a victim of an accident or succumbs to a heart attack or other life threatening event. Data from the WBAN in this case could be uplinked to the portable radio repeater network and sent to emergency physicians in a distant hospital, if the portable radio repeater network has been trunked to a digital satellite Internet gateway. In such a dire situations, the emergency physicians could further diagnose WBAN sensor data from the patient while instructing colleagues at the base camp or support vehicle what to do to stabilize the patient. This process could remain ongoing until emergency medical teams arrive or the decision is made to transport the patient using the camp support vehicle if no other emergency medical transport or medevac capabilities are available, such as in remote portions of Africa or Antarctica expeditions. In such a manner, the employment of a wearable system involving components such as WBAN, HTs, a wearable computer and an integrated radio repeater system with a trunked satellite uplink could provide a significant health and safety technology otherwise completely unavailable for back country fieldwork researchers.

3. Wearable Computer Systems for Offline Geoscience Applications

A scientific research team has utilized wearable computer systems (McGuire, et al., 2004) for investigating a "cyborg" approach (human geologists augmented with a computer vision system in which the geological targeting and investigational analysis was directed by the computer system). The objective of this research was to optimize a computer system capable of being integrated into a future autonomous Mars rover mission for NASA, which would be able to independently target geological features of interest and initiate an independent, algorithmic-based analysis without the direct need for human intervention. One of the goals of this research was to pair this wearable computer system including autonomous computer vision and sophisticated image analysis software with the mobility of a human to enable maneuvering to a site of geological interest.

The computer system would then provide recommendations on how to approach any observed geological outcrops after obtaining and analyzing mosaics of digital color images. The wearable computer system in this situation would provide these suggestions in real-time to investigate the most unusual regions detected in these mosaic image datasets. This is advantageous due to the radio signal time delay between the Earth and Mars as it could potentially be employed on an autonomous robot rover in a similar manner as was done with the human investigators.

Such research however underscores the potential suitability for the employment of wearable computer systems for fieldwork applications in the geosciences. McGuire's team utilized an advanced (at the time)

667 MHz wearable computer with 112 MB of on-board physical memory, and a head-mounted VGA display designed to be useable in daylight conditions. One particularly unique feature of their system was the use of the Transmeta "Crusoe" CPU unit (Klaiber, 2000), a low-power capable consumption central processing unit intended to maximize battery efficiency for field operations. This processor is capable of providing variable operating modes, due to its dynamic core design and is capable of varying voltage and frequency cycles dynamically under various load conditions. This allows for minimizing power consumption by the wearable computer central processor under field conditions if exploited correctly by the application software. What this and similar research has shown is that wearable technologies uniquely suited to earth science can be readily adopted for back country field work applications. Robust wearable PCs can be used not only for specific cybernetic geology application research like shown by the McGuire team, but can also be used to enable to field researcher to record data about phenomena observed such as geophysical characteristics or mineralogical attributes, depending upon the type of inquiries being undertaken.

Such examples could include induced polarization (geophysics) studies, biogeochemistry (Woolman, Yi 2013) or aeromagnetic surveys where key attributes could be recorded digitally during the initial data gathering phases, or about mineralogical observations such as sip and strike data when a mineral vein is located during a prospecting survey. Integrating a computer vision system or other type of interface with the physical world such as radiological or paramagnetic sensors for geophysical data gathering, along with traditional geological exploration methods such as biogeochemical samples, rock sampling or core drilling techniques employed by heavy equipment or teams of prospectors. Being able to visually record stratigraphic settings or outcrop locations can add significant supporting information to field data gathering, and assist in preparing reports. The wearable technology could thus be prove to be an extremely a useful aspect for gathering additional data for geological surveys by employing a combination of systems creating a hybrid multidimensional dataset (Woolman, 2013) to augment the traditional data gathering methods. This would allow a greater amount of research data to be collected during fieldwork expeditions that are both potentially expensive and of limited time window duration for the scientist. By gathering as much data as accurately as possible during these limited windows of research data gathering opportunity, the wearable computer system can potentially be of significant value.

4. Applications for Earth Science Education

The applicability of the use of this type of personal technology has immediate applications to science education. Because of noted potential field hazards, such safety, navigation and support equipment lends itself to enhancing outdoor field exercises that are a necessary and vital part of earth science curriculums. Being able to communicate and share information between classroom teams and instructors as well as being able to collaborate on data gathering exercises adds a previously unavailable dimension of learning and safety.

This further serves to emphasis the relationship between research and education. Geoscience teams conducting field research frequently collaborate during data gathering tasks. The use of wearable technologies in field classroom environments serves to reinforce the type of operational aspects that one would encounter in a field research setting. Related disciplines such as archaeology, botany, and paleontology may similarly benefit from the use of these systems in field education exercises. By using these wearable technologies as an aid to successful and more accurate data gathering in a learning environment, these systems can be an important multiplier in gathering useful data for training and future research. Likewise, the adoption of these wearable technologies in educational field exercises could lead to an increasing use by students and future researchers. As students become familiar with the capabilities and operations of these systems, their inherent advantages become greater as habituation increases.

Issues

Several recurring issues demonstrated in all four of the above referenced areas of application include the need to be able to operate productively in a sustained offline fashion. This requires not only the ability to continue functioning for extended periods without access to external power, it also requires software applications which are larger and more sophisticated because they need to operate on the wearable computer or as firmware on the worn technologies, and not rely on distributed network or cloud based applications performing the "heavy lifting" computations.

Other unique aspects to earth science wearable technology utilizations include robustness and reliability. Temperature extremes as well as significant humidity variances can be expected in many back country conditions which might not otherwise be experienced by the average wearable technology end user. The ability to survive abrasions and mild impacts associated with working in steep, forested terrain or other challenging technical environments would also be useful for these types of roles. Light weight is also a desired trait in the wearable components as is the ability of the various physical components of the worn technologies to integrate with each other both wirelessly (or with a minimum of wires) and on a software application level between the wearable computer and the integrated worn peripheral devices requiring a wearable computer to function (such as WBANs).

Problems

Problems noted with various wearable technologies include a current lack of specific earth science software applications designed for field research. While this hindrance can be partially addressed by developing custom applications for wearable mobile platforms, the cost and time involved to do this can be significant, although development of these applications has become simpler than in the past due to new CASE development tools. However, it can be expected open source software applications for wearable offline-capable mobile devices suited for geosciences will emerge over time. This is particularly true for applications suitable to both offline navigation and health and safety monitoring as they would have a potential market beyond just earth science field research. Additionally, WBAN systems and software applications are an especially emerging subset of wearable technologies and operate on a different network frequency and protocol than other types of wearable peripherals, as their intended use is primarily within hospitals and other types of medical facilities.

This is a unique requirement governed by the American FCC and has been regulated to address issues of bandwidth saturation from hundreds or even thousands of Wi-Fi devices in the local area of these same medical facilities. While it is anticipated WBANs may be usable as medical health and safety monitoring systems outside of the confines of hospitals including earth science fieldwork, such WBAN peripherals will need a wearable computer capable of communicating with them on the 40 MHz spectrum and running application software that functions with them.

RECOMMENDATIONS

One immediate recommendation would be to develop a wearable technology for earth science working group composed of geoscientist researchers, representatives of earth science industries such as mining, oil and gas exploration, government agencies such as the U.S. Geological Survey and engineering organizations such as the IEEE. Partnering with the open source development community (Von Krogh and Von Hippel, 2006) for the application software development may be an additional avenue worth pursuing, as well as enlisting the assistance of university computer science departments for assistance with creating open source offline mobile applications for this field. Expanding the availability of open source offline software applications for wearable computers may have the desirable effect of increasing the number of scientists who become interested in the field and also allow for improved amounts of data collection.

This could also influence the need for new and innovative wearable devices attempting to better exploit such offline wearable computers. Furthermore, next generation wearable offline navigation technology designers may wish to integrate inertial navigation capabilities into the devices to augment the use of GPS data. This is due to the fact UHF-band GPS data signals can be attenuated in bad weather conditions or under heavy tree canopies. Development of expanded WBAN applications specifically intended for health and safety monitoring in remote wilderness environments is recommended. To maximize the effectiveness of WBAN monitoring systems, it would be ideal if the wearable computer and related WBAN software could integrate with the HT autonomously and wirelessly to broadcast emergency signal telemetry from the WBAN. Such a system may require the development of custom wearable technology software applications, utilizing the employment of COTS (commercial off the shelf) wearable peripheral devices. Such an enhancement would allow WBANs to be employed in wilderness environments and turn them into medical distress and emergency field medicine diagnostic tools for remote telemedicine systems.

CONCLUSION

Earth science field research has several unique requirements representing both an extreme challenge for information technology systems, and in particular wearable technologies. Because of the challenges associated with extreme weather and terrain as well as unreliable (at best) access to any type of Internet connection, wearable technologies intended for earth science research must be extremely self-supporting while deployed. Being able to employ robust, light weight and reliable information technology systems into offline back country field environments can enhance health and safety, data gathering and field communications for the researcher.

Wearable systems can be beneficial to the field researcher not only by providing enhancements to data gathering, field communication and navigation but by enhancing physical safety and comfort. The ability to monitor the health status and in some cases even actively maintain the medical well-being of a user can be especially vital when performing operations in remote sites.

Offline navigation wearable technologies employing the ability to not only identify topographical terrain features of possible interest but also incorporate data from remote sensing systems will enable the field research teams to better make use of their limited data gathering times. This could enable an exploration staff to make better informed decisions in the field as to what features of interest to explore. This could be especially useful as weather or terrain conditions change making access to some locations more or less difficult than initially expected.

By improving the ability of the research team to remain in the field longer, with improved situational safety awareness and gather more data as well as share data amongst geographically dispersed team members, wearable technologies can become a vital piece of earth science equipment. It is foreseeable some of these wearable technology trends may have the capacity to become lifesaving research support technologies for this field.

REFERENCES

Andersen, J., & Hansen, F. (1977). Antennas for VHF/UHF personal radio: A theoretical and experimental study of characteristics and performance. *IEEE Transactions on Vehicular Technology*, *26*(4), 349–357. doi:10.1109/T-VT.1977.23707

Brown, W. C. (1984). The history of power transmission by radio waves. . *IEEE Transactions on* microwave theory and techniques, *32*(9), 1230–1242.

Chen, L.C., Lai Y.C., Yeh, Y.H., Lin, J.W., Lai, C.N. & Weng, H.C. (2013). Enhanced mechanisms for navigation and tracking services in smart phones. *Journal of applied research and technology*, 11(2), 272-282.

Darwish, A., & Hassanien, E. (2011). Wearable and implantable wireless sensor network solutions for healthcare monitoring. *Sensors (Basel)*, *11*(6), 5561–5595. doi:10.3390/s110605561 PMID:22163914

Gochfeld, M., Volz, C., Burger, J., Jewett, S., Powers, C., & Friedlander, B. (2006). Developing a Health and Safety Plan for Hazardous Field Work in Remote Areas. *Journal of Occupational and Environmental Hygiene*, *3*(12), 671–683. doi:10.1080/15459620601009201 PMID:17050349

Golden, A. (1978). A foliage penetration summary. Rome Air Development Center, Air Force Systems Command, Griffiss Air Force Base, New York.

Grant, I. S., & Phillips, W. R. (2008). Electromagnetism (2nd ed.). Manchester Physics, John Wiley & Sons.

Higgitt, D., & Bullard, J. (2010). Assessing fieldwork risk for undergraduate projects. *Journal of Geography in Higher Education*, *23*(3).

802.15 Working group for WPAN. (2015). IEEE. Retrieved from http://www.ieee802.org/15/

Jordan, E., & Balmain, K. G. (1968). *Electromagnetic waves and radiating systems* (2nd ed.). Prentice-Hall.

Kaivanto, E. K., Berg, M., Salonen, E., & de Maagt, P. (2011). Wearable circularly polarized antenna for personal satellite communication and navigation. *IEEE Transactions on Antennas and Propagation*, *59*(12), 4490–4496. doi:10.1109/TAP.2011.2165513

Karulf, E. (2008). Body area networks (BAN). Retrieved from http://www.cse.wustl.edu/~jain/cse574-08/ftp/ban.pdf

Klaiber, A. (2000). *The technology behind crusoe processors. Low-Power X86-Compatible Processors Implemented with Code Morphing Software*. Transmeta Technical Brief.

Komljenovica, D., Grovesb, W. A., & Kecojevic, V. J. (2008). Injuries in U.S. mining operations – A preliminary risk analysis. *Safety Science*, *46*(5), 792–801. doi:10.1016/j.ssci.2007.01.012

McGuire, P.C., Elvira, J.G., Manfredi, J.A., Martinez, E.S., Ormo, J., Martinez, E.D., Ritter, H., Oesker, H., Haschke, R. & Ontrup, J., (2004). Field Geology with a wearable computer: First Results of the Cyborg Astrobiologist System.

Mearian, L. (2012). 'Body area networks' should free hospital bandwidth, untether patients (at-home use of disposable wireless monitors could be the biggest market of all. *Computerworld*. Retrieved from http://www.computerworld.com/article/2503882/healthcare-it/-body-area-networks--should-free-hospital-bandwidth--untether-patients.html

Wearable technology to enter the Mining industry. (2014). *Mining.com*. Retrieved from http://www.mining.com/web/wearable-technology-to-enter-the-mining-industry/

Muhammad, M. A., & Hamida, E. B. (2014). Surveying wearable human assistive technology for life and safety critical applications: Standards, challenges and opportunities. Retrieved from http://www.mdpi.com/1424-8220/14/5/9153/htm

Newman, J. (2015, January 12). Wearables go niche and narrow in the search for consumer relevance. *PC World*.

Pielot, M., & Boll, S. (2010). Tactile wayfinder: Comparison of tactile waypoint navigation with commercial pedestrian navigation systems. *Proceedings of the 8th International Conference on Pervasive Computing* (pp. 76-93).

Polliack, A., & Scheinberg, S. (2006). A new technology for reducing shear and friction forces on the skin: Implications for blister care in the wilderness setting. *Wilderness & Environmental Medicine*, *17*(2), 109–119. doi:10.1580/PR30-05.1 PMID:16805147

Rosine, L. L. (1964). Advances in electronic circuit packaging. *Proceedings of the Fifth International Electronic Circuit Packaging Symposium, Boulder*, Colorado.

Sadiku, M. N. O. (2007). *Elements of Electromagnetics* (4th ed., pp. 380–390). New York, USA: Oxford University Press.

Serra, A., Carboni, D., & Marotto, V. (2010). Indoor pedestrian navigation system using a modern smartphone. *Proceedings of the 12th International Conference on Human Computer Interaction with Mobile Devices and Services*. doi:10.1145/1851600.1851683

Sheehy, A. (2014). *8 Mind-blowing uses of wearable technology (Seriously…). Generator Research.*

Tracy, K. W. (2012). Mobile application development experiences on Apple's iOS and Android OS. *Potentials, IEEE*, *31*(4), 30–34. doi:10.1109/MPOT.2011.2182571

Valtchev, S. S., Baikova, E. N. & Jorge, L. R. (2012). *Electromagnetic field as the wireless transporter of energy*. Facta Universitatis Ser. Electrical Engineering.

Von Krogh, G., & Von Hippel, E. (2006). The promise of research on open source software. *Management Science*, *52*(7), 975–983. doi:10.1287/mnsc.1060.0560

Woolman, T.A. (2013). Base metals prospecting in Kagera region, Tanzania using remote sensing and biogeochemistry analytics. *The Compass: Earth Science Journal of Sigma Gamma Epsilon*, *85*(4).

Woolman, T. A., & Yi, J. C. (2013). An Analysis of the Use of Predictive Modeling with Business Intelligence Systems for Exploration of Precious Metals Using Biogeochemical Data. *International Journal of Business Intelligence Research*, *4*(2), 39–53. doi:10.4018/jbir.2013040103

What are electromagnetic fields: Summary of health effects. (2015). World Health Organization.

KEY TERMS AND DEFINITIONS

Attenuation: A phenomenon also known as foliage attenuation or multipath induced fading. This is typically associated with UHF band frequencies including GPS signals under deciduous tree canopies or other obstruction layers. Deciduous tree needles are often close to ¼ wavelength in both size and shape for frequencies used by UHF band data transmissions, which makes them an ideal natural absorber (attenuator) of RF energy in that spectrum portion.

Dynamic Core: Also known as dynamic multicore processor architecture. A computer hardware architecture that provides the capability for individual central processing unit cores to adapt resources during executing to maximize processor capability and manage power consumption. This provides for a more efficient performance management capacity for both computational resources and power efficiency control for a wide range of mission demands.

HT: Handheld transceiver. A portable two-way radio transceiver, typically operating on bands such as 2 meters, 70 centimeters, or one of several UHF or VHF frequency allocations. Modulation schemes can include FM, AM, SSB, CW or digital modes such as PSK31 and even GPS position reporting.

Induced Polarization: A geophysical image data gathering technique intended to locate potential subsurface anomalies. An electric current is induced into subsurface material via the use of electrodes, and voltage variances or resistance inconsistencies are then plotted within the sampling area.

Inverse Square Law: Typically utilized in radio communications or electrical power transmission equations, the inverse square law is a physical law stating that a quantity is inversely proportional to the square of the distance from the source of the emission.

WBAN: Wireless Body Area Networks, also referred to as WSBNs or Wireless Sensor Body Networks. An interoperability standard classified by the U.S. Federal Communication Commission and the IEEE on a 40 MHz of radio frequency spectrum slice specifically intended for medical-related health and safety monitoring devices on an individual human being. This frequency spectrum allocation regulation dictates the use of low-power, wide-area radio links within the 2360 MHz to 2400 MHz range. One benefit of using this frequency range is not competing with or overlapping already congested Wi-Fi spectrums commonly employed around the globe.

WPT: Wireless Power Transfer. Also known as wireless energy transfer, it is the transmission of electrical power from a source to a consuming electrical device without the use of wires or other conducting physical material.

Chapter 14
Developing an Elementary Engineering Education Program through Problem-Based Wearable Technologies Activities

Bradley S. Barker
University of Nebraska-Lincoln, USA

Jennifer Keshwani
University of Nebraska-Lincoln, USA

Gwen Nugent
University of Nebraska-Lincoln, USA

Carl Nelson
University of Nebraska-Lincoln, USA

Neal Grandgenett
University of Nebraska at Omaha, USA

Ben Leduc-Mills
SparkFun Electronics, Inc., USA

ABSTRACT

This chapter describes the efforts and pilot study results of the Nebraska 4-H Wearable Technologies (WearTec) project, funded by the National Science Foundation's Innovative Technology Experiences for Students and Teachers (ITEST) program. The overall goal for the three-year WearTec project is to study a systematic set of learning experiences focused upon the use of wearable technologies to effectively support student comprehension of the engineering design process and to increase interest in STEM academics and careers by students in grades 4 to 6. The key components of the WearTec project include (a) utilizing small sewable electronic components; (b) a focus on engineering design; (c) computer programming via the LilyPad Arduino microcontroller, and (d) bridging the formal and non-formal learning environments in order to deliver the WearTec project.

DOI: 10.4018/978-1-5225-0069-8.ch014

INTRODUCTION

Wearable electronic textiles provide a unique opportunity to teach students engineering design and other important science, technology, engineering and mathematics (STEM) concepts through the creation of functional solutions to real problems. The artistic aspects of wearable technologies are particularly attractive and accessible to new and diverse audiences to STEM (Qiu, Buechley, Baafi, & Dubow, 2013). In this chapter, we describe our National Science Foundation (NSF) funded project to teach engineering design, circuitry, and programming to students in grades 4 to 6 through the design and creation of wearable technologies. Specifically, the chapter will explore the following: the current state of using wearable electronic textiles in education, theoretical underpinnings, developing a shared vocabulary, an overview of the current electronic textiles kits, a comparative analysis of electronic textiles with educational robotics, an argument for introducing engineering education at the elementary school level, and examination of the need to more effectively bridge formal and informal education. In addition, the chapter presents the Nebraska 4-H Wearable Technologies (WearTec) pilot study that investigated the delivery of a newly developed student curriculum and professional development training centered on wearable electronic textiles.

Context of the Nebraska 4-H Wearable Technologies Project

The Nebraska 4-H Wearable Technologies (WearTec) project was funded by the National Science Foundation's Innovative Technology Experiences for Students and Teachers (ITEST) program in October of 2014. The overall goal for the three-year WearTec project is to study a systematic set of learning experiences focused upon the use of wearable technologies to effectively support student comprehension of the engineering design process and to increase interest in science, technology, engineering and mathematics (STEM) academics and careers by students in grades 4 to 6. The project is also interested in exploring the best practices for educators to bridge formal instruction time with after-school programming using wearable technologies as one contextual model to provide additional opportunities for students to pursue STEM-related experiences. Each step of the curriculum is tied to *Next Generation Science Standards* (NGSS) standards to encourage the effective integration of the activities into the school curriculum. The after-school programming allows students time to build and to iterate their designs.

The engineering design process is woven throughout the 60-hour WearTec curriculum, providing experiences from simple circuits all the way through to building and programming an electronic textile. These experiences provide students with the skills to create a variety of soft circuits allowing them to create real products from interactive greeting cards to backpack embellishments to electronic quilt blocks.

The curriculum begins with students learning basic circuitry skills by designing greeting cards with embedded paper circuits. The paper circuits use a small 3-volt coin cell battery, copper tape, a small LED, and card stock. Within the activity, students are challenged to use the engineering design process to build a mechanical on/off switch to preserve the life of their battery in the greeting card circuit. As students progress through the curriculum, they are given design problems and material constraints to guide them through the engineering design process.

Students learn basic sewing skills next by sewing a circuit design onto the fabric material. Their sewn circuit is then attached to the front cover of their engineering design notebook. Students combine their sewing and circuitry skills in the following project to design and create a communication flag. In this activity students use the engineering design process to solve the problem of sending messages using only their flag. The flag design is limited to two LEDs and a power supply. The sewing, circuitry, and design skills developed in this activity are then used to design an illuminated gym bag helping them to travel safely at night. The gym bag includes an additional component (TinyLily) to introduce students to preprogrammed microcontrollers. Students then spend time learning how to write their own code for a microcontroller (LilyPad). In the capstone project, each student designs and constructs an electronic quilt block using what they've learned about wearable technologies – sewing, circuitry, computer programming, and the engineering design process—to tell a personal story. The quilt blocks are connected to form a large electronic quilt that can be displayed at their school. By including an engineering design problem in each project, students are able to move through the design steps several times to develop an understanding of the overall process.

In the final iteration of the project, in contrast to the piloting process, the WearTec educators receive professional development through three-day summer workshops. It should be noted that during the piloting process educators only received one day of professional development training for roughly 20 hours of curriculum. Participants are required to attend the training in teams made up of both a certified teacher and an after-school educator from the same school. The teams are given time during the workshop to plan how they will present the material in both formal and informal learning environments. The teams provide feedback while implementing the WearTec curriculum on how they are incorporating the material in both the formal and informal learning settings.

Overview of the Educational Potential of Wearable Technologies

In our evolving experiences with this project, we are finding wearable technologies indeed provide an ideal hands-on learning space for students. Wearable technologies are both exciting and personally relevant, especially for females who tend to be more interested in textiles and design than their male counterparts (Barker, Melander, Grandgenett, & Nugent, 2015). Smart textiles and wearable technologies bring together engineering and computing to make computers, which are "soft, colorful, approachable, and beautiful" (Buechley, Peppler, Eisenber, & Kafai, 2013, p.1). The approachability of wearables derives from the personal way students interact with their creations. For example, one of the students at our first pilot site designed and created a flower bracelet with an LED in the center of the flower. Wearing her bracelet allowed her to begin building a personal identity as a creator/maker. In addition, when her classmates saw her creation they also became interested in learning about wearable technologies and how to build them. Wearable technologies are not limited to aesthetic design. Sensors and other inputs/outputs can be added to create sophisticated "smart" objects capable of interacting and responding to their environment. Students learn how to implement these sophisticated components and create unique smart wearable objects in the WearTec curriculum. These capabilities empower students to see themselves as creators instead of just consumers of the technology they use every day.

Theoretical Underpinnings

The WearTec curriculum and its use in engineering education includes the design, development, and building of wearable computer-based electronic artifacts and the learning activities are firmly situated in the *constructivism* theory of learning (Piaget, 1972). This theory purports youths' new knowledge is actively constructed from the world around them through experiential practice and by independent/

guided research to integrate new knowledge and to arrive at a deeper understanding. This pedagogical approach has been shown to promote sensorimotor intelligence, which enables learners to invent new learning through deduction (Piaget, 1972; McGrath et al., 2012). Problem-based learning (PBL) is a constructivist learning model using ill-structured problems to invite multiple approaches for developing a solution (Husain, 2011), which is also consistent with wearable technology projects. Like constructivism, PBL is student-centered, allowing learners to determine what they need to solve the instructional challenge with minimal facilitator input (Klegeris & Hurren, 2011). Also, rather than focusing exclusively on problem-solving in a single discipline, PBL typically requires students to engage in more interdisciplinary thinking and problem solving that is seen as foundational to STEM learning and application (Carnevale, Smith, & Stoll, 2010; Nicholl & Lou, 2012; Wiznia et al., 2012).

From these two well-recognized theoretical perspectives, we believe wearable textiles in many ways can be thought of as cognitive tools (Jonassen, 2000) to enhance the learning process and allow students to creatively and collaboratively problem solve while drawing on engineering design, computer programming, mechanical and textile design, and other STEM content. In addition, wearable technology activities can be closely aligned with components of engineering design that have been shown to support the study of science and its collaborative problem solving (Riskowski et al., 2009). These components include:

1. **Interaction:** Collaboration between team members and groups to design/build/test artifacts;
2. **Artifact Development:** Fostering individual and team knowledge by creating useful artifacts;
3. **Critical Analysis:** A process of continual iterative design for artifact refinement.

Several studies support the combination of engineering design, PBL, and design-based science as effective means to increase understanding of STEM with long-term retention of learning, and to decrease the achievement gaps between groups (Hmelo et al., 2000; Mehalik et al., 2008).

We have hypothesized learning engineering design processes, especially among female and minority students, will be enhanced through the design and creation of wearable technology artifacts capable of solving real-world, personally relevant problems. As described earlier, wearable technologies provide a unique interface for learning engineering design, which has been shown to support the study of science and the development of collaborative problem-solving skills (Riskowski et al., 2009).

Through wearable technologies, students are able to independently create functional solutions to real-world problems. This student-centered exploration and thinking is a hallmark of PBL. Students are given the freedom to identify what they need to solve a given design problem with minimal facilitator input (Klegeris & Hurren, 2011). For example, students could create a variety of wearable artifacts to solve the problem of walking home safely at night. In order to solve this design problem, students draw on the circuitry and engineering design skills learned in earlier sections of the WearTec program. In addition, the students make connections to their prior personal knowledge to determine how they might wear lights or what configuration or color of lights would best solve the problem.

The WearTec curriculum encourages interdisciplinary thinking by bridging several subject areas in addition to general personal knowledge to promote STEM learning and application as described by PBL theory (Carnevale, Smith, & Stoll, 2010; Nicholl & Lou, 2012; Wiznia et al., 2012). By combining real world, personally accessible problems with technology capable of providing real solutions, we predict students will be more motivated to utilize the engineering design process, leading to enhanced learning outcomes.

Wearable Technologies and STEM Education

There are many different names (wearable technologies, computational textiles, electronic textiles, e-textiles, smart textiles) behind the idea of combining computational devices and textiles within the realm of STEM education. While conceptually similar, electronic textiles (e-textiles) in the broadest sense represents a field of engineering combining electronics and computing with textiles and design. Wearable technology refers to electronic textiles or electronic accessories that can be worn; such as watches, eyeglasses, or clothing like a shirt or jacket containing electronics and a computation device (Johnson et al., 2013). Electronic textiles (e-textiles) may or may not be worn, whereas wearable electronics are electronic textiles specifically made to be worn on the body.

Theoretically, these programs rely on the use of a computing device combined with textiles or another "soft" material and electrical connections to create personally relevant artifacts for the builder (Qiu, Buechley, Baafi, & Dubow, 2013). When used in educational settings, students create such artifacts by combining fabric with components connected through conductive fabric or conductive thread (Katterfeldt, Dittert, & Schelhowe, 2009). When students add sensors and programmable microcontrollers to their artifacts, they can be transformed into smart and connected e-textiles. The novelty of e-textiles is the combination of physical and digital media, and the integration of hands-on/minds-on experiences with technology normally hidden from the students. By constructing electronic textiles the students can literally *see* the connections between the electrical/technical components made by hand and relevant theories from physics, engineering, and computing (Buechley, Peppler, Eisenberg, & Kafai, 2013). In addition, e-textiles provide an alternative pathway into engineering by providing practices and materials traditionally in the domain of women and other underrepresented groups. They can also connect at a very personal level, since clothing items are typically personalized artifacts for both adults and students. E-textiles offer an avenue for challenging, personalized projects that solve real-world problems. In addition, e-textiles offer a window into the world of technology and provide an idea pathway to acquire technology literacy that may be more inclusive than other pathways like game development or educational robotics (Qiu, Buechley, Baafi, & Dubow, 2013).

In the past, these technologies were too difficult for nonprofessionals to manipulate and create (Buechley, Peppler, Eisenber, & Kafai, 2013). With the rise of the Maker and open-source cultures, the demand for accessible and easily manipulated technologies is rising (Tanenbaum, Williams, Desjardins, & Tanenbaum, 2013). For instance, Leah Buechley's team has developed the LilyPad sewable microcontroller purchased relatively inexpensively along with compatible conductive thread, LED lights, sensors, motors, and switches (Buechley, Peppler, Eisenber, & Kafai, 2013). The WearTec curriculum is written for the LilyPad system, although several comparable wearable technology platforms are available. All of these platforms make visible the electrical components and functional mechanics of wearable technologies that are normally hidden. However, this is accomplished without sacrificing the integrity of the technology. There exists great potential for using e-textiles to increase knowledge of and attitudes towards STEM fields in many different learning environments. Because e-textiles and their use as wearable technologies are relatively new, they are not widely used in the classroom. In addition, few formal curriculum products and professional development trainings are available to the educational community (Qiu, Buechley, Baafi, & Dubow, 2013). Likewise, little research exists on the use of e-textiles as an educational tool in either formal or informal learning environments (Katterfeldt, Dittert, & Schelhowe, 2009).

The existing literature does provide evidence that the use of similar computational manipulatives, like educational robotics, are, in general, an effective approach to increase interest in STEM, improve workplace and life skills, and inspire youth to enter STEM careers (Melchior et al., 2005; Nugent, Barker, & Grandgenett, 2012). Unlike standard robot kits, however, the e-textile kits like the LilyPad and associated electronic components are not hidden behind plastic coverings. The components are bare and must be treated carefully to avoid problems like shorting out the circuit. In addition, unlike entering STEM and computing through robotics experiences traditionally dominated by males, wearable technologies offer an alternative pathway to STEM through textiles and computing more closely aligned with the typical societal experiences of some females. Past research suggests using wearable textile projects helps students understand underlying electrical structures and processes normally hidden. Peppler and Glosson (2013) found middle school students in an out-of-school time (OST) club learned about electrical current, battery polarity, circuit conductivity, and diagramming series circuitry through the use of wearable textiles. In another study, Kafai, Fields, and Searle (2013) delivered an e-textile workshop to high school students, looking specifically at the intersection of coding, circuitry, and crafting. Like Peppler and Glosson (2013), they concluded making the technology visible to students provided a better understanding of the circuits and computer programming.

Creating wearable computation devices involves identifying, debugging, and solving problems in coding, circuitry, and crafting (Kafai, Fields, & Searle, 2013). These design problems help learners develop STEM inquiry skills, including observation, hypothesis generation, testing, and evaluation of solutions (Sullivan, 2008). When building artifacts, concepts like electrical voltage, current, resistance, and polarity are reinforced for students. When working with wearable technologies, getting the artifact to work is not enough. Peppler, Sharpe, and Glosson (2013) argue another important aspect of wearable technologies is the aesthetics reflecting the voice of the designer. The intersection of STEM and art in wearable technologies reinforces computing, circuitry, and engineering while also unlocking the creativity and artistry of design. Kafai, Fields, & Searle (2014) found that women and girls had a clear preference for the sewable LilyPad Arduino over the traditional Arduino microprocessor using soldered electronic connections. Qiu, Buechley, Baafi, and Dubow (2013) reported e-textiles are an effective way to introduce technology and computer science in STEM content areas.

Current Wearable Technologies for Education

Commonly, when we talk about wearable technology, we are referring to "hard" ware—recent inventions such as Google Glass, fitness trackers such as FitBit or Nike Fuel Band, or even virtual reality headsets such as the Oculus Rift—are wearable technologies, yes, but decidedly inflexible, and very much in the vein of traditional consumerist electronics, not necessarily tools for learning. In our case, we wish to discuss "soft" ware the world of electronic textiles, sewable circuitry, and conductive threads—and how they can be used as powerful, personally expressive tools for education. This section, in particular, will focus on the ecosystem of wearable and e-textile technologies most widely used in educational settings.

The electronic textiles toolbox is, unsurprisingly, composed of a varied array of textiles and electronics materials designed to work together gracefully. Typically, an e-textiles project consists of several common elements: a power source (most often a battery of some sort), a computational "brain" (e.g., a microcontroller), one or more sensors and actuators (e.g., LEDs, motors, tilt or light sensors), and some way to connect them electrically (e.g., conductive thread, conductive fabric, copper tape, conductive ink, etc.). One of the first instances of a formalized wearable electronics kit, the LilyPad Arduino, initially included such components as a fabric-mounted tilt sensor and vibrating motor, a stitchable battery holder, two fabric switches—one a pressure switch and the other a hook and loop (Velcro) on-off switch—and a spool of conductive thread (Buechley, Eisenberg, Catchen, & Crockett, 2008, p. 425).

The interest in using e-textiles and wearable electronics for education is, in fact, integral to the reasoning behind developing comprehensive e-textiles tools in the first place; namely, to provide an alternative pathway towards computer science, electrical engineering, and general STEM fields appealing to a more diverse audience—particularly around gender—by using materials to encourage hands-on exploration and individuality instead of plodding through didactic lessons that "teach to the test." One of the first curricula for e-textiles was a high school unit outlined in Buechley, Eisenberg, & Elumeze, (2007), including nine-lesson plans based on Lego® Robotics curriculum beginning with introductions to e-textiles, sewing, and electronics, before moving on to cover basic programming, and hardware, inputs, and outputs.

In the past eight years, the use of e-textiles for learning in (and outside of) the classroom has been on the rise. Examples of e-textile resources include instructional topics such as learning introductory computer science (Qiu et al., 2013), teaching basic electronic circuits (Peppler & Glosson, 2013), crafting e-fashion (Peppler, Gresalfi et al., 2014), and even interactive books with standards-aligned activities based around creating e-puppets for storytelling (Peppler, Salen et al., 2014) and. Yet, one commonality threads through all of these efforts—the continued use of the LilyPad Arduino ecosystem as the primary tool through which these efforts are centered.

The LilyPad Technology

The LilyPad, essentially an ecosystem of sewable electronic components, developed by Leah Buechley while at the Craft Technology Laboratory at the University of Colorado, Boulder. With the help of SparkFun Electronics, a local electronics company, LilyPad became the symbol and de facto tool of a new movement in wearable and electronic textiles. The LilyPad ecosystem is built upon several timely advances in do-it-yourself manufacturing: a) the open source hardware movement, and textiles science, specifically, b) the existence of free circuit design software and (relatively) affordable services for small production runs of printed circuit boards (PCBs), c) the popularity of the open-source Arduino microcontroller and d) software development environment, and the growing availability of conductive threads and fabrics suitable for wearable circuitry design. This section situates the LilyPad technology within the context of these factors, discussing both the technological and social achievements leading to the LilyPad becoming such a powerful tool for education.

The first LilyPad prototypes were "iron-on" stencils, whereby the circuit design was laser-etched and then adhered via an iron to a square-shaped fabric substrate. The microcontroller and components were plated through-hole components, meaning they were much larger in all three axes than surface-mounted components, and thus took up a significant footprint space both on a garment and above it. The initial tool chain used to program the LilyPad involved AVR microcontrollers, programming in C, compiling through the command line with the GNU Compiler Collection (GCC) compiler, and using an STK500 development board and UISP software to download programs onto the AVR chips—hardly a beginner-friendly setup, as the author acknowledges (Buechley, 2006). After running several workshops with children and getting feedback on their experiences, the LilyPad eventually moved to a circular design and smaller components with evenly distributed I/O "pads" made of conductive fabric, and formally adopted the name "LilyPad."

At this point, each LilyPad was still laser etched and hand assembled out of fabric and small electronic components, a time-consuming and non-scalable practice. This prompted the LilyPad's conversion to the printed circuit board design it has today, and facilitated the move towards a commercially available product. The design of the LilyPad circuit board maintained the circular shape, but replaced the conductive fabric pads with large, copper-plated holes; these holes were big enough to fit a needle and thread through, and thus allowed the PCB to be sewn onto a garment with conductive thread. These signature "pads" appear in the designs for the entire LilyPad ecosystem, from push buttons to blue LEDs, ensuring every component could be sewn on without extra work on the part of the user.

The other significant development was the adoption of the Arduino platform as the means to develop for and load programs onto the LilyPad microcontrollers. Arduino is an open-source hardware company maintaining and developing a line of beginner-friendly microcontrollers as well as a simplified integrated programming environment (IDE) brought an unprecedented level of accessibility and ease to the world of microcontroller programming. Users could now connect a USB cable from their LilyPad Arduino to their computer, write a program in the Arduino IDE using a simplified version of the C language, and hit a button to upload their program onto the LilyPad device—a much simpler and cheaper process than the previous LilyPad versions.

Based on academic and education community reactions, and our own evolving experiences across our funded projects, these two developments seem to have brought the design of electronic textile projects within the reach of educators and their students (Buechley & Hill, 2010, Kafai et al., 2010, Lovell and Buechley, 2010). As of the time of this writing, the LilyPad ecosystem displays 47 different products sold on the SparkFun Electronics website (sparkfun.com, 2015). This includes several types of Arduino-compatible LilyPad microcontrollers, seven different colors of sewable LEDs, an integrated MP3 player board, a light sensor, accelerometer, vibration motor, temperature sensor, and several different kinds of project-based development kits.

While the LilyPad ecosystem has arguably lowered the entry point for many within formal and informal educational environments there exists some limitations. Like most technology driven interventions the cost to implement it is an important consideration. For the WearTec program the individual electronic components are relatively inexpensive, generally costing less than a few dollars per item. However, components used in the WearTec program are generally not reusable since the products become very personal to students and with each learner built project the costs can escalate especially when more expensive items are introduced like sensors and the LilyPad microcontroller. Another limitation of the current ecosystem is the durability of the conductive thread. The LilyPad components are used for rapid prototyping and are not usually capable of being washed. Moreover, the thread is somewhat brittle and is prone to breaking and stretching causing problems with electrical circuits. Projects that are three dimensional, like sweatshirts or t-shirts that are worn on the body are especially prone to the stretching and breaking of the conductive thread.

WEARABLES AS AN EVOLUTION TO EDUCATIONAL ROBOTICS

Wearable e-textiles subscribe to the same learning theories of constructivism and PBL as many of the successful hands-on, minds-on educational robotics programs that have been used by formal and informal educators for many years. Educational robotics programs, organized as summer camps, after-school clubs, and competitions, have shown promise for building knowledge, problem-solving skills, and STEM self-efficacy in participating youth at the middle school and high school levels (Nugent, Barker, Grandgenett, & Adamchuk, 2010; Nugent, Barker, & Grandgenett, 2012; Benitti, 2012). Therefore, it makes sense to build complementary programs to engage younger students, steering them towards eventual involvement in robotics or other activities with similar benefits. Here we outline ways in which a program in wearable technologies, aimed at students in late elementary school, can prepare participants for later STEM involvement such as educational robotics programs.

First, let us examine the characteristics of educational robotics programs to see where connections can be made between wearable technologies and robotics. A typical robotics program may include hands-on activities founded in problem-based learning. These activities would teach various STEM skills, such as computer programming, mechanical design, and circuits/electronics in a scaffolded manner. Along the way, concepts of problem solving and engineering design would be introduced, requiring analysis, planning, and execution of solutions. The involvement of an informal educator is generally important throughout this process.

Similarly, the field of wearable technology can encompass STEM knowledge from various disciplines (e.g., electronics/circuits, programming, and design). Because wearable technology is all about producing a useful product, the problem-based, hands-on approach is very appropriate. Although introduced differently, perhaps, than would be done in an educational robotics program, concepts from engineering design and problem-solving approaches are very useful in the process of developing a wearable technology product. Mentoring from an informal educator along the way is likewise helpful in developing self-efficacy in such a program.

Based on these parallels, the progression of students from wearable technology as a step to something more advanced within STEM developmental environments seems to be reasonable. Not only can programs such as wearable technologies and robotics be partially modeled after each other, but specific successes and lessons learned from the educational robotics community can be leveraged to create a successful introductory STEM experience for younger students based on wearable technology.

BUILDING ON WHAT IS KNOWN FROM EDUCATIONAL ROBOTICS

Studies of educational robotics programs have generated some important conclusions about why they work and what their particular strengths are. Building on this knowledge will be important for creating successful educational programs in wearable technology.

One important characteristic is that robotics is an applied science. It draws on knowledge and skills from multiple disciplines and allows a relatively open-ended spectrum of problems to be posed to aspiring roboticists. This casts a wider net in terms of student interest; although an individual may not be highly interested or possess self-efficacy in all of the related sub-domains (circuits, programming, etc.), she or he is relatively likely to find interest and/or native ability in at least one of these sub-domains. Likewise, wearable technologies draw from a similar breadth of STEM sub-disciplines and can thus attract a wide audience. There is a further advantage to wearables, however. The textiles and artistic aspects of these activities can be leveraged to attract girls to STEM who are otherwise an underrepresented group (Kafai, Fields, & Searle, 2014; Weibert et al., 2014). This is a key distinction between the learning platforms of robotics and wearables and is a significant aspect of the educational approach described in this chapter.

The problem-based approach is also important for generating both interest and self-efficacy (Hmelo-Silver, 2004). Because it is exploratory in nature, it fuels curiosity (interest), and allows participants to experience successes (self-efficacy) as they solve problems. One cascade effect is the generation of student interest in these STEM topics affects learning in those topics as well as career orientation towards the associated fields (Maltese & Tai, 2011; Maltese & Tai, 2010). Therefore, building on this problem-based approach is a critical aspect of designing a successful youth program in wearable technology. Fortunately, because of the applied nature of wearable technology, it is also well suited to this approach.

Finally, robotics programs have demonstrated the importance of the informal learning environment and of the informal educator as a contributing mentor (Barker & Ansorge, 2007; Nugent, Barker, Grandgenett, & Welch, 2014). The informal educator plays an important role in a variety of ways, including contextualization of content learning through guided projects and serving as a role model for STEM interest and a potential career opportunity youth can emulate. Just as didactic learning in the formal learning environment is critical for laying a foundation in STEM, the informal learning environment provides the flexibility and opportunity for self-direction building beyond the basic foundation and stimulates increased interest while strengthening and solidifying content learning. The benefits of leveraging these strategies in parallel suggest educational programming should be designed with integration of the formal and informal learning environments in mind, with collaboration between formal and informal educators.

In addition to highlighting these common points between robotics and wearable technology programs, it is noteworthy to point out at least one difference from our experiences in both environments. The cultures of robotics programs tend to be centered around competition events, in which a team of youth succeeds together based on their combined effort. In wearable technologies, the objective is typically not to succeed in a competitive event, but rather to create a useful and personalized product. This can be a more enduring outcome with a lasting personal connection. Wearable technology tends to be more of an individual creation, although group or team-based projects are certainly possible (e.g., the quilt example mentioned previously).

In summary, we believe that wearable technology and robotics have enough similar aspects to make them complementary in steering youth towards STEM. Lessons gained through successful implementation of robotics programs show us the important characteristics of successful programs should be maintained as new wearable technology programs are developed.

Characteristics to Include in Wearable Technology Programs

- Applied nature of educational programming.
- Problem-based learning.
- Informal learning environment (and coupling with formal learning environment).
- Influence of educator as mentor.

By building around these principles, it seems likely wearable technology programs can be successfully developed to achieve similar positive results to robotics programs, while attracting students to STEM in a younger age range and across different demographics.

The Need for Innovative Engineering Education

It is important to note we also see wearable technologies as a possible instructional environment for supporting K-12 engineering education. Effective K-12 education in STEM continues to be critically important to supporting the U.S. economy, our national security, and our quality of life (NRC, 2010).

As described in numerous reports such as *Rising Above the Gathering Storm Revisited* (NRC, 2010), the United States is facing desperate challenges and growing international competition associated with STEM education. The report acknowledges that schools are not producing enough graduates in STEM areas, which is negatively impacting America's ability to compete for new industries, laboratories, and jobs (NRC, 2010, p.4). Engineering professions in particular are already experiencing a serious shortage of qualified individuals as many current engineers are entering retirement (Engineering: Short on fact, 2012). While STEM, in general, continues to be identified as critical for an effective K-12 education, the "E" of STEM or "engineering" has typically received the least attention in schools. Since the early 1990s less than one percent of K-12 students have received formal engineering curriculum during the school year (NRC, 2010).

The current K-12 system often produces students unprepared for higher education within engineering (AASCU task force report, 2012). High school students have little experience with innovative engineering activities and generally have negative feelings towards engineering as a future profession (Adams et al., 2011). College freshmen engineering students struggle to implement basic engineering design processes (Atman et al., 2007). But there is some hope for changing the situation by providing engineering experiences for students during early formative years when interest in science and engineering can more easily be developed (Brotman & Moore, 2008; Catsambis, 1995; Clewell & Braddock, 2000. Children are naturally creative problem solvers during play (Adams et al., 2011; Forman, 2010). These traits are beneficial to engineering design, and if these natural tendencies can be nurtured at a young age and developed throughout K-12 education, students will be better-prepared and more excited to engage in engineering education as well as future careers in STEM fields (NRC, 2010; Mann et al., 2011). For example, Lachapelle, Phadnis, Jocz, and Cunningham (2012) reported students in an engineering curriculum designed specifically for elementary children showed positive changes in their attitudes towards engineering.

The need for innovative engineering programs in K-12 STEM education is clear and the *Next Generation Science Standards* (NGSS) exhibit a commitment to fully integrate engineering and technology into science education, with engineering design identified as a major core idea (NRC, 2012). It makes sense in order to understand the engineering process, K-12 students need to learn engineering concepts and the core skills of design and redesign (NRC, 2010). According to the NGSS standards, students at any grade level should be able to ask questions, define the problem to be solved, and elicit ideas leading to the constraints and specifications for its solution (NRC, 2012, p. 56). Clearly, a quality hands-on, minds-on engineering curriculum will be of critical importance for students and teachers to meet the goals of NGSS and to increase the number of students in the STEM pipeline.

As previously mentioned, some initial research efforts combining wearable electronic textiles, engineering design, and computer programming suggest well-focused engineering activities appear to have the potential to help students understand underlying structures and processes normally hidden. Similar to engineering design, wearable technologies also appear to have the potential to contribute to the learning of computer programming constructs. Creating wearable computation devices involves identifying, debugging, and solving problems in coding, circuitry, and crafting (Kafai, Fields, & Searle, 2013). These design problems help learners develop inquiry skills, including observation, hypothesis generation, testing, and evaluation of solutions (Sullivan, 2008). When building wearable technology artifacts, concepts like electrical voltage, current, resistance, and polarity are easily reinforced for students.

Bridging Formal and Non-Formal Learning Environments

Currently, many of the more innovative engineering education experiences occur in informal settings like 4-H Robotics and the FIRST Robotics competition, and there is a need to effectively bridge and align these non-formal engineering programs with what is taught in the formal K-12 classroom. Current research indicates out-of-school programs can be truly effective avenues for promoting learning in STEM content areas (Barker & Ansorge, 2007; Barker, Nugent, & Grandgenett, 2008; Clarke, 2010; McGrath, Lowes, McKay, Sayres, & Lin, 2012; NRC, 2009). In Nebraska, for example, one in three, age eligible youth participate in 4-H informal education programs (135,000) about diverse topics like robotics, coding, agriculture, and animal science (NU Foundation, 2013). Certainly, there would seem to be educational potential in aligning such innovation in after-school programs with in-school experiences.

While OST programs can provide innovative learning experiences, these experiences typically do not align with what the child is learning during the school day unless directly focused for such alignment. When formal and informal learning goals are aligned and congruent, the young person's learning becomes more personally meaningful and relevant (Noam, Biancarosa, & Dechausay, 2002). In its 2009 report, *Learning Science in Informal Environments*, the National Research Council determined the goals between formal and informal educational environments complement and overlap each other (p. 296). Providing a congruent engineering experience from the in-school classroom to out-of-school time experience has the potential to increase learning and to perhaps reverse the trend of STEM interest loss at the late elementary school ages.

Across both formal and informal educational settings, there continues to be a lack of research on how students learn and understand engineering concepts, particularly when the focus is on engineering design. The comprehensive report on engineering education commissioned by the NRC (2009) identified three key strategies where engineering projects, and particularly those articulated across in-school and OST settings, might contribute to the research. The first is a focus on iterative, purposeful revisions of student designs in both settings. The new framework for science standards suggests, "it is the iterative cycle of design that offers the greatest potential for applying science knowledge in the classroom and engaging in engineering practices" (NRC, 2012, pp. 201-202). The second strategy is the need for immersion in extended design activities, where after-school programs can certainly bring additional immersion time possibilities. The third strategy is sequencing of instruction to build from the easiest to more difficult concepts. By coordinating engineering activities across both settings, there may well be a chance to better align instructional time with concept difficulty and exploration considerations. Wearable technology activities may be an excellent approach for building articulated engineering education, crossing instructional settings.

The use of an integrated instruction spanning both formal and informal learning time would seem to be able to provide opportunities for initial design in the classroom and more extensive iterative redesign in OST activities. The OST environment also provides the flexibility and freedom for immersive extended wearable technology activities where in-school opportunities may be limited. Finally, wearable technology instructional activities crossing these instructional day boundaries can involve learning progressions where students are provided with background knowledge within the school day to lay the foundation for developing a prototype, making iterative design revisions, and completing a final device or product in the OST setting. This complementary formal/informal context capitalizes on the strengths of both learning environments to promote the integration of engineering functionality and aesthetic design.

CASE STUDY: THE WearTEC PILOT

The pilot of the WearTec instructional program took place in after-school clubs in eight Nebraska elementary and middle schools during spring semester, 2015. The purpose of the pilot was to test the curriculum and teacher professional development, as well as the research instrumentation and procedures. The pilot curriculum covered 20 hours (of the 60 total hours planned) and consisted of four student lessons focusing on (a) electric circuits and engineering design, (b) understanding microcontrollers, (c) computer programming, and (d) using engineering design to complete a T-shirt project. The lesson infused content dealing with engineering design, circuitry, and computer programming within structured hands-on student activities resulting in student products.

The lessons began with using the LilyTiny, a small, preprogrammed microcontroller with four basic blink (output) patterns, LEDs, a 3-volt power supply, and conductive thread. The instruction provided an opportunity for students to integrate these components into a bracelet or a felt pennant with multi-colored fabric embellished with LEDs connected to the LilyTiny. The programming activity used the LilyPad ProtoSnap board, which consists of a microprocessor connected to multiple LEDs, sensors, and switches prewired on a single board. Students then made felt-based toys with LilyPad microcontrollers and LEDs. For the final activity, students used a stand-alone LilyPad Arduino microcontroller, LEDs, light, buzzers, and sensors to design and create a T-shirt.

The one-day professional development for the teachers consisted of an overview of the curriculum, presentation of engineering design process, and facilitated hands-on activities to develop a flag/pennant to incorporate wearable technology components. Formal and informal educators attended the professional development training in teams and began planning how they would jointly deliver the instruction.

Participants

All but one of the participating schools in the project were part of the 21st Century Community Learning Center (21st CCLC), which serves students attending high-need schools. The one exception was an after-school club choosing to focus on the computer programming elements of the curriculum. Students self-selected into the after-school clubs, varying in length and number of weekly sessions depending on the schools' club schedules. Students were in grades 4 thru 8, with most in the fourth or fifth grades. Fifty-eight percent of participants were female while 42% were male. The majority (58%) were Caucasian; the remaining percentage represented Native American (6%), Hispanic (16%), African-American (12%), and multiracial (8%). Data was obtained from four certified teachers and nine after-school educators who attended the professional development sessions to prepare educators to present the materials.

Research Design and Data Analysis

The research used a repeated measures pre-post design examining both teacher and student outcomes and included a split plot analyses examining changes exhibited by certified teachers versus after-school educators.

Instruments

Teachers. Teacher instruments were administered before and after their participation in the professional development. Questionnaires focused on their confidence in (a) their knowledge and use of wearable technologies, (b) promoting youth engagement in STEM, (c) helping students design and develop a wearable technology product, and (d) implementing the *WearTec* curriculum to include both formal and informal components. Teacher confidence or self-efficacy has been shown to be correlated with achievement and performance outcomes (Sorge, 2007). Teachers also participated in interviews to provide more in-depth qualitative feedback about their experiences.

Students. Student instruments were administered before and after their involvement in the *WearTec* club. The instruments included an assessment of their wearable technologies knowledge (circuitry, engineering design, and computer programming), their attitudes towards technology and engineering, and their self-efficacy with electric circuits and computer programming (5-point Likert scales). The circuitry and computer programming assessments were multiple-choice questions. The engineering design assessment presented a scenario in which students were asked to select activities appropriate at each step in the engineering design process. Instruments were based on ones developed and validated for a robotics NSF project (Nugent, Barker, Welch, Grandgenett, Wu, & Nelson, 2015) but were revised to reflect the wearable technologies content emphasis. Questions presented on the instrument with basic statistical descriptives are shown below in Table 1.

RESULTS OF THE PILOT STUDY

Teachers

Teachers entered the professional development with confidence levels ranging from 59% confidence ("I can differentiate between the engineering design process and the science inquiry process") to 94% confidence ("I can encourage positive attitudes towards science, technology, engineering, and mathematics"). In general, teachers were more confident in their pedagogical abilities (i.e., encouraging critical thinking, collaborative learning, and engagement) than their technological abilities (i.e., set up a LilyPad Protosnap to download and troubleshoot programs from the Arduino programming software). After completing the professional development, teachers showed increased confidence in all areas, including wearable technologies. Post confidence scores were significantly higher than pre ($t(12) = 4.51, p < .0001$). Overall, confidence scores of the certified teachers were considerably higher than those of the informal educators at baseline, but the gap was closed after completing the professional development ($\Lambda = 5.37$ (1, 11), $p < .05$). The pre and post PD confidence levels are shown below (Figure 1).

Interviews and feedback from participating educators highlighted the challenges of engaging teachers and informal educators to utilize and integrate STEM into the classroom and out-of-school time learning. All of the instruction was delivered in the after-school environment but teaching responsibilities were generally shared. Teachers emphasized the difficulty of integrating the curriculum into the school day, where they typically have to teach a certain curriculum addressing state standards.

Table 1. Descriptives for student survey items

Question	N	Pre Mean	Post Mean
Cognitive Assessment: Electric Circuits[1]			
Choose the image best showing a closed circuit (a complete path for electricity to flow):	40	43%	70%
In the picture above, identify which components (different colored LEDs) are wired in parallel.	40	83%	68%
Where should a switch be placed in the circuit below to control only the red LED?	40	57%	53%
In the image below, where should connections be sewn to control the LED with the LilyPad microcontroller?	40	41%	51%
A switch in a circuit can turn off an LED by:	40	48%	48%
In the picture above, which number would result in a short circuit?	40	25%	28%
A short circuit happens when electricity in a circuit:	40	18%	25%
Which of the circuits below shows the correct wiring for an LED?	40	35%	18%
Cognitive Assessment: Programming			
What would you change in the code above to blink the LED highlighted (LED 5) on the ProtoSnap board?	26	45%	50%
Which line in this code determines how fast the light will blink?	26	43%	50%
What line(s) of the code would need to be repeated and modified to blink two LEDs?	26	37%	48%
How would you expect the variable **LED 1** from the code below to behave?	26	36%	43%
Cognitive Assessment: Engineering Design			
Defining the Problem			
Learn what materials and machines are available to build the system*	24	63%	88%
Find out how many T-shirts should be processed in one day*	24	54%	71%
Identify the type of packaging needing to be used*	24	42%	67%
Go to the local store to see how much a typical T-shirt costs	24	58%	58%
Find out what labels should be placed on each T-shirt*	24	54%	54%
Conduct a customer survey of T-shirt preferences	24	63%	46%
Calculate the approximate distance between processing stations*	24	50%	42%
Count how many T-shirts are sold in your area	24	46%	29%
Develop A Solution			
Think about ways to label and sort things*	24	50%	79%
Look up patent information (inventor's right to make and sell his/her invention) on clothing equipment*	24	67%	67%
Go to the textile design department of a local college or university	24	71%	63%
Call an expert in fabric processing to help you with ideas*	24	67%	58%
Start a blog or Facebook page about T-shirts	24	42%	58%
Learn about the packaging and labeling process*	24	71%	54%
Do an Internet search for "T-shirt label"*	24	33%	42%
Look up "T-shirt" in an encyclopedia	24	29%	29%
Implement the Solution			
Test prototype systems*	23	78%	78%
Try out different materials and mechanisms in the system construction*	23	39%	70%

continued on following page

Table 1. Continued

Question	N	Pre Mean	Post Mean
Conduct an extensive economic analysis before building anything	23	57%	65%
Think about what can possibly go wrong with the design*	23	35%	65%
Get feedback from those who sort T-shirts on a regular basis*	23	70%	52%
Create computer simulations of the most favorable solutions*	23	57%	52%
Write a report about the design	23	35%	48%
Add a computer-controlled camera to the design	23	74%	39%
Attitudes towards Technology			
It is important for me to learn about wearable technologies.	39	3.77	4.08
It is important for me to learn how technology works.	39	4.31	4.08
It is important for me to learn how to program a computer.	39	4.1	3.97
It is important for me to learn how to make changes to improve a wearable technology product.	39	3.95	3.85
I like learning new technologies such as e-textiles.	39	3.74	3.79
Attitudes towards Engineering			
It is important for me to understand basic engineering concepts (e.g., design trade-offs, circuits, power) related to e-textiles	39	3.79	4
I like using the engineering design process to solve problems.	37	3.68	3.7
Self-Efficacy: Circuitry			
I am certain I can use a switch in a circuit to turn an LED on and off.	39	4.1	4.21
I am confident that I can assemble a working electric circuit using two lights, a power source, and a computer.	39	3.95	4.08
I am certain I can fix a broken circuit in an e-textile product.	38	3.42	3.55
Self-Efficacy: Programming			
I am confident I can use a computer to blink 3 LEDs in a pattern.	38	3.29	3.87
I am confident I can transfer/move a program I've written from a computer to a circuit.	39	3.62	3.72
I am confident that I can program a computer to change colors of an RGB (Red, Green, Blue) LED light.	7	3	3
I am confident that I can program a computer to solve a wearable technology problem	7	3.57	2.71

Students

Knowledge. Student results are shown in Table 2. Because the number of club sessions varied, not all teachers were able to present all lessons. This required limiting post-test assessments to questions dealing with material teachers had covered. The result was varying numbers of students completing certain sets of questions. As measured by scores on the multiple choice knowledge tests, students showed increases in their knowledge of circuitry, computer programming, and engineering design. The only *significant* increase was for computer programming, but the increases for circuitry and engineering design were approaching significance ($p = 0.28$ and $p = 0.12$).

Attitudes. There were increases in students' attitudes towards engineering but not technology. Students also showed increases in the programming self-efficacy but not their circuitry.

Figure 1. Pre and post PD confidence levels

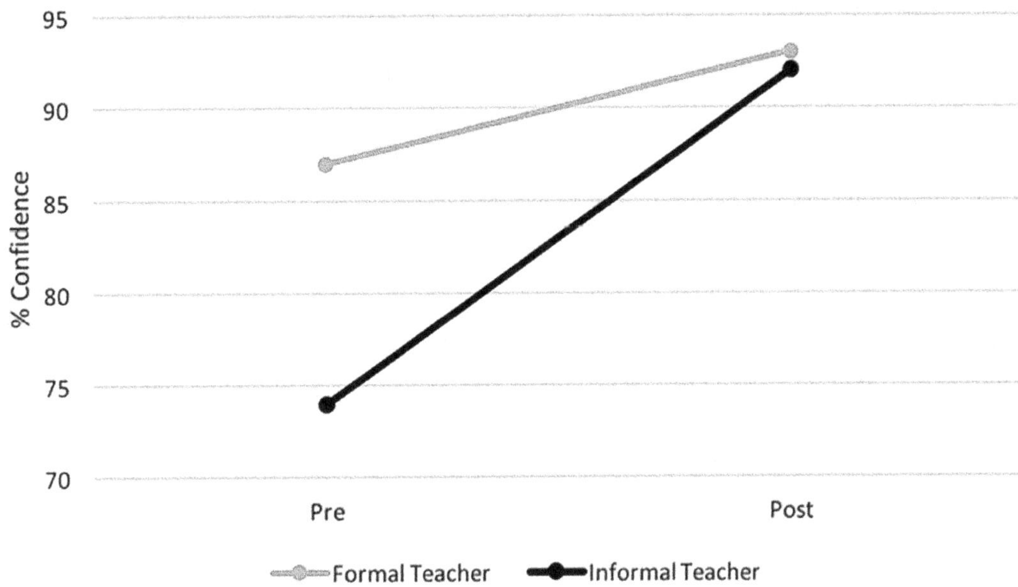

Table 2. Student results

Outcome	n	Mean		t(df)	p	Observed Power	Effect Size dz
		Pre (%, SD)	Post (%, SD)				
Knowledge							
Circuitry	40	3.28 (47%, 1.36)	3.58 (51%, 1.36)	1.09 (39)	0.28	.186	.172
Eng. Design	23	8.43 (56%, 3.11)	9.61 (64%, 2.39)	1.60 (22)	0.12	.333	.333
Programming	26	.81 (21%, .80)	1.81 (45%, 1.20)	4.37 (25)	.000	.987	.858
Attitudes							
Technology	40	3.98 (.76)	3.95 (.81)	.26 (39)	0.80	.058	.0267
Engineering	40	3.74 (.94)	3.84 (.91)	.61 (39)	0.55	.091	.096
Self-efficacy							
Circuitry	39	3.83 (.93)	3.94 (.81)	.68 (38)	0.50	.102	0.109
Programming	39	3.50 (.95)	3.79 (1.01)	1.50(38)	0.14	.310	.240

Summary and Discussion

The Nebraska 4-H Wearable Technologies (WearTec) project is centered on utilizing electronic textiles to support teachers and students in the understanding of the engineering design process and to increase student interest in STEM careers and academics. When complete, the project will result in 60 hours of hands-on, carefully piloted and refined curriculum for students and a three-day, face-to-face professional development workshop and teachers guide mapped to national and Nebraska state standards. Because of the nature of electronic textiles, which combines computing, circuitry, and engineering design with textiles, students are able to construct personally relevant artifacts. By utilizing the LilyPad ecosystem of sewable electronic components, including conductive thread and programmable microcontrollers, students are able to experiment with the technology that is normally hidden from them. This combination of technology and computing with softer materials like textiles is an ideal medium to encourage female and other underrepresented groups of students to participate in STEM activities and over time to pursue additional academic opportunities and careers in engineering and other related STEM fields. From an educational learning theory perspective, wearable electronic textiles are similar to educational robotics and are situated in the learning theories of constructivism and problem-based learning. Both educational robotics and wearable electronic textiles draw on applied science and knowledge from multiple disciplines to solve open-ended engineering problems/challenges. While educational robotics is arguably an effective tool for attracting students into STEM, its appeal has not always been well received by females.

The purpose of the pilot was to test the curriculum and teacher professional development, as well as the research instrumentation and procedures. The pilot curriculum covered 20 hours (of the 60 total hours planned) and consisted of four student lessons that focused on (a) electric circuits and engineering design, (b) understanding microcontrollers, (c) computer programming, and (d) using engineering design to complete a T-shirt project. The lessons infused content dealing with engineering design, circuitry, and computer programming within structured hands-on student activities resulting in student products.

In an effort to determine the effectiveness of the Nebraska 4-H WearTec project's early curriculum and professional development efforts and to also examine the research instruments and procedures, a pilot study was conducted on the first 20 hours of the planned 60-hour curriculum in 2015. Overall, the project attracted a majority number of females (58%) in contrast to other STEM clubs focusing on robotics, which typically report a ratio of male to female students of 2:1 or greater (Martin et al., 2011) or 30% female participation (Nugent, Barker, & Grandgenett, 2014). The results of this study also support the assumption wearable technologies offer an alternative pathway to STEM through textiles and computing and may be more closely aligned with the typical societal experiences of some females.

Results show teachers can develop confidence in delivering wearable technologies curriculum, even with limited professional development. Certified teachers entered the professional development more confident in their abilities to implement a wearable technologies project than did the after-school educators; however, the confidence levels equalized by the end of the training. Results also underscore the challenges of presenting wearable technologies curriculum during the school day because of the need for formal teachers to cover specified local curriculum and state standards.

Due to the varying lengths of the after-school clubs, not all students completed all four lessons of the curriculum. In addition, attendance in after-school clubs is optional and students could drop in and out at will. These inconsistencies resulted in pre and post questionnaire matches of only 40 of the students participating in the *WearTec* clubs. These small numbers considerably reduce the power of the results. As shown in Table 1, the observed power of all but the computer programming knowledge outcome was below the desired .80.

The student attitudes results showed no difference in student pre and post perceptions of the value and importance of technology, but closer examination of the pre and post means for these questions showed distinct differences, depending on the specific aspect of technology. In particular, their perceptions of the value of computer programming and learning how technology works declined from pre to post; however, attitudes towards learning about wearable technologies increased (see Table 1). It should also be noted student pre scores for the two general technology questions were both over 4.0 on a 5-point scale, making it difficult to show increases.

There was also an increase in student perception of the value and importance of engineering. The later result may be due to the fact the project used an engineering design framework as the basis of the individual projects. This result may also reflect the fact engineering is not part of the K-12 curriculum in Nebraska, and this project may well have been a student's first introduction to this discipline. Student self-efficacy results showed increases in student confidence regarding circuitry and their programming skills. As with engineering design, computer programming is not well established at the elementary and middle school levels. The significant increase in student knowledge of computer programming and their increased self-confidence may again be related to their first exposure to such material. Despite these increases, however, the pre and post knowledge scores dealing with computer programming were also considerably lower than those of the other content areas.

While pre-post comparisons generally were not statistically significant, all but one of the seven outcomes showed increases as a result of student participation in the project. Results provide evidence of the promise of the intervention and support other research showing the positive effects of a wearable technologies program in increasing students' self-efficacy in working with wearable technologies and producing e-textile products (Barker, Melander, Grandgenett, & Nugent, 2015). These results also extend previous research by providing evidence of impacts on student knowledge of electric circuitry, engineering design, and programming.

Pilot Study Limitations

The main goal of the pilot study was to test the curricular activities and teacher professional development of the WearTec project. The results for the teacher professional development were promising with significant increases from pre to post test. However, while educators reported increases in confidence in the content and delivery it is not known how much of the curriculum teachers and after school educators actually implemented during the pilot. In an effort to document and better understand what lessons of the curricula will be delivered in future iterations of the project a project implementation log will be developed and collected by project staff immediately after the program completion. In addition, during the professional development workshops teachers and educators will have time to develop implementation plans to determine when and how they will deliver the WearTec program during the school year and how they will partner to bridge the formal and informal learning environments. Another limitation of the study was the small sample size and predominance of informal educators (four certified teachers and nine informal educators). The results may not be generalized to the population of teachers and afterschool educators as a whole because of the skewed participation rates for the pilot. In future iterations of the program it is expected additional representation from certified teachers and a large sample of both classroom teachers and informal educators will participate in the project and corresponding research.

The student results did show positive gains from pre to posttest however there were unexpected decreases on some of the cognitive and attitudinal survey questions. In particular questions concerning circuitry and attitudes towards technology (importance of programming a computer) actually decreased from pre to post test. One question in particular "In the picture above, identify which components different colored LEDs are wired in parallel" fell from 83% on the pretest to 68% on the posttest. It is not clear why the pretest scores for this question were elevated and additional analysis of the individual questions is needed. In addition, the overall reading level for the instruments which was calculated at the 8th grade level. This lack of alignment with the 4 – 6th grade target audience may explain some of the score results due to students purely guessing on questions directly attributed to their lack of reading comprehension.

In addition, a decrease in scores could be explained by the lack of content coverage by the teachers either due to time constraints to implement the program fully or, like in the case of programming, they were not comfortable teaching the content. As previously mentioned in future iterations of the program the personal implementation plans will be used as a way to examine fidelity of implementation.

The pilot was implemented near the end of the third quarter of the school year, which resulted in many students transferring from the WearTec club program to another afterschool activity in the fourth quarter of the school year. This caused a decrease in the number of matched repeated measures scores (pre to post) and reduced the overall power of the study. Many students that transferred into the project in the fourth quarter may not have had the opportunity to build the early circuit activities thereby not receiving vital background information needed for advanced activities. Individual student participation was not tracked during the pilot, which muddled our understanding of the instructional dosage each student received. This could explain low posttest scores since students may have missed important lessons or entire units of instruction leading to the low posttest scores. To help better understand dosage and student attendance future iterations of the project will ask teachers to track and report student attendance. Another limitation of the pilot study was the small student sample size, which led to an underpowered pilot study.

FUTURE RESEARCH DIRECTIONS

Additional studies with larger samples and appropriate statistical power, as well as more complete teacher fidelity and student attendance data, are needed to better understand the effect of the intervention on student learning and attitudes towards STEM. Future research for the project will focus on teacher and educator preparation to implement programs involving wearable technologies looking specifically at implementation strategies, student to teacher ratios, efficacy and how such an intervention could fit into the school day to support STEM learning. Finally, the WearTec project is particularly interested in a model extending the school day from the formal learning environment led by certified teachers into the informal learning environment led by afterschool educators. While there were teams of teachers/educators attending the professional development workshops it appears most of the implementation for the pilot was conducted with informal educators in the after school-learning environment. In future iterations of the program participating schools will be strongly encouraged to form teaching teams providing a mechanism to better study the bridging of the formal and non-formal learning environments through the use of wearable electronic textiles.

CONCLUSION

The WearTec pilot study was an attempt to examine a preliminary set of learning experiences focused upon the use of wearable electronic technologies to effectively support student comprehension of the engineering design process, circuitry, and programming and to increase interest in STEM by students in grades 4 to 6. In addition, the project explored ways to bridge formal instruction time with after-school programming using wearable electronic technologies as one contextual model to provide additional opportunities for students to pursue STEM-related experiences. While the results on student learning and attitudes were mixed the program does show promise in providing an exciting and personally relevant learning space for students. Like previous studies (Barker, Melander, Grandgenett, & Nugent, 2015;

Kafai, Fields, & Searle (2014) more females than males participated in the program, which provides additional evidence that wearable electronic textiles may represent an alternative pathway into STEM for traditionally underrepresented groups especially females. Once fully developed the WearTec program may be a useful tool for schools and other youth-serving organizations to attract student into STEM programming. In addition the program may provide teachers and educators much needed experience with integrated STEM projects to help increase their efficacy around STEM-based content.

Overall this chapter described the initial development and delivery of the Nebraska WearTec program and results of the associated pilot study examining learning and attitudes towards STEM. Wearable electronic textiles were chosen as the foundation of the WearTec program because of the potential to attract a new and diverse audience to STEM. From a theoretical standpoint electronic textiles are similar to educational robotics and both are situated in the constructivism learning theory whereby new knowledge is constructed through experimental practice in the world surrounding the learner (Piaget, 1971). The Nebraska WearTec program also relies on engineering centric problem-based learning where students solve interdisciplinary instructional challenges. The project elected to use the LilyPad ecosystem of components including the LilyPad Arduino-compatible microcontroller along with conductive thread due to their aesthetic appeal to younger students and the open-source programming environment. As the wearable electronic textiles industry as a whole matures and we continue to see an abundance of new and innovative products hit the marketplace it is our expectation that the WearTec project will develop a cohort of youth that are interested in developing their own inventions and become the next generation of STEM innovators.

ACKNOWLEDGMENT

This material is based upon work supported by the National Science Foundation under Grant No. ITEST #1433822 awarded to the University of Nebraska-Lincoln. Any opinions, findings, and conclusions or recommendations expressed in this material are those of the author(s) and do not necessarily reflect the views of the National Science Foundation.

REFERENCES

A framework for K-12 science education: Practices, cross- cutting concepts, and core ideas. (2012National Research Council. Washington, DC: National Academies Press.

Adams, R., Evangelou, D., English, L., Dias de Figueiredo, A., Mousoulides, N., Pawley, A. L., & Wilson, D. M. et al. (2011). Multiple perspectives on engaging future engineers. *The Journal of Engineering Education*, *100*(1), 48–88. doi:10.1002/j.2168-9830.2011.tb00004.x

Serving America's Future: Increasing College Readiness. (2012). American Association of State Colleges and Universities (AASCU), Task Force on College Readiness. Retrieved from http://www.aascu.org/CollegeReadiness/ExecutiveSummary/

Atman, C. J., Adams, R. S., Cardella, M. E., Turns, J., Mosborg, S., & Saleem, J. (2007). Engineering design processes: A comparison of students and expert practitioners. *The Journal of Engineering Education, 96*(4), 359–379. doi:10.1002/j.2168-9830.2007.tb00945.x

Barker, B., Melander, J., Grandgenett, N., & Nugent, G. (2015). Utilizing wearable technologies as a pathway to STEM. In D. Slykhuis & G. Marks (Eds.), *Proceedings of Society for Information Technology & Teacher Education International Conference 2015* (pp. 1770-1776). Chesapeake, VA: Association for the Advancement of Computing in Education (AACE).

Barker, B. S., & Ansorge, J. (2007). Robotics as means to increase achievement scores in an informal learning environment. *Journal of Research on Technology in Education, 39*(3), 229–243. doi:10.1080/15391523.2007.10782481

Barker, B. S., Nugent, G., & Grandgenett, N. F. (2008). Examining 4-H robotics and geospatial technologies in the learning of science, technology, engineering, and mathematics topics. Publication in the *Journal of Extension*, 46(3). Retrieved from http://www.joe.org/joe/2008june/rb7.shtml

Benitti, F. B. V. (2012). Exploring the educational potential of robotics in schools: A systematic review. *Computers & Education, 58*(3), 978–988. doi:10.1016/j.compedu.2011.10.006

Bevan, B., & Semper, R. J. (2006). Mapping informal science institutions onto the science education landscape. *Center for Informal Learning and Schools.* Retrieved from http://cils.exploratorium.edu/resource_shared/downloads/4229/Mapping%20Informal%20Science%20Instituions%20onto%20the%20Science%20Eductaion%20Landscape.pdf

Brotman, J. S., & Moore, F. M. (2008). Girls and science: A review of four themes in the science education literature. *Journal of Research in Science Teaching, 45*(9), 971–1002. doi:10.1002/tea.20241

Buechley, L. (2006). A construction kit for electronic textiles. *Proceedings of the 2006 10th IEEE International Symposium on Wearable Computers* (pp. 83-90). IEEE. doi:10.1109/ISWC.2006.286348

Buechley, L., Eisenberg, M., Catchen, J., & Crockett, A. (2008). The LilyPad Arduino: using computational textiles to investigate engagement, aesthetics, and diversity in computer science education. *Proceedings of the SIGCHI conference on Human factors in computing systems* (pp. 423-432). ACM. doi:10.1145/1357054.1357123

Buechley, L., Eisenberg, M., & Elumeze, N. (2007, June) Towards a curriculum for electronic textiles in the high school classroom. *Proceedings of the Conference on Innovation and Technology in Computer Science Education (ITiCSE)*, Dundee, Scotland. doi:10.1145/1268784.1268795

Buechley, L., & Hill, B. M. (2010, August). LilyPad in the wild: how hardware's long tail is supporting new engineering and design communities. *Proceedings of the 8th ACM Conference on Designing Interactive Systems* (pp. 199-207). doi:10.1145/1858171.1858206

Buechley, L., Peppler, K., Eisenberg, M., & Kafai, Y. (2013). *Textile Messages: Dispatches from the world for e-textiles and education.* New York, NY: Peter Lang.

Carnevale, A. P., Smith, N., & Stoll, J. (2010). Help wanted: Projections of jobs and education requirements through 2018. Washington, DC: Georgetown University, Center on Education and the Workforce. Retrieved from http://files.eric.ed.gov/fulltext/ED524310.pdf

Catsambis, S. (1995). Gender, race, ethnicity, and science education in the middle grades. *Journal of Research in Science Teaching*, *32*(3), 243–257. doi:10.1002/tea.3660320305

Clarke, K. (2010). A science, engineering and technology (SET) approach improves science process skills in 4-H animal science participants. *Journal of Extension*, *48*(1). Retrieved from http://www.joe.org/joe/2010february/iw3.php

Clewell, B. C., & Braddock, J. H. (2000). Influences on minority participation in mathematics, science, and engineering. In G. Campbell Jr, R. Denes, & C. Morrison (Eds.), *Access denied: Race, ethnicity, and the scientific enterprise* (pp. 89–136). New York, NY: Oxford University Press.

Engineering: Short on fact? [Editorial]. (2012, February). *Quality Progress*, 45(2), 14-15. Retrieved from http://asq.org/quality-progress/2012/02/keeping-current.html

Fallik, O., Rosenfeld, S., & Eylon, B. S. (2013). School and out-of-school science: A model for bridging the gap. *Studies in Science Education*, *49*(1), 69–91. doi:10.1080/03057267.2013.822166

Forman, G. E. (2010). When 2-year-olds and 3-year-olds think like scientists. Early Childhood Research & Practice, 12 (2). Retrieved from http://ecrp.uiuc.edu/v12n2/forman.html

Support the IANR Extension division. (2013). *NU Foundation*. Retrieved from https://nufoundation.org/unl/institute-of-agriculture-and-natural-resources/colleges-and-areas/unl/extension-division

Hmelo, C. E., Holton, D., & Kolodner, J. L. (2000). Designing to learn about complex systems. *Journal of the Learning Sciences*, *9*(3), 247–298. doi:10.1207/S15327809JLS0903_2

Husain, A. (2011). Problem-based learning: A current model of education. *Oman Medical Journal*, *26*, 295. doi:10.5001/omj.2011.74 PMID:22043442

Johnson, L., Adams, S., Cummins, M., Estrada, V., Freeman, A., & Ludgate, H. (2013). *The NMC Horizon Report: 2013 Higher Education Edition*. New Media Consortium.

Jonassen, D. (2000). *Computers as mindtools for schools. Engaging critical thinking* (2nd ed.). Saddle River, NJ: Prentice-Hall.

Kafai, Y., Fields, D., & Searle, K. (2013). Making connections across disciplines in high school e-textiles workshops. In L. Buechley, K. Peppler, M. Eisenberg, & Y. Kafai (Eds.), *Textile Messages: Dispatches from the World for e-Textiles and Education*. New York, NY: Peter Lang.

Kafai, Y. B., Peppler, K. A., Burke, Q., Moore, M., & Glosson, D. (2010, June). Fröbel's forgotten gift: textile construction kits as pathways into play, design and computation. *Proceedings of the 9th International Conference on Interaction Design and Children* (pp. 214-217). doi:10.1145/1810543.1810574

Katterfeldt, E., Dittert, N., & Schelhowe, H. (2009). EduWear: smart textiles as ways of relating computing technology to everyday life. In *Proceedings of the 8th International Conference on Interaction Design and Children* (IDC '09) (pp. 9-17). ACM, New York. DOI: http://doi.acm.org/10.1145/1551788.155179110.1145/1551788.1551791

Klegeris, A., & Hurren, H. (2011). Impact of problem-based learning in a large classroom setting: Student perception and problem-solving skills. *Advances in Physiology Education*, *35*(4), 408–415. doi:10.1152/advan.00046.2011 PMID:22139779

Lachapelle, C. P., Phadnis, P., Jocz, J., & Cunningham, C. M. (2012). The impact of engineering curriculum units on students' interest in engineering and science. *Paper presented at the NARST Annual International Conference*, Indianapolis, IN, USA.

Learning science in informal environments: People, places, and pursuits. (2009National Research Council. Washington, DC: The National Academies Press.

Lovell, E., & Buechley, L. (2010, June). An e-sewing tutorial for DIY learning. *Proceedings of the 9th International Conference on Interaction Design and Children* (pp. 230-233). ACM. doi:10.1145/1810543.1810578

Maltese, A. V., & Tai, R. H. (2010). Eyeballs in the fridge: Sources of early interest in science. *International Journal of Science Education*, *32*(5), 669–685. doi:10.1080/09500690902792385

Maltese, A. V., & Tai, R. H. (2011). Pipeline persistence: Examining the association of educational experiences with earned degrees in STEM among US students. *Science Education*, *95*(5), 877–907. doi:10.1002/sce.20441

Mann, E. L., Mann, R. L., Strutz, M., Duncan, D., & Yoon, S. Y. (2011). Integrating engineering into K-6 curriculum: Developing talent in the STEM disciplines. *Journal of Advanced Academics*, *22*(4), 639–658. doi:10.1177/1932202X11415007

McGrath, E., Lowes, S., McKay, M., Sayres, M., & Lin, P. (2012). Robots underwater! Learning science, engineering, and 21st century skills: The evolution of curricula, professional development and research in formal and informal contexts. In B. S. Barker, G. Nugent, N. Grandgenett, & V. I. Adamchuk (Eds.), *Robots in K-12 education: A new technology for learning* (pp. 141–167). Hershey, PA, USA: IGI Global. doi:10.4018/978-1-4666-0182-6.ch007

Mehalik, M. M., Doppelt, Y., & Schunn, C. D. (2008). Middle-school science through design-based learning versus scripted inquiry: Better overall science concept learning and equity gap reduction. *The Journal of Engineering Education*, *97*(1), 71–85. doi:10.1002/j.2168-9830.2008.tb00955.x

Melchoir, A., & Cohen, F. (2005). *More than Robots: An evaluation of the FIRST robotics competition participant and institutional impacts.* Waltham, MA: Center of Youth and Communities, Heller School for Social Policy and Management, Brandeis University.

Milestones in Computer Development. (n. d.) Retrieved from http://www.cis.usouthal.edu/faculty/daigle/project1/timeline.htm

Nicholl, T. A., & Lou, K. (2012). A model for small-group problem-based learning in a large class facilitated by one instructor. *American Journal of Pharmaceutical Education*, *76*(6), 117. doi:10.5688/ajpe766117 PMID:22919093

Noam, G. G., Biancarosa, G., & Dechausay, N. (2002). *Learning to bridge – bridging to learn: A model and action plan to increase engagement between schools and afterschool programs in Boston (Report commissioned by Boston's After-School for All Partnership Learning Goal Research)*. Cambridge, MA: Harvard Education Press.

Nugent, G., Barker, B., & Grandgenett, N. (2012). The impact of educational robotics on student STEM learning, attitudes and workplace skills. In B. S. Barker, G. Nugent, N. Grandgenett, & V. I. Adamchuk (Eds.), *Robotics in K–12 education: A new technology for learning*. Hershey, PA, USA: IGI Global. doi:10.4018/978-1-4666-0182-6.ch009

Nugent, G., Barker, B., Grandgenett, N., & Welch, G. (2014). Robotics camps, clubs, and competitions: Results from a US robotics project. *Proceedings of 4th International Workshop Teaching Robotics, Teaching with Robotics &5th International Conference Robotics in Education Padova (Italy) July* (Vol. 18, No. 2014, pp. 11-18).

Nugent, G., Barker, B., Grandgenett, N. F., & Adamchuk, V. (2010). Impact of robotics and geospatial technology interventions on youth STEM learning and attitudes. *Journal of Research on Technology in Education*, *42*(4), 391–408. doi:10.1080/15391523.2010.10782557

Nugent, G., Barker, B., Welch, G., Grandgenett, N., Wu, C., & Nelson, C. (2015). A model of factors contributing to STEM learning and career orientation. *International Journal of Science Education*, *37*(7), 1067–1088. doi:10.1080/09500693.2015.1017863

Peppler, K., & Glosson, D. (2013). Stitching circuits: Learning about circuitry through e-textile materials. *Journal of Science Education and Technology*, *22*(5), 751–763. doi:10.1007/s10956-012-9428-2

Peppler, K., Gresalfi, M., Salen, K., & Santo, R. (2014). *Soft circuits: Crafting e-fashion with DIY electronics*. Cambridge, MA: MIT Press.

Peppler, K., Salen, K., Gresalfi, M., & Santo, R. (2014). *Short circuits: Crafting e-puppets with DIY electronics*. Cambridge, MA: MIT Press.

Peppler, K., Sharpe, L., & Glosson, D. (2013). E-Textiles and the new fundamentals of fine arts. In L. Buechley, K. Peppler, M. Eisenberg, & Y. Kafai (Eds.), *Textile Messages: Dispatches from the World for e-Textiles and Education*. New York, NY: Peter Lang.

Piaget, J. (1972). *The child's conception of the world*. Towota, NJ: Littlefield Adams.

Qiu, K., Buechley, L., Baafi, E., & Dubow, W. (2013, June). A curriculum for teaching computer science through computational textiles. *Proceedings of the 12th International Conference on Interaction Design and Children* (pp. 20-27). ACM. doi:10.1145/2485760.2485787

*Rising Above the Gathering Storm, Revisited: Rapidly Approaching Category 5*National Research Council. (2010). Washington, DC: The National Academies Press.

Riskowski, J., Todd, C., Wee, B., Dark, M., & Harbor, J. (2009). Exploring the effectiveness of an interdisciplinary water resources engineering module in an eighth grade science course. *International Journal of Engineering Education*, *25*(1), 181–195.

Sorge, C. (2007). What happens: Relationship of age and gender with science attitudes from elementary to middle school. *Science Educator*, *16*, 33–37.

Sullivan, F. R. (2008). Robotics and science literacy: Thinking skills, science process skills and systems understanding. *Journal of Research in Science Teaching*, *45*(3), 373–394. doi:10.1002/tea.20238

Tanenbaum, J. G., Williams, A. M., Desjardins, A., & Tanenbaum, K. (2013). Democratizing technology: pleasure, utility and expressiveness in DIY and maker practice. *Proc. of CHI'13* (pp. 2603-2612). doi:10.1145/2470654.2481360

Weibert, A., Marshall, A., Aal, K., Schubert, K., & Rode, J. (2014). Sewing interest in E-textiles: Analyzing making from a gendered perspective. In *Proceedings of the 2014 conference on Designing interactive systems* (DIS '14), New York, NY, USA (pp. 15-24). Doi: http://doi.acm.org/10.1145/2598510.260088610.1145/2598510.2600886

Wiznia, D., Korom, R., Marzuk, P., Safdieh, J., & Grafstein, B. (2012). PBL 2.0: Enhancing problem-based learning through increased student participation. *Medical Education Online*, *17*(0). doi:10.3402/meo.v17i0.17375 PMID:22778540

KEY TERMS AND DEFINITIONS

Arduino: An open-source hardware company maintaining and developing a line of beginner-friendly microcontrollers as well as a simplified integrated programming environment (IDE).

Constructivism: A theory of learning that purports youths' new knowledge is actively constructed from the world around them through experiential practice and by independent/guided research to integrate new knowledge and to arrive at a deeper understanding.

Electronic Textiles: A field of engineering combining electronics and computing with textiles and design.

LilyPad: An ecosystem of sewable electronic components, developed by Leah Buechley while at the Craft Technology Laboratory at the University of Colorado, Boulder.

Microcontroller: An computation device that contains input and output for peripherals.

Problem-Based Learning: A student-centered learning model that uses ill-structured problems to encourage multiple approaches to develop a solution.

STEM: An acronym that stands for Science, Technology, Engineering and Mathematics.

Wearable Textiles: Electronic textiles or electronic accessories that can be worn; such as watches, eyeglasses, or clothing like a shirt or jacket containing electronics and a computation device (Johnson et al., 2013).

Chapter 15
Smart Device Clickers: Learning Basic Sciences and Biotechnology

Érika Bertozzi de Aquino Mattos
Universidade Federal Fluminense, Brazil

Alexander Gonçalves da Silva
Universidade Federal do Rio de Janeiro, Brazil

Isabelle Mazza Guimarães
Universidade Federal Fluminense, Brazil

Claudia Marcia Borges Barreto
Universidade Federal Fluminense, Brazil

Gerlinde Agate Platais Brasil Teixeira
Universidade Federal Fluminense, Brazil

ABSTRACT

In the traditional instructional paradigm, faculty members act like actors on a stage. They memorize their speech and deliver it to the audience, many times with very little to no interaction at all with the audience. On the other hand, in the student-centered learning paradigm, faculty members act like coaches interacting full time with their team. This chapter is based on a study conducted at a Brazilian Federal University. The study depicts the distance between science production and teaching, and reports on experiences using smart phone clickers to track and analyze students' content acquisition. The objective is to improve the interactive quality of teaching and learning, thus promoting steps to shift towards a student-centered instructional paradigm. Although smartphones were used in this study, with wearable technologies continuing to grow, other wearables such as smart glasses and smart watches could be used instead.

INTRODUCTION

Humankind has been improving technology since antiquity, enabling advances in knowledge beyond what is visible. The exploration of events invisible to the naked eye began with the vastness of the sky, diving into the cellular and molecular world, giving rise to Biotechnology. Within the technological advances, Biotechnology has made it possible to understand complex systems such as Molecular Genetics, Cellular Biochemistry, Immunology, Physiology, Pathology, etc. Therefore, the advancement of Science, Technology and more specifically Biotechnology has made it possible to comprehend the assemblage

DOI: 10.4018/978-1-5225-0069-8.ch015

of survival strategies in the array of living systems on this planet. (Flood, 1993) On the one hand, this context enabled the elucidation of a variety of biological "black boxes", and on the other, the amount of new abstract knowledge developed on a daily basis makes the study of molecules, cells, systems and organisms increasingly more complex.

Studying complex systems is not an easy task due to the number of interconnected facts to be known in order to properly understand the phenomenon. (Flood, 1993) This difficulty is many times intensified because basic Science subject matter is still mostly taught using traditional methodology, in other words, teacher-centered where memorization is the main evaluated aspect (Bradforth et al., 2015). In the teacher-centered scenario, students generally address learning in a passive form. They focus on the speech given by the teacher, memorize, and then repeat facts on tests, used to classify or rank students without being "assessed" by providing necessary feedback to congratulate achievement or indicate more or better learning strategies are required. Consequently, quite often information is quickly lost. On the other hand in the student centered paradigm students more often establish connections with what is said, understand the phenomenon reaching the "bottom line" (essence) of the content. Consequently, new knowledge is gained permitting students to draw their own conclusions (Chin & Brown, 2000) (Figure 1).

Analyzing the annual outcome report of the Immunology Department at a Brazilian federal university, the authors observed the need to initiate a change in the teaching paradigm of its faculty members due to a high student failure index. The initial study compared faculty teaching strategies with student outcomes. In summary, the study revealed the intentional use of strategies to overcome long learned misconceptions by one faculty members was fundamental to a better outcome of the first year medical students (Barreto & Teixeira, 2013). Based on this finding and after reading Joel Michael's (2007) paper "What makes physiology hard for students to learn? Results of a faculty survey." the researchers asked for authorization to adapt Michael's survey questionnaire. The survey was translated to Portuguese and adapted to identify the main factors influencing the teaching-learning processes of Immunology. Brazilian students were asked to answer the question: "What makes immunology hard for students to

Figure 1. Comparison between aspects of deep learning and surface learning
(Adapted from Chin & Brown, 2000)

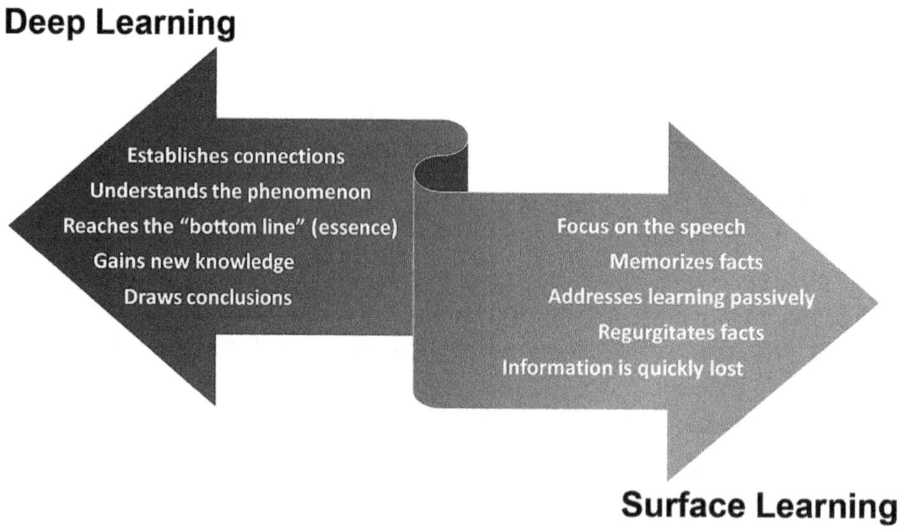

learn?" The main response from students' opinion was Immunology is hard to learn mainly due to the complexity of the content. As in Michael's results, in the Brazilian students' opinion teaching strategies were not the main problem.

Because of these results, the next step was to develop a specific strategy to further simplify learning of Immunology for faculty already using the student-centered learning paradigm and those still in the traditional instructional paradigm. The chosen strategy was the use of an Audience Response System (ARS) or clickers. Knowing the learning objectives for basic Immunology and that in most of the health courses, Immunology is a first year discipline; the authors prepared a group of questions to be used by the immunology department faculty during their classes. They were designed to be projected with short intervals to allow for interactions with classmates and answer the questions using an ARS. As soon as the class responds, the results of the poll is projected in a graphical format and the professor justifies why each answer is correct or incorrect (Bertozzi, 2015). Undergraduate students and attendees of an outreach immunology course positively evaluated the preliminary strategy. Thus, the aim of this chapter is to present clickers as a successful strategy to spark change from a lecture-centered (traditional) to a student-centered learning paradigm.

BACKGROUND

According to previous educational research, performed by Barreto and Teixeira (2013) the intentional and active clarification of concepts used in the presentation and discussions of scientific topics are very important. Thus, this chapter will start with the etymology of some central words important for understanding this work, as presented in the Concise Oxford Dictionary (Dictionary, 2002).

- **Technology**
    - From the Greek *tekhne* (technical, craft) + *logy* (study knowledge).
- **Biotechnology**
    - From the Greek *bio* (life) + *tekhne* (technical, craft) + *logy* (study knowledge).
- **Immunology**
    - From the Latin *imunis* (exemption of a service or duty by individual) + the Greek *logy* (study knowledge).
- **Methodology**
    - From the Greek *methodos* (goal, objective or purpose) + *hodos* (path or intermediary) + *logy* (study knowledge).
- **Education**
    - From the Latin *in* (on) + *signare* (mark or flag, designate and show things).
- **Clicker**
    - Derived from the word to click, meaning a sharp, short sound. The word clicker was created to indicate, among other things, the person standing on a store front counting the number of people entering. In scientific circles, the word was borrowed to refer to a diagnostic tool capable of revealing the thinking of students (Biological Concepts Instrument)(Klymkowsky, Underwood, & Garvin-Doxas, 2010).

- **ARS**
    - Acronym for Audience Response System. One of the most striking factors of the ARS is the speed of feedback. The results are immediately calculated and displayed in a graphical form as soon as the audience answers the question. It was first introduced at Stanford University in 1966 by Louis Abrahamson.

Let us start out by asking a few questions. After all, what is Technology? And Biotechnology? What is their importance to society? How long have these subject areas been present in human history? Why relate basic sciences with biotechnology? Furthermore, why study biotechnology in high school?

Currently, the term biotechnology is highly valued and emphasized, mainly due to its unlimited potential to serve and benefit humanity, in matters such as those relating to health, food and the environment. It can potentially promote a better quality of life and sustainability. Information on Biotechnology and events on the subject are increasingly frequent in the daily news (Verma, Agrahari, Rastogi & Singh, 2011). Thus, adequate press releases in academic publications for the community in general is vital since misconceptions may constrain the use of biotechnology in addressing imperatives of development and escalating challenges from nature, especially in less-industrialized nations.

The word Biotechnology was first used in 1919 by the Hungarian engineer named Károly Ereky, in his book "Biotechnologie", defined as "the application of principles of engineering and life sciences to create new products derived from biological raw materials, such as in the production of vaccines or nutritional improvement of foods " (Fári & Kralovánszky, 2006). Another and more recent concept of biotechnology is "the application of scientific and engineering principles to produce valuable substances through biological agents" (Kumar, 2015).

Although the word biotechnology was only coined in the early twentieth century (Pele & Cimpeanu, 2012), when considering the idea of using living organisms to generate products or services, we can consider biotechnological processes have been developed from ancient times to the present day. According to Verma et al. (2011) biotechnological developments can be systematically divided into three categories:

- **Ancient Biotechnology:** Marked by empirical observations (from antiquity to the end of the XVII century);
- **Classical Biotechnology:** Characterized by early scientific work (of the late seventeenth century to almost half of the twentieth century);
- **Modern Biotechnology:** Characterized by molecular biology (from the mid-twentieth century to present).

Using this framework, the first biotechnological discoveries of antiquity were based on observations of nature and had direct consequences on the improvement of human life, especially with regard to food, clothing and housing. Human beings, like their ancestors, initially fed on the wild biological diversity surrounding them and eventually started to plant crops and domesticate animals. In this early period of farm development, farmers unconsciously began to perform artificial selections of plants and animals. The crops, close to the houses, allowed the gradual improvement of plants. Crop variety increased as farmers started to perform the selection of desirable characteristics such as shorter growth periods; increased resistance to diseases and pests, larger seeds and fruits, changes in nutritional content, longer shelf life, crops better adapted to the different ecological conditions where they were planted, etc. (Pele

& Cimpeanu, 2012; Wieczorek & Wright, 2012). Likewise, the domestication of wild animals caused a gradual variation in their characteristics in order to be suitable for humans. Among others, larger chickens and smaller wild cattle were selected, sheep with better wool quality and mammals with greater and better milk production and / or meat were bred, and horses with greater physical force were raised to assist in the preparation of land for cultivation (Diamond, 2002; Verma et al., 2011).

Roughly, 3500 years ago, once plants and animals with potential food properties were "domesticated", produced and stored, humankind found time for new observations. One of these observations was fruits and grains, when left covered, altered over time their state resulting in the production of different types of beverages such as wines and beers. The word fermentation is derived from the Latin *fervere*, which means "to boil" due to the fact that crushed grapes set in tanks, overtime, produced bubbles as if the mixture was boiling (Alba-Lois & Segal-Kischinevzky, 2010). Even without understanding the way the recipes work, systematic observations culminated in the discovery of the importance of temperature and air exposure for success in the production of these brews, resulting in the improvement of fermentation (Alba-Lois & Segal-Kischinevzky, 2010).The discovery of a variety of fermentation processes, using microorganisms led to a new phase in human feeding style (Pele & Cimpeanu, 2012).

The invention of the microscope enabled scientific discoveries associated with the study of living unicellular organisms invisible to the naked eye as well as the study of the cells of multicellular organisms. One of the main actors at this stage was Robert Hooke who studied many very small animals and plants and illustrated them in his book Micrographia, published in 1665 (Hooke, 1961). A few years after Hooke's publication, a Dutch tradesman called Antony van Leeuwenhoek developed his own microscope with high quality lenses, leading to the discovery of a new world yet smaller. The invention and the improvement of the microscope led to the visualization and description of different microscopic beings such as fungi, bacteria and protozoa, allowing the development of microbiology (Martins & Pereira, 2012; Pele & Cimpeanu, 2012). This, in turn, permitted solving a diversity of Ancient Biotechnological puzzles and consequently opened the path for new discoveries (Verma et al., 2011).

It was believed the particles floating in the wort (the liquid obtained during the initial brewing process of beer and whiskey) were starch grains until, according to Leeuwenhoek's observations, it was discovered these particles were in fact cells (Van Leeuwenhoek, 1722). Still in the XVIII century the substance placed in the preparation of drinks to make it bubble or into dough to lighten and puff the breads was called "leaven". Thus, yeasts were seen as organic chemicals needed only for fermentation, not realizing they were living organisms and fermentation was the result of their metabolism. The scientific study of fermentation began in the eighteenth century, when chemists began to decipher the nature of the alcoholic fermentation by analytical chemistry and chemical nomenclature (Alba-Lois & Segal-Kischinevzky, 2010).

Antoine Lavoisier a French chemist (1743 – 1794) was interested in the basic theoretical questions on the transformation of substances, such as the transformation of sugar into alcohol and carbon dioxide during fermentation. To answer this question, Lavoisier measured the proportion of sugars and water early in the chemical reaction, added yeast (yeast paste) and compared the proportions of alcohol and carbon dioxide obtained at the end of the experiment. From his results, Lavoisier concluded that sugars were broken down and formed two-thirds alcohol and the other third was oxidized to form carbon dioxide (the source of the bubbles observed during fermentation). Lavoisier noted in a closed system the mass of the total end product equals the total mass of reagents, creating the principle of mass conservation (Alba-Lois & Segal-Kischinevzky, 2010). At that time, although the experiments provided a clear vision of the basic chemical reactions necessary for the production of ethanol, the role of yeast in the process had not yet been understood. The chemists assumed the yeast remained unchanged during the chemical reactions and they would be responsible for the start of the fermentation process.

However, in 1815, the French Joseph-Louis Gay-Lussac took a step further in the elucidation of the phenomenon. The experiment was projected to first prevent fermentation of grape juice and then add yeast to prove its necessity. Thus, immediately after boiling the juice, it was put in a hermetically sealed container. The juice remained unfermented while it stayed without air contact. At the end of the experiment, yeast was added, which led to a fermented drink and the conclusion that yeast is fundamental for the fermentation process (Alba-Lois & Segal-Kischinevzky, 2010). However, the fermentation process was only better understood in 1858 when Louis Pasteur another French scientist performed experiments to find out what led to "milk sickness". Pasteur concluded milk became sour, because microbes transformed sugar into lactic acid, for which he gave the name lactic fermentation. Following this line of thought, the term "wine disease" was used to explain the fermentation of grape juice by microbes. As a consequence, a hypothesis was established relating microbes responsible for the fermentation process to infectious diseases (Gillen & Sherwin III, 2008).

Another scientific black box was how biological characteristics descend from one generation to the other. Gregor Mendel, published in one of the most prestigious scientific magazines of the time (1865), the Brunn Society of Natural History his important discovery, proposing the transmission of biological characteristics occurring through invisible heritage units, designated as factors (Mendel, 1996). Even though the journal had a significant impact on science, Mendel's work was ignored until about 1900 when three different scientists, de Vries, Correns and Tschermak-Seysenegg, obtained independently the same results confirming Mendel's findings (Freire-Maia, 1995).

This was one of the greatest discoveries in the history of science and a milestone of genetics as we know it today (Verma et al., 2011). Another very significant event for biotechnology and biological science in general, was the elucidation of the DNA structure by Watson and Crick (1953), the hallmark of modern biotechnology and advances in scientific research, such as the production of new drugs, new treatments and modern biotechnological techniques (Goeddel et al., 1979). In synthesis, since antiquity, biotechnology has enabled advances in knowledge beyond what is visible and currently allows the understanding of living beings at the systemic, cellular and molecular levels. On one hand, these advances have allowed the clarification of several black boxes of biology; on the other, it produced a colossal quantity of abstract knowledge making the study of cells, biological systems and whole organisms a lot more complex.

This short history of science was provided to demonstrate the time lapses between one important discovery and another are getting shorter and demonstrate when discoveries do not fit into the mainstream paradigm, or the community is not ready for change, new ideas are often set a side. In parallel to the evolution of Biotechnology a realm of technological developments have happened including new technologies in information and communication (ICT). During the XVIII and XIX centuries critical discoveries were made during a lifetime and today important discoveries occur on a daily basis, some leading to important paradigm shifts. In addition, contrasts between today's speed of basic Sciences and Biotechnology progress, the development if ICT, and the stability of Science Education, which is a lot slower in incorporating pedagogical innovations and appropriate methods of student learning assessments including the use of e-learning (electronic-learning), m-learning (mobile learning) and u-learning (ubiquitous-learning). Agreeing with Morita (2003) we believe that mobile phones, and in special smart phones, are the most wearable technology diffused in today's society thus m-learning could provide very flexible learning opportunities to learners. However, in the Brazilian scenario, the use of teaching strategies that include wearable devices is not frequently used in the K-12 or college classroom. This scientific scenario reflects in the learning challenges basic school and university students have to deal with.

SCIENCE TEACHING OR SCIENCE LEARNING? WHICH IS THE AIM OF SCHOOL?

Science Teaching in Brazil

Science teaching has undergone numerous changes in recent years. Due to the scientific and technological advances during the last century, information is generated and distributed faster every day. This context leads to two situations: the first a globalized world where information is instantly transmitted to any party at any time, and second to the impossibility of expecting students to memorize all details in their curricula. It is now clear science and technology are important ingredients for economic, cultural and social development, thus within the educational context they become the object of many changes in teaching. As a result, it is necessary to develop critical citizens competent in finding information that is adequate and discarding what is not (Krasilchik, 2000).

Influenced by the Cold War in the 1960s, the United States invested in the production of educational projects in Physics, Chemistry, Biology and Mathematics, aiming to encourage young people to follow scientific careers, thus ensuring the hegemony through winning the space race. In Brazil, the need to prepare students for the enhancement of national scientific and technological development was important to cease absolute dependence on other countries. Moreover, as political changes took place in the country, the basic school became responsible for training all Brazilian citizens (Krasilchik, 2000). With the promulgation of the 1961 educational bill (Lei de Diretrizes e Bases n.4.024), a shift occurs and science education starts at the 5th grade and not only in high school. At that time, all teaching was lecture based, with student assignments designed to mechanically absorb all information. The quality of the courses were evaluated by the amount of content delivered and not by the actual learning of students. In addition, to aggravate the scenario, scientific knowledge was taken as neutral and scientific truths were not questioned (Brasil, 1997). In the 1970s a new educational bill (lei n.5.692/1971) set Natural Science as a mandatory discipline as of the first grade. Since then, the energy crisis, the environmental problems and health issues gained a major role in the science education curriculum. The issues were addressed at different depths and relevance according to the educational level of the students (Brasil, 1997).

In short, until the 1970's science education had a passive character. The student's prior knowledge was not taken into consideration, the teachers gave classes mechanically and students memorized the subject matter. As of the 1980's, there was growing concern with the teaching-learning process and basic education took on a constructivist trend, considering, for example, the existence of misconceptions, intuitive and alternative concepts previously not recognized. This new educational scenario brought to the school context the concern over significant learning, students motivation in the classroom, the contextualization of the contents to daily life in society and the search for educational tools to facilitate learning (Bellanti, 1978). In the official documents of the 1980s, there was the caution to establish a transition from teaching based on memorization to a more reflective teaching. However, after 25 years, this form of teaching-learning process has not yet been fully established in schools and thus is still far from the ideal of 100% of schools, public and private, encouraging reflective, critical and scientific thinking.

Contradictions in Teaching Methods

The traditional teaching methods are based on the theory that teaching consists of using logical strategies allowing everyone to learn everything. The teacher plays a major role and uses logic typical for adults, fully matured and developed intelligence, even when learners are children and adolescents (Veiga, 1991). On the other hand, some of the pillars sustaining the active teaching methodology respects individual differences, such as the student's rhythms and potential. Further development of the students learning abilities are triggered not only by activities (learning by doing, experiencing and observing) but also by the integration of content by both teachers and students. Using this concept, students becomes the center of the educational / training process and teaching focuses on the process of acquiring attitudes, skills, and competencies to encourage interactions between the subject and the elements involved in the educational process (Veiga, 1991, Barr & Tagg, 1995). In table 1 the author demonstrates the duality in teaching paradigms based on a synthesis performed by Barr and Tagg and published in 1995.

The social interactionist perspective proposed by Vygotsky seeks to explain how individuals develop knowledge emphasizing the role of social interactions between the developing child and linguistically knowledgeable adults. According to this view, the higher mental processes (thinking, language, choice behavior, reasoning, abstraction, representation, etc.) are learned during these social interactions mediated by the tools and symbols built on social, historical and cultural aspects (Moreira, 1999, Mortimer & de Carvalho, 2013). Thus, the constitution of the characteristics of each individual, such as personality, habits, ways of acting and mental capacity, are influenced by the social environment they pertain to (Rego, 2000).

Active teaching methods are still a challenge in the Brazilian formal educational institutions (Andrade, 2011). This is the reality for K-12 education and unfortunately worsens in higher education. To be a

Table 1. Duality in teaching strategies (Bertozzi, 2015 adapted from Barr & Tagg, 1995)

Emphasis on Teaching	Criteria	Emphasis on Learning
What do faculty want to teach?	Standing Questions	What do students need to learn?
Deliver instructions	Aims	Stimulate learning
How can faculty cover it all?	Designated learning objectives	How can students complete course material?
Transfer knowledge to students	Teacher's role	Elicit student discovery and knowledge construction
Classify and sort students	Assessment	Feedback on acquired competencies
Quality of lectures	Success criteria	Quality of learning
Quality and quantity of teaching resources	Assets	Quality and quantity of student outcomes
Accumulated credit hours	Degree	Demonstrated knowledge and skills
Any expert can teach	Assumption about teachers	Teaching is complex and requires considerable training
Independent/department based on collaboration	Disciplines	Independent/department collaborate

faculty member at a Brazilian university, the mandatory requirement is to have a graduate course. All graduate courses in Brazil have a compulsory teaching training program. However, since many of the professors have never had formal training in different teaching methods, their graduate students continue to reproduce the traditional lecture based teaching procedures. Thus, without a theoretical basis to guide the use of alternative methods appropriate for the target audience (students) they repeat old methods, which were used during their formal education (Veiga, 1991).

Teacher-student interactions affect students in some form, consequently a scenario should be constructed by the teacher to not make learners feel intimidated and apprehensive. When the teachers' attitudes and strategies are the desired ones such as having clarity of exposure within a respectful student-teacher interactions, learning is facilitated. Although, this is not the mainstream attitude found for many educational settings (Tassoni, 2000). Setting the stage for the advancement of Science and the educational system the author presents four questions to stimulate teachers to align educational development with scientific development.

- How can you stimulate students' interest in participating in class?
- How quickly and effectively can you evaluate whether a particular subject needs further explanations before proceeding to the next?
- How can you make sure students learned what was expected for the program content?
- How can you determine if strategy changes are needed to achieve the learning objectives before the end of the course?

In regards to the questions, the use of the traditional lecture structure as the single method is dominant in many universities worldwide, especially when classes are large. The correct answer is - Use a diversity of active teaching methods available in the literature most adequately aligning to students, goals and objectives. Traditional lecture style classes often fail to promote meaningful learning and intellectual engagement of students (Smith et al., 2005). However, when faculty actively seek new information, organize it in a meaningful way, and explain it to students (Huba & Freed, 2000), they are the ones who learn most because they have reserved for themselves the conditions necessary to promote learning. Therefore, both active involvement and efficient feedback constitutes equally important elements in the learning processes (Trees & Jackson, 2007). This being said, teachers and faculty should promote learning settings allowing students to achieve what they (the teacher) have already achieved in regards to increasing subject matter knowledge.

Active Learning Strategies

There is an array of different active learning strategies to use for different scenarios. However, most present better results in classes with a small number of students because it is easier for the teacher or faculty to maintain student control. As the scope of this chapter is to discuss the use of ARS, this section will be limited to the presentation of a few active learning strategies most closely related to the use of clickers.

Encouraging classroom discussions with peers either in small groups or in conversation circles is an interesting active teaching method. However, these may not result in the expected learning outcomes if some students overwhelm those who have trouble with the subject. Some students may be too shy to participate because of a fear of embarrassment. Therefore, an atmosphere of trust and support allowing students feel they can give their opinion without being intimidated is required to engage them in an open discussion. When the teacher is able to construct an atmosphere of confidence and conduct profitable discussions even shy students will put forward their ideas providing moments of reflection balancing out those who always come forward and give their opinion. This stimulus is extremely important since students learn best when they are actively engaged. Thus, the teacher's role in this context is to provide the appropriate stimulus to generate and mediate the discussion. (de Barros, Remold, da Silva & Tagliati, 2004; Tenreiro-Vieira, 2004; Berbel, 2012).

Active teaching methods may be useful to disrupt strongly rooted misconceptions (Michael, 2006). When the teacher / faculty knows the current conceptions in a population and encourages students to present their beliefs these may be put into discussions. The different opinions may be analyzed and the current scientific concepts introduced and learned (Barreto & Teixeira, 2013). Once convinced the information learned up to this point is not always correct, students tend to critically question the information given, its origins and possible biases contributing to the learning potential of the class as a whole. Another important aspect of this scenario is in regards to self-confidence students acquire to present their doubts and criticisms in the classroom. In classes where open discussions are not yet the best alternative, the teacher can use a variety of strategies to get answers to questions made by students. Below you will find some instructional examples.

- Set a question or a small problem. ➔ Ask students to write a short answer to the problem. ➔ Collect all answers and redistribute them so students do not know whose answer is in their hands. ➔ Ask a student to read aloud the reply given to him or her. ➔ Repeat the process by asking another student to read the reply in their hands until all answers have been read. ➔ Count the frequency of each type of answer. ➔ Discuss all of the alternative answers justifying why they are correct or not.
    - In this setting, criticism to responses with misconceptions usually occur with a lot of tranquility
- A variation of this strategy is to first ask a student to read aloud the reply given to him or her. ➔ Then ask how many have the same answer as the first one read. ➔ If all agree, then discuss the concepts the students have by justifying the appropriateness or not of the answer. ➔ If not all agree with the first reply, ask another student to read an alternative answer. ➔ Repeat the process by asking if there are other answers and ask another student to read the reply in their hands until all alternative answers have been read. ➔ Count the frequency of each type of answer. ➔ Discuss all types of answers justifying why the answers are correct or not.
    - In this setting, a greater maturity of the topic or concept being studied is needed. Students will need to interpret whether peer written responses are adequate or not, i.e. they need to perceive misconceptions and incomplete answers. In this scenario students usually place their critics with a lot of tranquility.
- A third variation of this strategy is, after writing the answer, ask students to read their own answers. The teacher can coordinate which students will answer or let the students set themselves forward and read their answer. ➔ After the first opinion is placed, the strategies are the same as the two previously described.
    - In this situation, a more mature relationship with the technique and maturity is required, for the critic placed by the peers is not anonymous.

Yet another interesting strategy is to perform polls, for example, asking "how many in the class believe...". This usually yields a quick overview of the student's opinion on the subject matter; however, peers can influence those students not having a firm opinion on the answer. In one example, of having students raise their hands Bonaiuti, Calvani and Piazza (2015) showed that 18% of the students never raised their hands during the lessons, 23% participated only once, 18% twice and 15% three times or more. To circumvent the problem there are a few poll strategies to be used from the most simple to the most sophisticated, from the technological point of view. For example:

- One of the simplest variations of the poll strategy is the use of multiple-choice questions in which letters or numbers, corresponding to the answers, are substituted with colored symbols. Each student receives a set of cards with the matching colors. In this strategy, a question is exposed to the students, after a brief discussion with a classmate, the teacher projects the alternative answers. Each student chooses independently the corresponding colored card for the answer he or she believes is correct. When prompted, all student lift the chosen card and the teacher or students count how many chose each alternative. The last step is to discuss each appropriate and inappropriate alternative permitting all students to understand why and what is accurate or not in each alternative.
    - The teacher should intentionally choose how to discuss the topic. He or she may give the answers directly or request one of the participants, who gave a certain response, to justify the chosen answer. The order of the alternatives, from the most appropriate to the most inadequate, the most frequent to the least frequent, or other criteria the teacher feels is the most appropriate for the specific group of students should be intentionally chosen.
- A more sophisticated system is the use of remote handsets connected to a computer system and associated with the teachers' presentation. Using the same strategies as the colored cards the teacher can set a question, reveal immediately the answers to be chosen or give students time to discuss with classmates what is being asked and then show the answers. In this setting, there is no need to count the number of students electing each answer. The system does this automatically and the results are then projected on the screen. From this point on, the teacher again decides how to deal with correct and incorrect answers.
- A further evolution for polling is the use of wearable technology such as smartphones, smart glasses and smart watches. The strategies are essentially the same with more possibilities for interpretation of the data and the type of questions being posed to students.

The Audience Response System (ARS) as a Didactic Resource

As described by Louis Abrahamson (2006), ARSs were first introduced at Stanford University in 1966, but were expensive at that time and did not function well because they were difficult to use. Two decades later, Abrahamson himself with two colleagues, Fred Hartline and Milton Fabert, developed and tested a new prototype called Classtalk I. They built the first prototype from surplus Atari keypads, which included an additional communication circuit board, a LED display, and was connected to the teacher's computer by a special-purpose digital multiplexer. In general, students and teachers received the new version well. Although the cost was prohibitive for widespread distribution, ARSs became commercially available in the 1990's. At the turn of the century, a new generation of more affordable, infrared ARS became available stimulating their use and as a consequence of their success, today, numerous secondary schools, colleges and universities use this tool (Abrahamson, 2006). With the introduction of smartphones and the availability of a number of different apps in the last decade this strategy has given another turn.

One of the most striking factors of ARS, from the beginning, is the speed of feedback. The results are immediately calculated and displayed in a graphical form as soon as the audience answers the question. Another important factor is the feedback can be both anonymous or, depending on the app, it is possible to identify each student for teaching intervention purposes (Kay & LeSage, 2009). As one of the main advantages of this feature is to obtain immediate feedback on how the class understands the concepts presented by the teacher, it is possible to immediately modify the course of instruction and students can work on misconceptions by discussing in large or small groups in class. Thus, the ARS is an interesting didactic resource permitting the diagnosis of doubts and issues students are having difficulty dealing with during class. This technology has positively impacted students from several universities in the United States, Australia and England (Yourstone, 2008) even though students to not necessarily realize its effectiveness (Lantz, 2010).

There is a growing search in higher education for teaching strategies less teacher-centered and more student-centered, thus interactive teaching resources are very welcome. The known limitations of traditional lecture classes, such as attention span and the inefficient acquisition of knowledge for many listeners are some of the reasons for this change (Biggs & Tang, 2011). However, as pointed out by Judson (2002), the key to student comprehension is not the incorporation of the technology but rather the pedagogical practices of the instructor. ARS are viewed as tools holding the promise of facilitating earnest discussions, thus, smart clickers may be a good strategy used in a variety of ways, to increase interactions among students, knowledge acquisition awareness, promote discussions, assess students' knowledge, and teachers can adjust the pace of teaching and their didactic strategies (Kay & LeSage, 2009). Most groups report that the use of smart clickers can promote interactive learning and student engagement, particularly with a large number of students in class, associated with a better acquisition of both short and long-term knowledge (Ten Cate & Durning, 2007; Kay & LeSage, 2009). Faced with the versatility of this teaching resource, the need for pedagogical innovations and the importance Biotechnology has had in recent decades, led to another question: What are the pros and cons of using ARS? To answer this question, Immunology was used as an example.

In Brazil the handheld remote devices are still too expensive, thus smart phone ARS (applications or websites) were selected for our educational needs. The following criteria was used for this search: Availability for Android, iOS and Windows Phone operating systems, free download, free use, availability in Portuguese or English, and versatility (if the application had different functions and allowed the development of both open and closed questions).

The starting point for the study was searching for existing ARS applications for all smartphone operating systems. The search was conducted in the App Store, Google Play and Marketplace with the keyword "clicker", using "Education" as a filter for the App Store and "Applications" in Google Play. When an application was found for only one of the operating systems, a targeted search was conducted to determine if it is available for additional platforms. For this, a new search was performed using the applications' name. After the first search, 21 applications were found at the App Store, 4 applications at Google Play and only 2 applications at Marketplace. All the applications found on Google Play and the Marketplace were also found in the App Store. Thus, after performing the second survey using the specific names of the applications found in the App Store, 4 more applications were found at Google Play. As the only two applications available for all the three operating systems did not meet the pre-established needs, a search for sites of ARS was held for Internet sites.

In the search for sites offering this feature, the keywords "clicker" and "audience response system" were used in Google, which returned approximately 15.700 results. Browsing through the titles of the first page, the authors came across a site presenting a comparison between five ARS (http://www.polleverywhere.com/ars-comparison). Although a comparison was not performed for other ARS, the following list was provided: SMART Senteo, Promethean/Logical Choice Activote, Quick Tally, Pearson/Scantron Engage, TI-Navigator, iRespond, Ortek, Interactive Horizons IRIS, iRespond, GeneeUs, EduGame, Dolphin Inteactive Presenter, CLiKAPAD, iTech Keypads, Sun-Tech, or XTOL. After performing the comparisons and preliminary tests, the authors decided that Poll Everywhere (http://www.polleverywhere.com/) attended to the needs and no further search was conducted. This ARS application presents a free plan for audiences or classes of up to 25 participants, is extremely versatile, intuitive, and has a good design. An extra advantage of this site is the ability to use the resource directly in the slide viewer through the installation of a Power Point ® add-in. Moreover, this site is very user friendly and offers videos and detailed step-by-step instructions to help users to become familiar with it quickly.

Having chosen the application, a set of multiple-choice questions were prepared to be used by the faculty of the Immunology Department during their classes. The questions were validated during a Basic Immunology Outreach Course offered for K-12 teachers and health professionals. These were projected and participants had a short interval to interact with a classmate and then answer the questions using their Smartphone as an ARS. As soon as the class responded, the results of the poll was projected in a graphical format and the faculty justified why each answer was correct or incorrect (Bertozzi, 2015).

Preparing Multiple Choice Questions Used With ARS

As a starting point, the authors used a questionnaire prepared by Professor Claudia Barreto, which is the result of a five-year survey applied to first year health-professional students on the first day of the basic immunology course. She asked her students to present their concepts for the seven words considered central to immunology, pertaining to the vocabulary high school students should acquire as expected by the Brazilian National High School Standards. These standards include aspects such as understanding the importance of decision making related to their personal and family health issues, such as taking or not vaccines, the use of anti-inflammatory drugs, the need for the correct use of antibiotics, understanding and respecting medical recommendations.

Using content analysis as described by Stemler (2001) she came to the four most frequent categories for each concept. The answers reflect the immunological concepts taught at the high school level. Once the multiple-choice questions were prepared they were used to evaluate the prior knowledge of the students substituting the open-ended questionnaire and as a progress test at the end of the discipline of Immunology at a Brazilian Federal University (Barreto, 2011). The seven keywords are Antigen, Antibody, Inflammation, Infection, Immunity, Hypersensitivity and Vaccine. Four questions examples are shown depicted in Figure 2.

The Didactic Strategy Used in the Study

On the first day of the outreach course in basic Immunology, a brief explanation on the use of the clicker was given to attendees and after clearing all doubts, accessing the web or downloading the app the course started with the lectures "Immunology what is it all about" and "History of Immunology". Typically, the organizing group is composed by the person to give the lecture and one or two assistants who are in

Figure 2. Example of four questions presented during a basic immunology outreach course
(Source: Bertozzi, 2015, adapted from Barreto, 2011)

Which of the following is the most appropriate definition for "antigen"

a) Part of the infectious agent recognized by the antibody
b) A foreign body or invader of the body
c) A molecule that interacts with lymphocyte receptors
d) Action of an offending agent in an organism

Which of the following is the most appropriate definition for "vaccine"

a) Product that generates passive humoral response
b) Active immunization for the production of antibodies
c) An antigen preparation to elicit primary immune response
d) A method to actively immunize an organism

Which of the following is the most appropriate definition for "antibody"

a) A defense cell of the body specific for an antigen
b) A molecule secreted by plasma cells that mediates the elimination of antigens
c) A molecule whose production is induced by an antigen, and is able to react with it
d) A protein produced by B lymphocytes

Which of the following is the most appropriate definition for "immunity"

a) When the body presents memory and produces antibody against a particular antigen
b) Specific resistance to infectious diseases
c) Body's natural defense against infectious agents
d) Defense mechanism promoted by lymphoid cells and their products

charge of delivering the questions to be answered by the attendees with the ARS. The central topic of the first lecture was the maintenance of the body's homeostasis and then the reasons many researchers approach the Immune System dividing it as Innate and Adaptive systems were given. The major role of the Immune System's function, as believed by today's scientific community and the concepts of antigen, antibody and vaccine were addressed as closing topics.

The assistants presented questions to be answered at strategic moments. The decision was taken every time they needed to know if attendees adequately understood the concepts just delivered or if the prior concept of the next topic was to be documented. The aim of the outreach course was to update high school teachers and health professionals not only on biotechnology and immunology but also on active learning strategies. Thus, the organizing group used different ARS strategies. For example, the question covering the concept of immunity was applied before it was introduced in the lecture to evaluate the attendees' preconceptions, and thus perform the necessary interventions, while the other questions were inserted into the lecture slides after the topic was delivered to analyze whether the subject was adequately learned. After providing the answers and projecting the resulting graphs all possible answers were discussed with the group.

Although most participants responded appropriately that the body's homeostasis is a phenomenon occurring due to the interaction of the endocrine, nervous and immune systems one third of the participants did not include the immune system as responsible for the maintenance of homeostasis. The idea the immune system is a unique defense system can be related to the origin of the study of the system in medical microbiology, leading to the interpretation of immunological phenomena in defensive terms related to illnesses (Silverstein, 1982). These results agree with the paper "Immunology Teaching in Medical Education: Lessons from Akira Kurosawa," in which the authors show that Immunology textbooks from 1999 to 2006 have fragments written in a martial paradigm stressing the defense point of view. Therefore, it can be expected that teachers now acting in basic education are teaching according to their higher education training, where they had contact with books reinforcing this martial perspective of the immunological system. Consequently, a vicious cycle is established. After the explanation of the reasons of the adequate and for each of the inadequate options the lecture continued. At the appropriate moment, the second question covering the functions of the immune system was shown. All participants answered the question adequately, showing that those that did not include the immune system in the maintenance of the organism were able to replace the martial vision of the immune system with a physiological vision. After the individual questions were discussed a correlated discussion of answers to questions about the systems responsible for homeostasis and on the function of the immune system was then established.

When asked about the appropriateness of the separation of the innate and acquired immunological system 90% of the participants responded the division is inadequate. Among these, 40% adequately responded there is no dichotomy in the body and the cellular and molecular components are interdependent, although 10% still believe in the dichotomy. When learned concepts make sense, even if they are misconceptions, they are hard to remove or to substitute. Therefore, it is very important for faculty to intentionally break the vicious cycle approaching common misconceptions independently of the subject matter (Siqueira-Batista et al., 2009). According to Hoebe, Janssen and Beutler (2004), this dichotomy is not limited to only the didactic approach. It is also seen in the research scenario, dividing immunologists studying the so-called innate and adaptive systems using methods to justify their concepts. There are even international meetings for each of the research groups reinforcing these concepts. How can one expect the predominance of the integrated view of the immunological system if those producing new research data are still in a dichotomized paradigm? Some participants explicitly demonstrated their difficulty in viewing the immunological system in an integrated way. Thus, approaches and teaching resources used to facilitate the integration are needed. As for the Immune system, other content matters have similar problems needing to be addressed with intentional intervention through active learning strategies (Figure 3).

To identify the preconceptions of the participants and exemplify the didactic strategy, the next question (on the concept of immunity) was applied before the explanation (Figure 4). The most appropriate response among the options was "a specific resistance to infectious diseases", which was not chosen by any of the participants in the course. When such feedback is obtained, the teacher and the students can work on the misconceptions during the lecture discussion (Kay & LeSage, 2009). However, the change of convictions and updating of scientific concepts, from scientific and reflective thinking depend on the pedagogical competence of the teacher (Barreto & Teixeira, 2013). As discussed above, Immunology needs to be understood in the context of the function and interaction of its components and therefore, the appropriate understanding of particular concepts will certainly influence the understanding for others. For example, the concepts of immunity and vaccination are intimately related since the principle of vaccination is to generate immunity.

However, the term vaccine is commonly used for "desensitization" of allergic individuals, only because antigens are inoculated usually subcutaneously. Performing a search in the Pubmed database using the keywords desensitization, allergy and vaccine, over 500 papers are retrieved. An example of this can be observed in a paper published by Soyka, van de Veen, Akdis and Akdis (2014) in Chest, a medical journal. The paper "Scientific Foundations of Allergen-Specific Immunotherapy for Allergic Disease" presents de following statement: "Increasing safety while maintaining or even augmenting efficiency is the main goal of research for novel vaccine development and improvement of treatment

Figure 3. Three examples of questions used to determine participants' perceptions of the Immunological systems function
(Source: Bertozzi, 2015)

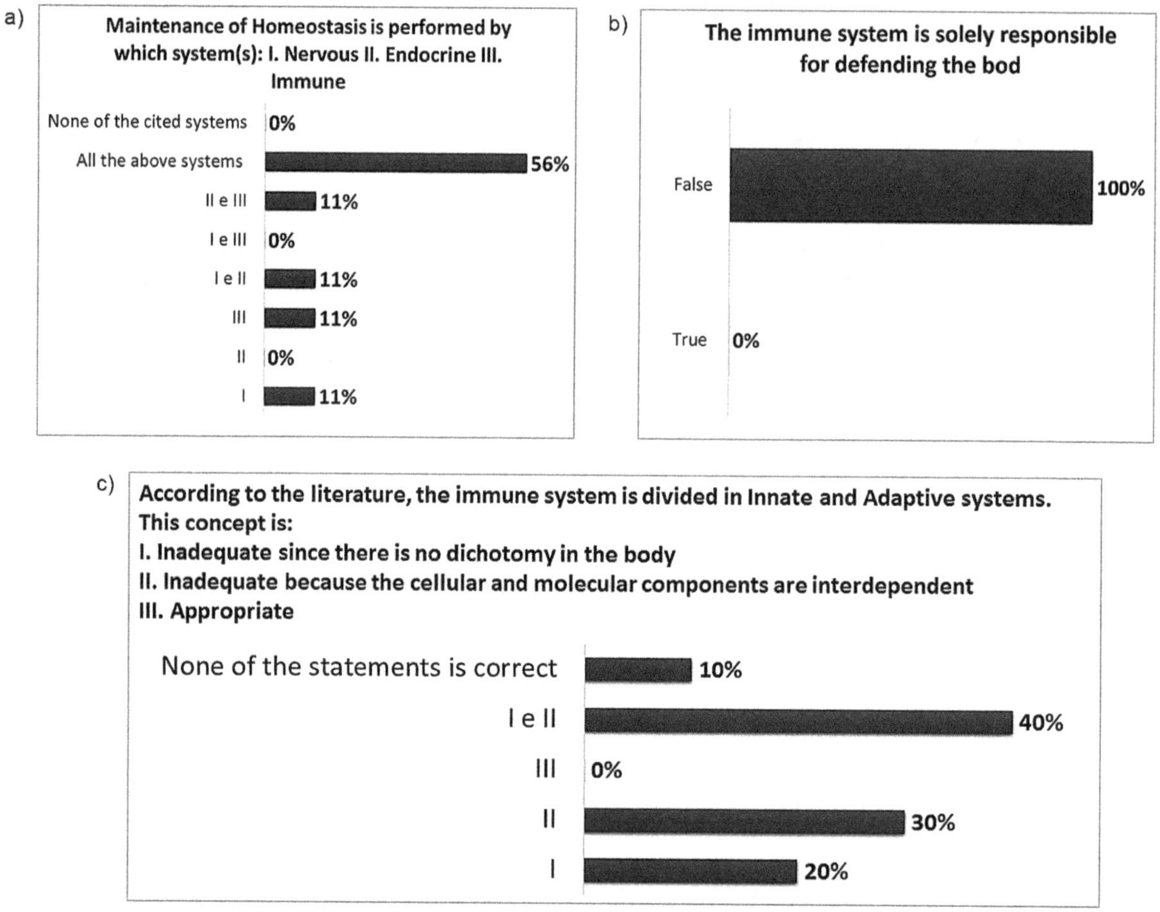

schemes in AIT", where AIT means Allergen-specific immunotherapy. However, no immunity to any microorganisms is being stimulated. Instead, the intended reaction is the opposite, to down-regulate the inflammatory reactions in such a manner that when the allergen interacts with the immunological system a non-inflammatory reaction is developed. Thus, if we return to the interdependence between basic sciences and biotechnology, it is possible to perceive that understanding basic Immunology will certainly influence the understanding and development of biotechnology tools.

The remaining questions referring to the concepts of vaccine, antigen and antibody were applied during the lecture on the history of Immunology. Since immunity was not a well-understood concept, it was extensively discussed in the previous lecture. The fact almost all attendees chose the correct alternative for the concept "vaccine" - a preparation of antigens to produce a primary immunological response" shows there was an adequate incorporation of the new knowledge. As the attendees of the outreach course were mainly high-school teachers the adequate understanding and dissemination of this concept is bound to have a major impact on the population. For example, during 2008, many Brazilians underwent improper re-vaccination for yellow fever, while in 2010 many people in risk groups failed to vaccinate for the H1N1

Figure 4. Examples of questions used to determine participants' perceptions of four terms central to understanding the Immunological system
(Source: Bertozzi, 2015, adapted from Barreto, 2011)

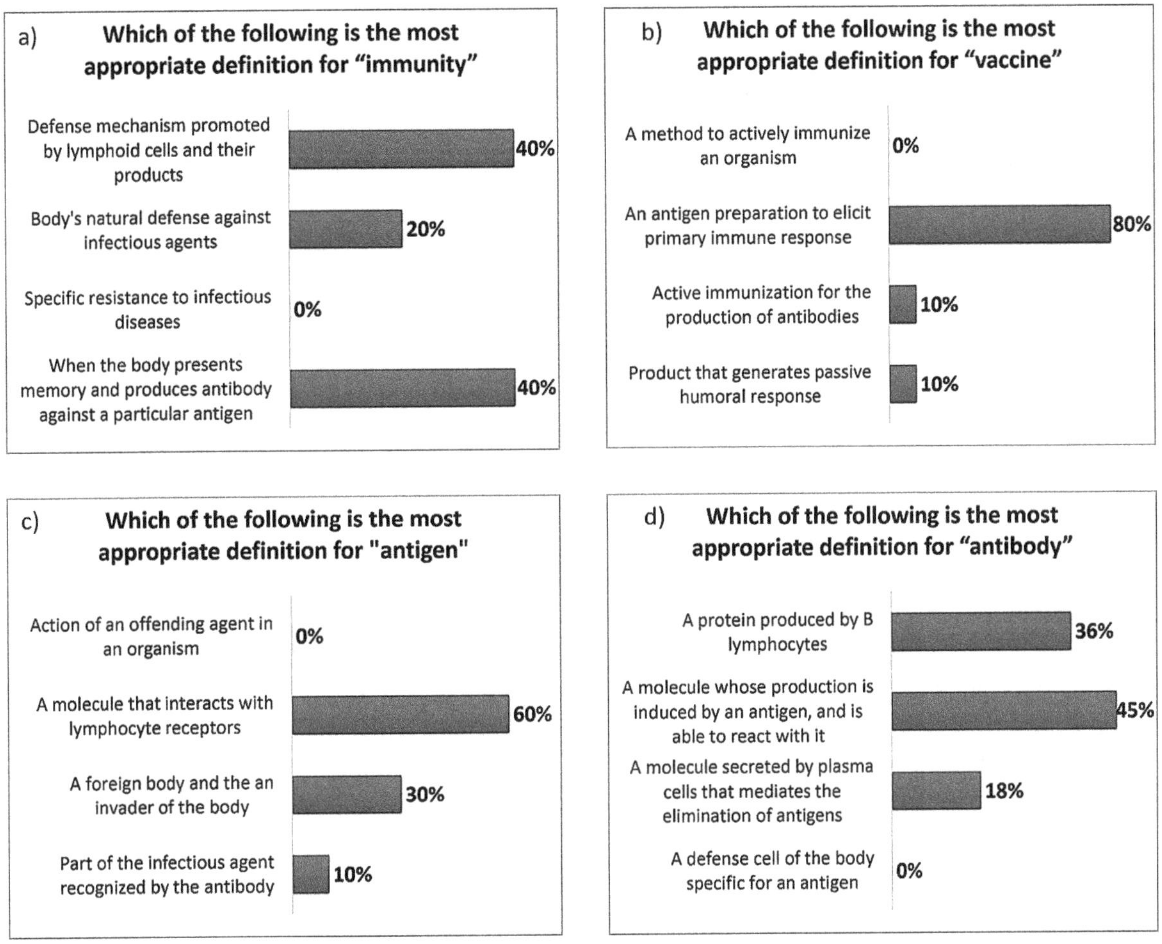

virus, as reported in newspapers of general circulation at the time. While the former wanted to ensure health at any cost, the latter believed they would get sick if they took the vaccine (Barreto & Teixeira, 2013). A recent paper, by Stukus, Green, Montandon, and Wada, (2015) shows an alarming deficit in allergy knowledge among physicians at academic medical centers worldwide.

Most participants answered correctly on the question related to antigens as "a molecule that interacts with clonal cell receptors". However, those who answered incorrectly still believe an antigen is something external to the body (an intruder or a foreign body) and that it interacts with antibodies. This view may hinder understanding that the immunological system consistently produces specific antibodies to proteins of its own body, including anti-antibodies. This misconception can be traced to the origin of the Immunological studies in medicine and more specifically to medical microbiology. Although advancement in understanding the physiology of the immunological system has shown this system does not exclusively defend organism by eliminating microorganisms, the emergence of HIV and AIDS reinforced the alternative view that antigens are microorganism or parts thereof. Thus, improper subsumed concepts can impair the understanding of a newly related knowledge (Barreto & Teixeira, 2013).

David P. Ausubel (1918-2008) is one of the strongest critics for the information learning-processing model since it does not account for discrepancies in the effort necessary to acquire knowledge of different content types or by different learners. He dedicated much of his professional career to define and promote the idea of meaningful cognitive learning. His theory was first presented in a paper submitted to the Journal of General Psychology entitled "Subsumption Theory of Meaningful Learning and Retention" (Ausubel, 1962) and the following year in the book "The Psychology of Meaningful Verbal Learning (Ausubel, 1963). Based on his continuous research and feedback from both students and colleagues Ausubel presented in 1978 a revision of his ideas. In the revision he refers to the assimilation theory formally emphasizing the major characteristics involved in learning new content is its interaction with existing cognitive structures (Ausubel, 1978). To contextualize his theory, Ausubel distinguished between two types of learning, rote and meaningful.

Given this context, ARS is a teaching tool enabling the interaction between teachers and students and can stimulate the desired educational reactions, if used strategically. A review of the literature related to undergraduate teaching and learning shows the use of ARS has interested faculty in all subject matter areas. Using "clicker" as a key word in the ERIC database to search the literature we found 206 peer reviewed papers in the last 20 years of which 131 were published in the last 5 years and 35 in the last year demonstrating a continuous interest. A variety of positive outcomes have been reported with the use of different ARS. As presented for other poll strategies, ARS is an excellent approach to provide immediate feedback for both faculty and learners (Connor, 2009). When faculty use ARS to assess students comprehension of key course concepts they can immediately determine at which pace the class sessions should be delivered, giving more or less details for the specific group (Kolikant, Drane & Calkins, 2010). Another important aspect is allowing shy and or insecure students to confidently participate because it is an approach permitting student to stay anonymous. However, ARS can foster peer discussions as students gain confidence (Chen, Whittinghill & Kadlowec, 2010) leading to an increased participative learning environment (Dallimore, Hertenstein & Platt, 2012), narrowing knowledge gaps (FitzPatrick, Finn & Campisi, 2011) and increasing academic achievement (Mayer et al., 2009). The use of ARS also allows faculty to measure students' reactions to problems or particular situations (Moss & Crowley, 2011) and in large classes it can reduce the distance between the faculty members and students (Cole & Kosc, 2010; Gebru, Phelps & Wulfsberg, 2012)

The viability of the strategic use of the clicker was evaluated based on criteria found in Table 2. With regards to the accessibility of the site, some problems were experienced with the older participants who are not digital natives. Not all participants installed the app on their smartphone since it is possible to respond using the webpage, thus it was not accessed. The use of the program on the webpage is feasible, since there is no need to install any software. All participants had free access to a stable Internet network with wireless routing and were able to connect and answer the questions. Most of the participants informed us the time to submit the responses did not affect the progress of the class nor the dynamics, since the discussion of the alternatives was previously scheduled. All participants agreed faculty discussed all the alternatives, justifying both appropriate and inadequate options and how the explanations sparked many discussions among students and the teacher, demonstrating the need for further learning in basic concepts of immunology, although all were trained in life sciences.

Complementing the data shown in Table 2, the K-12 teachers attending the Basic Immunology course stated the use of the ARS was a new and exciting experience and they were motivated to use this system, believing will increase interaction and discussions in their classes. These results corroborate those found in the article' "*An Audience Response System Strategy to Improve Student Motivation, Attention,*

Table 2. Evaluation criteria of the viability of the clicker strategy (Bertozzi, 2015)

Evaluation Criteria	Y	N	NA*
Did participants find the site accessible?		×	
Was the application installed properly in all wearables?			×
Was the use of the program on the webpage a viable alternative?	×		
Was everyone able to connect?	×		
Was the handling of the responses intuitive?		×	
Did the answers interfere with the progress of the class?		×	
Did the professor explain each alternative?	×		
Did the explanation spark discussion?	×		
Did all participants voice their opinions?	×		

and Feedback" (Cain, Black & Rohr, 2009). Since faculty members still mostly teach using traditional methodology, the use of smart ARS proved to be a very interesting resource for the transition from traditional to active learning methodology, letting both teachers and students feel comfortable encouraging active engagement in the classroom.

With a successful outcome with the use of the ARS during the immunology course, the researchers/authors believe this technology should be further made-known at the two ends of the educational process. In other words, faculty members already familiar with the technology should offer workshops meant for faculty members unfamiliar with ARS and use it during their undergraduate teacher training classes.

Since acquiring the technology is not enough, the programed activities include the use of alternative strategies, to stimulate active learning. Taking into account many faculty members are not digital natives using strategies less technology-dependent is important, and should be further encouraged in universities, especially in Brazil, where most teachers and faculty still use traditional methods in class. Given this scenario, the strategic use of ICT was introduced in an undergraduate biology-teacher training course. In order to encourage interaction and exchange, previous experiences between participants, were placed using ARS demonstrating its potential at the same time:

- Do you know what ICT means?
- During your university training, did the faculty use ICT?
- Were ICTs used in order to encourage participation and interaction between students and faculty?
- Do you believe it is possible to apply ICT in your future classroom?
- Would you use ICTs in your classes in the future?
- Would you facilitate classes using ARS?

Most of the undergraduate students did not know what ICT means although all have smartphones, computers etc. During the discussions, participants said some technological resources are frequently used in the classroom by faculty to teach their content area. Although they stated the main resource used is the projection of slides. Students have divergent opinions about how often ICTs were used during their university training (almost always, sometimes or almost never were the three alternative and these were evenly distributed). This fact demonstrates students perceptions of faculty trying to engage students is very weak. In fact, most of them answered ICTs weren't used to encourage participation and interaction between students and faculty, it was just used to deliver more content in a more rapid fashion. These results suggest faculty need to be familiarized with the strategic use of ICTs to stimulate active learning. The students believed it is possible to apply ICT in classroom and they would like to facilitate classes using ICTs and ARS.

SOLUTIONS AND RECOMMENDATIONS

The resource usage during class was relatively easy; however, there are some issues needing to be taken into account. Some of the teachers reported they would use the clickers in the high school Science courses they teach in the public schools. However, despite the enthusiasm with the resources presented during the course, some teachers reported they would have difficulties in applying the technology, as the necessary infrastructure is not always available. Considering the setting of many Brazilian public schools, they do not have the necessary infrastructure or the financial support to implement this feature (Internet based or handheld remote clickers) and the benefits of using the poll strategies we suggested teaches try to use the less technological strategies. As reported by Stowell and Nelson (2007) regardless of the feedback technique used in class, real-time review questions increases both student participation and reduces tediousness. In Stowell's experience the higher the technological resource used the higher participation in class, however, as there are many low-tech active learning alternatives teachers should not use as an excuse to return to traditional teaching paradigm when high-tech solutions and strategies cannot be performed. Thus, since the need for Internet access can be a restraint, the teacher should be familiar with alternative strategies to encourage participation and student discussions.

Another observation was individuals who are not digital natives had initial difficulties using the technology at two stages: to access and to respond to the question. To answer the questions participants needed to access the webpage directly in the browser. If the address is inserted directly in the search area, access to the questions is denied. At the answering stage, some participants did not immediately understand it was necessary to "scroll" the page to view all options. However, after the necessary aid was provided by the teaching assistants or from peers, the initial difficulties were overcome and did not affect the development of the class. Thus, it is important to save some time to account for possible difficulties especially when introducing a new technology. In regards to the limits for the free use of the Poll Everywhere software, it allows for 25 students to use it, the suggested strategy for bigger classes was using pairs or small groups enabling both the implementation of the resource and the stimulation of interaction between students. One last objection for some teachers was the fact the answers were anonymous and therefore it would not be possible to perform a more precise intervention or perform summative and or formative evaluations. However, the teachers were informed of other software enabling both summative and or formative evaluations. Therefore, the choice of the ARS software has to be aligned to the pre-established educational needs.

FUTURE RESEARCH DIRECTIONS

Our aim is to disseminate the use of ARS within the Brazilian universities and investigate the impact and changes in both students and faculty's attitudes towards learning. Another direction is applying ARS

Figure 5. Teacher-centered learning vs. Student-centered learning
(Source: Bertozzi, 2015)

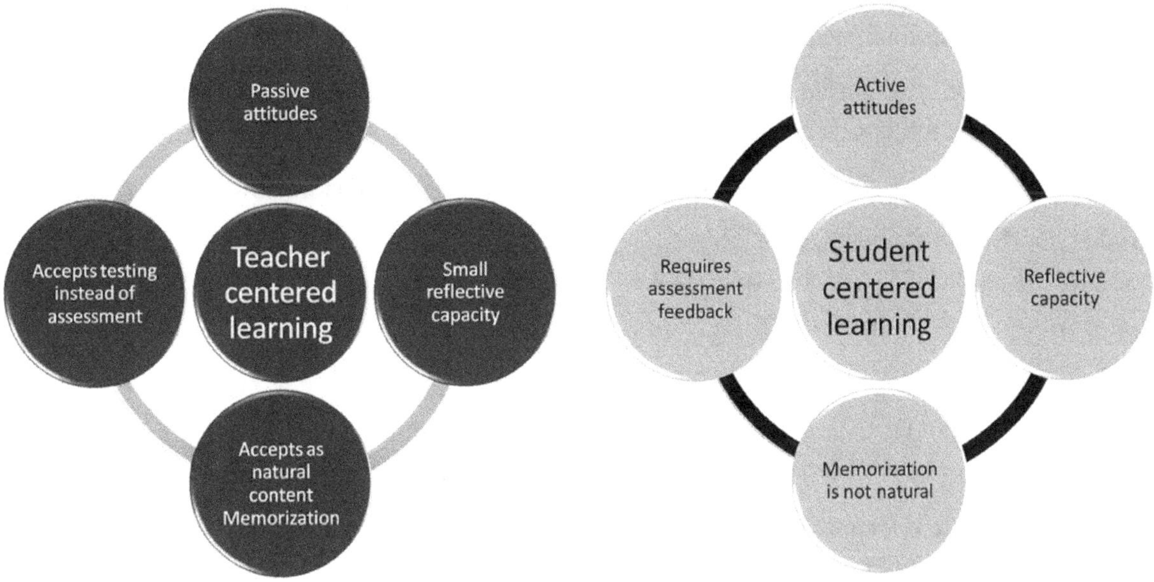

in a gamification scenario. Blending collaborative and competitive strategies with the formal learning process seems to be an interesting approach to further stimulate future high school teachers in the use of active learning strategies breaking at the university level the vicious cycle depicted in Figure 5.

CONCLUSION

In conclusion, the researchers/authors agree with Barr and Tagg (1995) that the structures reflecting an old paradigm can frustrate the best ideas and innovations. Instead of fixing the means – such as lectures and courses – the learning paradigm fixes the ends, the learning results, allowing the means to vary in its constant search for the most effective and efficient paths for learning.

In addition, more exciting courses can help participants decide on their career choices, for example choosing complex knowledge areas such as the Biotechnology. For example, the understanding of immunological phenomena and biotechnological techniques can stimulate young people to pursue higher education courses in these areas. As a result, this demand can meet the requirement for qualified professionals to work in these fields, as shown in the News and Views section of Nature Biotechnology "Appetite for cheaper biologics spreads across the map" (Senior, 2013).

REFERENCES

Abrahamson, L. (2006). A brief history of networked classrooms: Effects, cases, pedagogy, and implications. In D. A. Banks (Ed.), *Audience Response Systems in Higher Education: Applications and Cases* (pp. 78–100). Hershey, PA, USA: IGI Global. doi:10.4018/978-1-59140-947-2.ch001

Alba-Lois, L., & Segal-Kischinevzky, C. (2010). Beer & Wine Makers. *Nature Education*, 3(17).

Andrade, V. A. (2011). *Imunostase: uma atividade lúdica para o ensino de imunologia.* (Master Dissertation). Retrieved from Arca Fiocruz database. (Number Order A553)

Ausubel, D. P. (1962). A subsumption theory of meaningful verbal learning and retention. *The Journal of General Psychology*, *66*(2), 213–224. doi:10.1080/00221309.1962.9711837 PMID:13863333

Ausubel, D. P. (1963). *The psychology of meaningful verbal learning*. Oxford, UK: Grune & Stratton.

Ausubel, D. P. (1978). In defense of advance organizers: A reply to the critics. *Review of Educational Research*, *48*(2), 251–257. doi:10.3102/00346543048002251

Barr, R. B., & Tagg, J. (1995). From teaching to learning—A new paradigm for undergraduate education. *Change: The magazine of higher learning, 27*(6), 12-26.

Barreto, C. M. B. (2011). *Avaliação das relações entre ensino e aprendizado de imunologia: diagnóstico para mudança* [Doctoral Dissertation].

Barreto, C. M. B., & Teixeira, G. A. P. B. (2013). Concepções prévias de universitários sobre o sistema imunológico. *Revista Brasileira de Ensino de Ciência e Tecnologia*, *6*(1), 1–18.

Bellanti, J. A. (1978). Allergy and clinical immunology: An interdisciplinary concept. *Annals of Allergy*, *41*(3), 129–135. PMID:686506

Berbel, N. A. N. (2012). As metodologias ativas e a promoção da autonomia de estudantes. *Semina: Ciências Sociais e Humanas*, *32*(1), 25–40.

Bertozzi, E. (2015). *Imunologia e Biotecnologia: O clicker como recurso didático estratégico*. (Masters Dissertation). Available from Universidade Federal Fluminemse database. (Number Order M444)

Biggs, J., & Tang, C. (2011). *Teaching for quality learning at university: What the student does*. London, UK: McGraw-Hill Education.

Bonaiuti, G., Calvani, A., & Piazza, D. (2015). Increasing classroom engagement and student comprehension through the use of clickers: An Italian secondary school experience. *REM–Research on Education and Media*, *5*(1), 95–108.

Bradforth, S. E., Miller, E. R., Dichtel, W. R., Leibovich, A. K., Feig, A. L., Martin, J. D., & Smith, T. L. et al. (2015). University learning: Improve undergraduate science education. *Nature*, *523*(7560), 282–284. doi:10.1038/523282a PMID:26178951

Cain, J., Black, E. P., & Rohr, J. (2009). An audience response system strategy to improve student motivation, attention, and feedback. *American Journal of Pharmaceutical Education*, *73*(2), 1–7. doi:10.5688/aj730221 PMID:19513159

Chen, J. C., Whittinghill, D. C., & Kadlowec, J. A. (2010). Classes that click: Fast, rich feedback to enhance student learning and satisfaction. *The Journal of Engineering Education*, *99*(2), 159–168. doi:10.1002/j.2168-9830.2010.tb01052.x

Chin, C., & Brown, D. E. (2000). Learning in science: A comparison of deep and surface approaches. *Journal of Research in Science Teaching*, *37*(2), 109–138. doi:10.1002/(SICI)1098-2736(200002)37:2<109::AID-TEA3>3.0.CO;2-7

Cole, S., & Kosc, G. (2010). Quit surfing and start "clicking": One professor's effort to combat the problems of teaching the US survey in a large lecture hall. *The History Teacher*, *43*(3), 397–410.

Connor, E. (2009). Perceptions and uses of clicker technology. *Journal of Electronic Resources in Medical Libraries*, *6*(1), 19–32. doi:10.1080/15424060802705145

Dallimore, E. J., Hertenstein, J. H., & Platt, M. B. (2012). Impact of cold-calling on student voluntary participation. *Journal of Management Education*, *37*(3), 305–341. doi:10.1177/1052562912446067

de Barros, J. A., Remold, J., da Silva, G. S., & Tagliati, J. (2004). Engajamento interativo no curso de Física I da UFJF. *Revista Brasileira de Ensino de Física*, *26*(1), 63–69.

Diamond, J. (2002). Evolution, consequences and future of plant and animal domestication. *Nature*, *418*(6898), 700–707. doi:10.1038/nature01019 PMID:12167878

Dictionary, O. E. (2002). *Concise Oxford English Dictionary*. Oxford, UK: Oxford University Press.

Fári, M. G., & Kralovánszky, U. P. (2006). The founding father of biotechnology: Károly (Karl) Ereky. *International Journal of Horticultural Science*, *12*(1), 9–12.

FitzPatrick, K. A., Finn, K. E., & Campisi, J. (2011). Effect of personal response systems on student perception and academic performance in courses in a health sciences curriculum. *Advances in Physiology Education*, *35*(3), 280–289. doi:10.1152/advan.00036.2011 PMID:21908838

Flood, R. L. (1993). *Dealing with complexity: an introduction to the theory and application of systems science*. New York, NY: Springer Science & Business Media. doi:10.1007/978-1-4757-2235-2

Freire-Maia, N. (1995). *Gregor Mendel - Vida e Obra*. São Paulo, SP: TAQ.

Gebru, M. T., Phelps, A. J., & Wulfsberg, G. (2012). Effect of clickers versus online homework on students' long-term retention of general chemistry course material. *Chemistry Education Research and Practice*, *13*(3), 325–329. doi:10.1039/C2RP20033C

Gillen, A. L., & Sherwin, F. J. III. (2008). Louis Pasteur's views on creation, evolution, and the genesis of germs. *Answers Research Journal*, *1*, 43–52.

Goeddel, D. V., Kleid, D. G., Bolivar, F., Heyneker, H. L., Yansura, D. G., Crea, R., & Riggs, A. D. et al. (1979). Expression in Escherichia coli of chemically synthesized genes for human insulin. *Proceedings of the National Academy of Sciences of the United States of America*, *76*(1), 106–110. doi:10.1073/pnas.76.1.106 PMID:85300

Hoebe, K., Janssen, E., & Beutler, B. (2004). The interface between innate and adaptive immunity. *Nature Immunology*, *5*(10), 971–974. doi:10.1038/ni1004-971 PMID:15454919

Hooke, R. (1961). Micrographia. London, UK.

Huba, M. E., & Freed, J. E. (2000). *Learner-Centered Assessment on College Campuses: Shifting the Focus from Teaching to Learning*. Boston, MA: Allyn and Bacon.

Judson, E. (2002). Learning from past and present: Electronic response systems in college lecture halls. *Journal of Computers in Mathematics and Science Teaching*, *21*(2), 167–181.

Kay, R. H., & LeSage, A. (2009). A strategic assessment of audience response systems used in higher education. *Australasian Journal of Educational Technology*, *25*(2), 235–249.

Kay, R. H., & LeSage, A. (2009). Examining the benefits and challenges of using audience response systems: A review of the literature. *Computers & Education*, *53*(3), 819–827. doi:10.1016/j.compedu.2009.05.001

Klymkowsky, M. W., Underwood, S. M., & Garvin-Doxas, R. K. (2010). Biological Concepts Instrument (BCI): A diagnostic tool for revealing student thinking.

Kolikant, Y. B. D., Drane, D., & Calkins, S. (2010). "Clickers" as Catalysts for Transformation of Teachers. *College Teaching*, *58*(4), 127–135. doi:10.1080/87567551003774894

Krasilchik, M. (2000). Reformas e realidade: O caso do ensino das ciências. *São Paulo em Perspectiva*, *14*(1), 85–93. doi:10.1590/S0102-88392000000100010

Kumar, S. (2015). Biosafety and Biosecurity Issues in Biotechnology Research. *Biosafety*, *4*(01), 153. doi:10.4172/2167-0331.1000e153

Lantz, M. E. (2010). The use of 'Clickers' in the classroom: Teaching innovation or merely an amusing novelty? *Computers in Human Behavior*, *26*(4), 556–561. doi:10.1016/j.chb.2010.02.014

Martins, F. P. S., & Pereira, F. L. (2012). *O Ensino de Imunologia: Análise dos Principais Métodos e Recursos Didáticos Utilizados em Universidades Brasileiras*. Uberlândia: II Fórum Internacional Sobre Prática Docente Universitária.

Mayer, R. E., Stull, A., DeLeeuw, K., Almeroth, K., Bimber, B., Chun, D., & Zhang, H. et al. (2009). Clickers in college classrooms: Fostering learning with questioning methods in large lecture classes. *Contemporary Educational Psychology*, *34*(1), 51–57. doi:10.1016/j.cedpsych.2008.04.002

Mendel, G. (1996). Experiments in plant hybridization (1865). Verhandlungen des Naturforschenden Vereins Brünn.

Michael, J. (2007). What makes physiology hard for students to learn? Results of a faculty survey. *Advances in Physiology Education*, *31*(1), 34–40. doi:10.1152/advan.00057.2006 PMID:17327580

Moreira, M. A. (1999). Teorias de aprendizagem. São Paulo, SP: Editora pedagógica e universitária.

Morita, M. (2003). The mobile-based learning (MBL) in Japan. *Proceedings of the First Conference on Creating, Connecting and Collaborating through Computing* (C5'03) (pp. 128-129).

Mortimer, E. F. & de Carvalho, A. M. P. (2013). Referenciais teóricos para análise do processo de ensino de ciências. *Cadernos de Pesquisa,* 96, 05-14.

Moss, K., & Crowley, M. (2011). Effective learning in science: The use of personal response systems with a wide range of audiences. *Computers & Education*, *56*(1), 36–43. doi:10.1016/j.compedu.2010.03.021

Parâmetros Curriculares Nacionais: Ciências Naturais. (1997). *Secretaria de Educação Fundamental*, Brasil.

Pele, M., & Cimpeanu, C. (2012). *Biotechnology: An Introduction*. Boston, MA: Wit Press.

Rego, M. C. (2000). *Vygotsky: uma perspectiva histórico cultural da educação*. Petrópolis, RJ: Editora Vozes Limitada.

Senior, M. (2013). Appetite for cheaper biologics spreads across the map. *Nature Biotechnology*, *31*(12), 1059–1061. doi:10.1038/nbt1213-1059 PMID:24316622

Silverstein, A. M. (1982). History of immunology. Development of the concept of immunologic specificity, I. *Cellular Immunology*, *67*(2), 396–409. doi:10.1016/0008-8749(82)90231-3 PMID:6177425

Siqueira-Batista, R., Gomes, A. P., Albuquerque, V. S., Madalon-Fraga, R., Aleksandrowicz, A. M. C., & Geller, M. (2009). Ensino de imunologia na educação médica: Lições de Akira Kurosawa. *Revista Brasileira de Educação Médica*, *33*(2), 186–190. doi:10.1590/S0100-55022009000200004

Smith, A. C., Stewart, R., Shields, P., Hayes-Klosteridis, J., Robinson, P., & Yuan, R. (2005). Introductory biology courses: A framework to support active learning in large enrollment introductory science courses. *Cell Biology Education*, *4*(2), 143–156. doi:10.1187/cbe.04-08-0048 PMID:15917873

Soyka, M. B., van de Veen, W., Holzmann, D., Akdis, M., & Akdis, C. A. (2014). Scientific foundations of allergen-specific immunotherapy for allergic disease. *Chest*, *146*(5), 1347–1357. doi:10.1378/chest.14-0049 PMID:25367471

Stemler, S. (2001). An overview of content analysis. *Practical Assessment, Research & Evaluation*, *7*(17), 137–146.

Stowell, J. R., & Nelson, J. M. (2007). Benefits of electronic audience response systems on student participation, learning, and emotion. *Teaching of Psychology*, *34*(4), 253–258. doi:10.1080/00986280701700391

Stukus, D. R., Green, T., Montandon, S. V., & Wada, K. J. (2015). Deficits in allergy knowledge among physicians at academic medical centers. *Annals of Allergy, Asthma & Immunology*, *115*, 51–55. PMID:26024806

Tassoni, E. C. M. (2000). *Afetividade e aprendizagem: a relação professor-aluno. Psicologia, análise e crítica da prática educacional*. Campinas: ANPED.

Ten Cate, O., & Durning, S. (2007). Peer teaching in medical education: Twelve reasons to move from theory to practice. *Medical Teacher*, *29*(6), 591–599. doi:10.1080/01421590701606799 PMID:17922354

Tenreiro-Vieira, C. (2004). Formação em pensamento crítico de professores de ciências: Impacte nas práticas de sala de aula e no nível de pensamento crítico dos alunos. *Revista Electrónica de Enseñanza de las Ciencias*, *3*(3), 228–256.

Trees, A. R., & Jackson, M. H. (2007). The learning environment in clicker classrooms: Student processes of learning and involvement in large university-level courses using student response systems. *Learning, Media and Technology*, *32*(1), 21–40. doi:10.1080/17439880601141179

Van Leeuwenhoek, A. (1722). *Arcana naturae detecta*. Langerak.

Veiga, I. P. A. (1991). *Repensando a didáctica*. Campinas, SP: Papirus Editora.

Verma, A. S., Rastogi, S., Agrahari, S., & Singh, A. (2011). Biotechnology in the realm of history. *Journal of Pharmacy and Bioallied Sciences*, *3*(3), 321–323. doi:10.4103/0975-7406.84430 PMID:21966150

Watson, J. D., & Crick, F. H. (1953). Molecular structure of nucleic acids; a structure for deoxyribose nucleic acid. *Nature*, *171*(4356), 737–738. doi:10.1038/171737a0 PMID:13054692

Wieczorek, A., & Wright, M. (2012). History of agricultural biotechnology: How crop development has evolved. *Nature Education Knowledge*, *3*(10), 1–9.

Yourstone, S. A., Kraye, H. S., & Albaum, G. (2008). Classroom Questioning with Immediate Electronic Response: Do Clickers Improve Learning? *Decision Sciences Journal of Innovative Education*, *6*(1), 75–88. doi:10.1111/j.1540-4609.2007.00166.x

KEY TERMS AND DEFINITIONS

Audience Response System (ARS): A technological device that permits the audience to interact with the speaker, teacher or anyone conducting a show giving immediate feedback. The results are calculated and displayed in a graphical form as soon as the audience answers the question.

Clicker: A synonymous of Audience Response System.

Education: An action that involves the teaching and learning process.

Information and Communication Technology (ICT): An umbrella term for any communication device or application e.g.: radio, television, computers, satellite systems, cell phones, tablets, wearable devices, network hardware and software.

Student-Centered Learning Paradigm: A teaching method in which the students actively learn and the faculty members act as coaches interacting full time with their team.

Technology: Tools helping to solve problems.

Traditional Instructional Paradigm: A teaching method in which the teachers/faculty members act like actors on the stage. They memorize their speech and deliver it to the students, many times with very little to no interaction at all with them.

Compilation of References

7 things you should know about wearable technology. (2013). *EDUCAUSE*. Retrieved from https://net.educause.edu/ir/library/pdf/eli7102.pdf

7 things you should know about wearable technology. (2013, November 8). *Educause Learning Initiative*. Retrieved from http://www.educause.edu/library/resources/7-things-you-should-know-about-wearable-technology

7 Things You Should Know: Wearable Technology. (2013). *EDUCAUSE*. Retrieved from https://net.educause.edu/ir/library/pdf/ELI7102.pdf

802.15 Working group for WPAN. (2015). IEEE. Retrieved from http://www.ieee802.org/15/

A framework for K-12 science education: Practices, cross- cutting concepts, and core ideas. (2012National Research Council. Washington, DC: National Academies Press.

Abernathy, W., & Clark, K. B. (2007). Innovation: Mapping the winds of creative destruction. *Research Policy*, *14*(1), 3–22. doi:10.1016/0048-7333(85)90021-6

Abrahamson, L. (2006). A brief history of networked classrooms: Effects, cases, pedagogy, and implications. In D. A. Banks (Ed.), *Audience Response Systems in Higher Education: Applications and Cases* (pp. 78–100). Hershey, PA, USA: IGI Global. doi:10.4018/978-1-59140-947-2.ch001

Abramovich, S., Schunn, C., & Higashi, R. (2013). Are badges useful in education? It depends upon the type of badge and expertise of learner. *Education Technology Research Development*.

Adams, R., Evangelou, D., English, L., Dias de Figueiredo, A., Mousoulides, N., Pawley, A. L., & Wilson, D. M. et al. (2011). Multiple perspectives on engaging future engineers. *The Journal of Engineering Education*, *100*(1), 48–88. doi:10.1002/j.2168-9830.2011.tb00004.x

AdSense. (2015, May 22). *Wikipedia*. Retrieved from http://en.wikipedia.org/wiki/AdSense.

Afshar, V., & Peterson, D. (2014). Wearable technology: The coming revolution in healthcare. *Huffington Post*. Retrieved from http://www.huffingtonpost.com/vala-afshar/wearable-technology-the-c_b_5263547.html

Ahuja, A. K., Dorn, J. D., Caspi, A., McMahon, M. J., Dagnelie, G., Dacruz, L., & Greenberg, R. J. et al. (2011). Blind subjects implanted with the Argus II retinal prosthesis are able to improve performance in a spatial-motor task. *The British Journal of Ophthalmology*, *95*(4), 539–543. doi:10.1136/bjo.2010.179622 PMID:20881025

Alba-Lois, L., & Segal-Kischinevzky, C. (2010). Beer & Wine Makers. *Nature Education*, 3(17).

Alford, E. (2105). Wearables to heighten demand for local search: prepare now! *Search Engine Watch*. Retrieved from http://searchenginewatch.com/sew/news/2396194/wearables-to-heighten-demand-for-local-search-prepare-now#

Allen, A. L. (2008). Dredging up the past: Lifelogging, memory, and surveillance. *The University of Chicago Law Review. University of Chicago. Law School*, 2008, 47–74.

Alvarez, C., Brown, C., & Nussbaum, M. (2011). Comparative study of netbooks and tablet PCs for fostering face-to-face collaborative learning. *Computers in Human Behavior*, *27*(2), 834–844. doi:10.1016/j.chb.2010.11.008

Amhag, L. (2012). High school students' argument patterns in online peer feedback. E. Favaron, P. M. Pumilia-Gnarini, E. Pacetti, J. Bishop & L. Guerra (Ed.), *Handbook of Research on Didactic Strategies and Technologies for Education: Incorporating Advancements* (Vol. 2, Ch. 62, pp. 711-723). Retrieved from http://www.igi-global.com/chapter/high-school-students-argument- patterns/72113

Amhag, L. (2011). *Students´Argument Patterns in Asynchronous Dialogues for Learning*.

Amhag, L. (2013). Creativity in and between Collaborative Peer Assessment Processes in Higher Distance Education. In *Creative Education* (pp. 94–104). Special Issue on Higher Education; doi:10.4236/ce.2013.47A2011

Amhag, L., & Jakobsson, A. (2009). Collaborative Learning as a Collective Competence when Students Use the Potential of Meaning in Asynchronous Dialogues. *Computers & Education*, *52*(3), 656–667. doi:10.1016/j.compedu.2008.11.012

Andersen, J., & Hansen, F. (1977). Antennas for VHF/UHF personal radio: A theoretical and experimental study of characteristics and performance. *IEEE Transactions on Vehicular Technology*, *26*(4), 349–357. doi:10.1109/T-VT.1977.23707

Andrade, V. A. (2011). *Imunostase: uma atividade lúdica para o ensino de imunologia*. (Master Dissertation). Retrieved from Arca Fiocruz database. (Number Order A553)

Ariyatum, B., Holland, R., Harrison, D., & Kazi, T. (2005). The Future Design Direction of Smart Clothing Development. *Journal of the Textile Institute*, *96*(4), 199–212. doi:10.1533/joti.2004.0071

Armstrong, N., & Bray, S. (1991). Physical activity patterns defined by continuous heart rate monitoring. *Archives of Disease in Childhood*, *66*(2), 245–247. doi:10.1136/adc.66.2.245 PMID:2001112

Arpaci, I. (2015). A comparative study of the effects of cultural differences on the adoption of mobile learning. *British Journal of Educational Technology*, *46*(4), 699–712. doi:10.1111/bjet.12160

Atman, C. J., Adams, R. S., Cardella, M. E., Turns, J., Mosborg, S., & Saleem, J. (2007). Engineering design processes: A comparison of students and expert practitioners. *The Journal of Engineering Education*, *96*(4), 359–379. doi:10.1002/j.2168-9830.2007.tb00945.x

Ausubel, D. P. (1962). A subsumption theory of meaningful verbal learning and retention. *The Journal of General Psychology*, *66*(2), 213–224. doi:10.1080/00221309.1962.9711837 PMID:13863333

Ausubel, D. P. (1963). *The psychology of meaningful verbal learning*. Oxford, UK: Grune & Stratton.

Ausubel, D. P. (1978). In defense of advance organizers: A reply to the critics. *Review of Educational Research*, *48*(2), 251–257. doi:10.3102/00346543048002251

Avidov-Ungar, O., & Eshet-Alkakay, Y. (2011). Teachers in a world of change: Teachers' knowledge and attitudes towards the implementation of innovative technologies in schools. *Interdisciplinary Journal of E-Learning and Learning Objects*, *7*, 291–303.

Azim, R., & Hassan, A. (2013a, June 20-22). Understanding Recent Wireless and Mobile Technological Changes for Business Management Practices. Proceedings of the 6th International Conference on Business Market Management (BMM). The University of Bamberg, Bamberg, Germany. Retrieved from http://bit.ly/Ns5u1C

Azim, R., & Hassan, A. (2013b, June 20-22). Analysing the impact of mobile and wireless technology on Business Management Strategies. Proceedings of the he 6th International Conference on Business Market Management (BMM). The University of Bamberg, Bamberg, Germany. Retrieved from http://bit.ly/Ns5u1C

Bächlin, M., & Tröster, G. (2012). Swimming performance and technique evaluation with wearable acceleration sensors. *Pervasive and Mobile Computing*, *8*(1), 68–81. doi:10.1016/j.pmcj.2011.05.003

Baepler, P., Walker, J. D., & Driessen, M. (2014). It's not about seat time: Blending, flipping, and efficiency in active learning classrooms. *Computers & Education*, *78*, 227–236. doi:10.1016/j.compedu.2014.06.006

Bandura, A. (1984). Recycling Misconceptions of Perceived Self-efficacy. *Cognitive Therapy and Research*, *8*(3), 231–255. doi:10.1007/BF01172995

Bandura, A. (2002). Growing primacy of human agency in adaption and change in the electronic era. *European Psychologist*, *7*(1), 2–16. doi:10.1027//1016-9040.7.1.2

Barker, B. S., Nugent, G., & Grandgenett, N. F. (2008). Examining 4-H robotics and geospatial technologies in the learning of science, technology, engineering, and mathematics topics. Publication in the *Journal of Extension*, 46(3). Retrieved from http://www.joe.org/joe/2008june/rb7.shtml

Barker, B., Melander, J., Grandgenett, N., & Nugent, G. (2015). Utilizing wearable technologies as a pathway to STEM. In D. Slykhuis & G. Marks (Eds.), *Proceedings of Society for Information Technology & Teacher Education International Conference 2015* (pp. 1770-1776). Chesapeake, VA: Association for the Advancement of Computing in Education (AACE).

Barker, B. S., & Ansorge, J. (2007). Robotics as means to increase achievement scores in an informal learning environment. *Journal of Research on Technology in Education*, *39*(3), 229–243. doi:10.1080/15391523.2007.10782481

Barr, R. B., & Tagg, J. (1995). From teaching to learning—A new paradigm for undergraduate education. *Change: The magazine of higher learning, 27*(6), 12-26.

Barreto, C. M. B. (2011). *Avaliação das relações entre ensino e aprendizado de imunologia: diagnóstico para mudança* [Doctoral Dissertation].

Barreto, C. M. B., & Teixeira, G. A. P. B. (2013). Concepções prévias de universitários sobre o sistema imunológico. *Revista Brasileira de Ensino de Ciência e Tecnologia*, *6*(1), 1–18.

Bart, D., & Zach, M. (2014). *Make money from your site*. Retrieved from http://AdSensebrowse.blogspot.co.uk/

Basas, C. G. (2013). Universally designing the public sector workplace: Technology as disability access. *Working USA*, *16*(1), 69–86. doi:10.1111/wusa.12024

Baurley, S. (2003). Interactive and Experiential Design in Smart Textile Products and Applications.*Proceeding of International Textile Design and Engineering Conference (INTEDEC): Fibrous Assemblies at the Design and Engineering Interface,*Edinburgh (pp. 22–24).

Baya'a, N., & Daher, W. (2009). Learning mathematics in an authentic mobile environment: The Perceptions of Students. *International Journal of Interactive Mobile Technologies*, *3*, 6–14.

Bebell, D., O'Dwyer, L. M., Russell, M., & Hoffmann, T. (2010). Concerns, considerations, and new ideas for data collection and research in educational technology studies. *Journal of Research on Technology in Education*, *43*(1), 29–52. doi:10.1080/15391523.2010.10782560

Belkin, N. J. (1990). The cognitive viewpoint in information science. *Journal of Information Science*. Retrieved from http://jis.sagepub.com/content/16/1/11.abstract

Belkin, N. J. (1980). Anomalous states of knowledge as a basis for information retrieval. *The Canadian Journal of Information Science*, *5*, 133–143.

Bell, D. (2014). Smart watches in the library? *CSSIS.org*. Retrieved from http://cssis.org/news/blog/2015/01/smart watches-library

Bellamy, A. (2007). Exploring the influence of new technology planning and implementation on the perceptions of new technology effectiveness. *Journal of Technology Studies*, *33*(1), 32–40.

Bellanti, J. A. (1978). Allergy and clinical immunology: An interdisciplinary concept. *Annals of Allergy*, *41*(3), 129–135. PMID:686506

Benitti, F. B. V. (2012). Exploring the educational potential of robotics in schools: A systematic review. *Computers & Education*, *58*(3), 978–988. doi:10.1016/j.compedu.2011.10.006

Berbel, N. A. N. (2012). As metodologias ativas e a promoção da autonomia de estudantes. *Semina: Ciências Sociais e Humanas*, *32*(1), 25–40.

Bergamini, E., Picerno, P., Pillet, H., Natta, F., Thoreux, P., & Camomilla, V. (2012). Estimation of temporal parameters during sprint running using a trunk-mounted inertial measurement unit. *Journal of Biomechanics*, *45*(6), 1123–1126. doi:10.1016/j.jbiomech.2011.12.020 PMID:22325976

Berlin, J. E., Storti, K. L., & Brach, J. S. (2006). Using activity monitors to measure physical activity in free-living conditions. *Physical Therapy*, *86*(8), 1137–1145. PMID:16879047

Bertoni, D., & US Government Accountability, O. (2012). Employment for People with Disabilities: Little Is Known about the Effectiveness of Fragmented and Overlapping Programs. Report to Congressional Committees. GAO-12-677. *US Government Accountability Office*.

Bertozzi, E. (2015). *Imunologia e Biotecnologia: O clicker como recurso didático estratégico*. (Masters Dissertation). Available from Universidade Federal Fluminemse database. (Number Order M444)

Besch, E. L., Varosi, S. M., & Brigmon, R. L. (1992). Technical note: A noninvasive procedure for measuring goat heart rates. *Journal of Animal Science*, *70*(11), 3371–3375. PMID:1459897

Bevan, B., & Semper, R. J. (2006). Mapping informal science institutions onto the science education landscape. *Center for Informal Learning and Schools*. Retrieved from http://cils.exploratorium.edu/resource_shared/downloads/4229/Mapping%20Informal%20Science%20Instituions%20onto%20the%20Science%20Eductaion%20Landscape.pdf

Big data, wearable devices create anxiety about storage practices. (2013). *Trend Micro*. Retrieved from http://blog.trendmicro.com/big-data-wearable-devices-create-anxiety-storage-practices/

Biggs, J., & Tang, C. (2011). *Teaching for quality learning at university: What the student does*. London, UK: McGraw-Hill Education.

Bighiu, G., Manolic, A., & Roman, C. T. (2015). Compulsive buying behavior on the Internet. *Procedia. Economics and Finance*, *20*, 72–79.

Bilton, N. (2015) Why Google Glass broke. *The New York Times*. Retrieved from http://www.nytimes.com/2015/02/05/style/why-google-glass-broke.html?_r=0

Binboğa, E., & Korhan, O. (2014). Posture, Musculoskeletal Activities, and Possible Musculoskeletal Discomfort among Children Using Laptops or Tablet Computers for Educational Purposes: A Literature Review. *Journal of Science Education and Technology*, *23*(5), 605–616. doi:10.1007/s10956-014-9505-9

Bishop, J. L., & Verleger, M. A. (2013). The Flipped Classroom: A Survey of the Research. *Paper presented at the120th ASEE Conference & Exposition.*

Blow, D., & McConnell, S. (2012). Classroom technology rollouts. *Leadership, 42*(2), 32.

Blume, H. (2014). Head of L.A. schools' technology division resigns. *LA Times.* Retrieved from http://www.latimes.com/local/lanow/la-me-ln-lausd-technology-resign-20141031-story.html

Bonaiuti, G., Calvani, A., & Piazza, D. (2015). Increasing classroom engagement and student comprehension through the use of clickers: An Italian secondary school experience. *REM–Research on Education and Media, 5*(1), 95–108.

Bonato, P. (2005). Advances in wearable technology and applications in physical medicine and rehabilitation. *Journal of NeuroEngineering and rehabilitation.* Retrieved from http://www.jneuroengrehab.com/content/2/1/2

Booth, C., & Bretcher, D. (2015). Ok, library: Implications and opportunities for Google Glass. *College and Research Libraries.* Retrieved from http://crln.acrl.org/content/75/5/234.full

Borg, G. (1970). Perceived exertion as an indicator of somatic stress. *Scandinavian Journal of Rehabilitation Medicine, 2*(2), 92–98. Retrieved from http://www.ncbi.nlm.nih.gov/pubmed/5523831 PMID:5523831

Boser, U. (2013). Are schools getting a big enough bang for their education technology buck? *Center for American Progress.* Retrieved from http://www.americanprogress.org/issues/education/report/2013/06/14/66485/are-schools-getting-a-big-enough-bang-for-their-education-technology-buck

Botelho, S. (2014). IT centralization: Institutional hesitation. *University Business, 17*(9).

Bottani, N. (1986). *La ricreazione è finita: dibattito sulla qualità dell'istruzione.* Firenze, Italy: Il Mulino.

Boulos, M. N. K., & Yang, S. P. (2013). Exergames for health and fitness: The roles of GPS and geosocial apps. *International Journal of Health Geographics, 12*(1), 18. doi:10.1186/1476-072X-12-18 PMID:23561306

Bouten, C. V. C., Sauren, A. A. H. J., Verduin, M., & Janssen, J. D. (1997). Effects of placement and orientation of body-fixed accelerometers on the assessment of energy expenditure during walking. *Medical & Biological Engineering & Computing, 35*(1), 50–56. doi:10.1007/BF02510392 PMID:9136191

Boyce, C. J., Mishra, C., Halverson, K. L., & Thomas, A. K. (2014). Getting Students Outside: Using Technology as a Way to Stimulate Engagement. *Journal of Science Education and Technology, 23*(6), 815–826. doi:10.1007/s10956-014-9514-8

Bradforth, S. E., Miller, E. R., Dichtel, W. R., Leibovich, A. K., Feig, A. L., Martin, J. D., & Smith, T. L. et al. (2015). University learning: Improve undergraduate science education. *Nature, 523*(7560), 282–284. doi:10.1038/523282a PMID:26178951

Brage, S., Brage, N., Ekelund, U., Luan, J., Franks, P. W., Froberg, K., & Wareham, N. J. (2006). Effect of combined movement and heart rate monitor placement on physical activity estimates during treadmill locomotion and free-living. *European Journal of Applied Physiology, 96*(5), 517–524. doi:10.1007/s00421-005-0112-6 PMID:16344938

Branch, R., & Merrill, M. D. (2012). Characteristics of instructional design models. In R. Reiser & J. Dempsey (Eds.), *Trends and issues in instructional design and technology* (3rd ed.). Boston, MA: Pearson Education.

Brault, M. (2012). *Health and disability statistics branch.* United States Census Bureau of the Census.

Bravo, F. (2012). Giornalismo investigativo, vittime e diritto di cronaca *Rivista di Criminologia, Vittimologia e Sicurezza,* 6(2). Bologna: IT.

Brinck, T., Gergle, D., & Wood, S. D. (2002). *Designing Web sites that work: Usability for the Web.* San Francisco: Morgan Kaufmann Publishing.

Brooks, C. (2008). Alberta education technology policy analysis: Implications and future directions. *The International Journal of Learning, 15*, 139–146.

Brotman, J. S., & Moore, F. M. (2008). Girls and science: A review of four themes in the science education literature. *Journal of Research in Science Teaching, 45*(9), 971–1002. doi:10.1002/tea.20241

Broussard, M. S. (2011). A spoonful of sugar: Instructional games in libraries. *Games in libraries.* Retrieved from http://gamesinlibraries.blogspot.com/2011/12/lemontree-game-at-university-of.html

Brown, W. C. (1984). The history of power transmission by radio waves. . *IEEE Transactions on* microwave theory and techniques, *32*(9), 1230–1242.

Bruckman, A. (2004). Co-evolution of technological design and pedagogy in an online learning community. In S. Barab (Ed.), Designing for Virtual Communities in the Service of Learning (pp. 239-255). Cambridge University Press. doi:10.1017/CBO9780511805080.013

Brusseau, T., & Hannon, J. C. (2013). Pedometer-Determined Physical Activity of Youth while Attending School: A Review. *Sport Science Review, XXII*(5-6), 329–342. doi:10.2478/ssr-2013-0016

Buechley, L. (2006). A construction kit for electronic textiles. *Proceedings of the 2006 10th IEEE International Symposium on Wearable Computers* (pp. 83-90). IEEE. doi:10.1109/ISWC.2006.286348

Buechley, L., Eisenberg, M., Catchen, J., & Crockett, A. (2008). The LilyPad Arduino: using computational textiles to investigate engagement, aesthetics, and diversity in computer science education.*Proceedings of the SIGCHI conference on Human factors in computing systems* (pp. 423-432). ACM. doi:10.1145/1357054.1357123

Buechley, L., Eisenberg, M., & Elumeze, N. (2007, June) Towards a curriculum for electronic textiles in the high school classroom.*Proceedings of the Conference on Innovation and Technology in Computer Science Education (ITiCSE)*, Dundee, Scotland. doi:10.1145/1268784.1268795

Buechley, L., & Hill, B. M. (2010, August). LilyPad in the wild: how hardware's long tail is supporting new engineering and design communities.*Proceedings of the 8th ACM Conference on Designing Interactive Systems* (pp. 199-207). doi:10.1145/1858171.1858206

Buechley, L., Peppler, K., Eisenberg, M., & Kafai, Y. (2013). *Textile Messages: Dispatches from the world for e-textiles and education.* New York, NY: Peter Lang.

Buhalis, D., Leung, D., & Law, R. (2011). E-Tourism: Critical Information and Communication Technologies for Tourism Destinations. In Y. Wang & A. Pizam (Eds.), *Destination Marketing and Management* (pp. 205–224). CAB International. doi:10.1079/9781845937621.0205

Buxton, E. C., Burns, E. C., & De Muth, J. E. (2012). Instructional Design and Assessment. Professional Development Webinars for Pharmacists. *American Journal of Pharmaceutical Education, 76*(8), 1–7. doi:10.5688/ajpe768155 PMID:23129854

Cain, J., Black, E. P., & Rohr, J. (2009). An audience response system strategy to improve student motivation, attention, and feedback. *American Journal of Pharmaceutical Education, 73*(2), 1–7. doi:10.5688/aj730221 PMID:19513159

Calvani, A. (2007). *Tecnologia, scuola, processi cognitivi. Per una ecologia dell'apprendere.* Milano, Italy: Franco Angeli.

Carnevale, A. P., Smith, N., & Stoll, J. (2010). Help wanted: Projections of jobs and education requirements through 2018. Washington, DC: Georgetown University, Center on Education and the Workforce. Retrieved from http://files.eric.ed.gov/fulltext/ED524310.pdf

Cartelli, A. (2006). *Un esempio di progettazione di attività formativa mirata*. Udine, Italy: Idea Stampa.

Cascio, J. (2005). The rise of the participatory panopticon. *World Changing*. Retrieved from http://www.worldchanging.com/archives/002651.html

Catalano, F., & Doucet, K. (2013). *Digital badges emerge as part of credentialing's future*. New York, NY: Professional Examination Service.

Catsambis, S. (1995). Gender, race, ethnicity, and science education in the middle grades. *Journal of Research in Science Teaching*, *32*(3), 243–257. doi:10.1002/tea.3660320305

Cavanaugh, J. T., Coleman, K. L., Gaines, J. M., Laing, L., & Morey, M. C. (2007). Using step activity monitoring to characterize ambulatory activity in community-dwelling older adults. *Journal of the American Geriatrics Society*, *55*(1), 120–124. doi:10.1111/j.1532-5415.2006.00997.x PMID:17233695

Centro Europeo dell'Educazione., (1971). L'alternativa tecnologica. Il rapporto Perkins Mc Murrin. Frascati, Italy: Centro Europeo dell'Educazione

Chapanis, A. (1991). Evaluating usability. In B. Shackel & S. J. Richardson (Eds.), *Human factors for informatics usability* (pp. 359–395). Cambridge: Cambridge University.

Chen, L.C., Lai Y.C., Yeh, Y.H., Lin, J.W., Lai, C.N. & Weng, H.C. (2013). Enhanced mechanisms for navigation and tracking services in smart phones. *Journal of applied research and technology*, 11(2), 272-282.

Chen, C. M., & Chung, C. J. (2008). Personalized mobile English vocabulary learning system based on item response theory and learning memory cycle. *Computers & Education*, *51*(2), 624–645. doi:10.1016/j.compedu.2007.06.011

Chen, C. M., & Hsu, S. H. (2008). Personalized intelligent mobile learning system for supporting effective English learning. *Journal of Educational Technology & Society*, *11*(3), 153–180.

Chen, J. C., Whittinghill, D. C., & Kadlowec, J. A. (2010). Classes that click: Fast, rich feedback to enhance student learning and satisfaction. *The Journal of Engineering Education*, *99*(2), 159–168. doi:10.1002/j.2168-9830.2010.tb01052.x

Chen, J. D., Falkmer, T., Parsons, R., Buzzard, J., & Ciccarelli, M. (2014). Impact of experience when using the Rapid Upper Limb Assessment to assess postural risk in children using information and communication technologies. *Applied Ergonomics*, *45*(3), 398–405. doi:10.1016/j.apergo.2013.05.004 PMID:23735750

Childress, D. S. (1985). Historical aspects of powered limb prostheses. *Clinical Prosthetics & Orthotics*, *9*(1), 2–13.

Chin, C., & Brown, D. E. (2000). Learning in science: A comparison of deep and surface approaches. *Journal of Research in Science Teaching*, *37*(2), 109–138. doi:10.1002/(SICI)1098-2736(200002)37:2<109::AID-TEA3>3.0.CO;2-7

Chu, H. C., Hwang, G. J., & Tsai, C. C. (2010). A knowledge engineering approach to developing mind tools for context-aware ubiquitous learning. *Computers & Education*, *54*(1), 289–297. doi:10.1016/j.compedu.2009.08.023

Clarke, K. (2010). A science, engineering and technology (SET) approach improves science process skills in 4-H animal science participants. *Journal of Extension*, *48*(1). Retrieved from http://www.joe.org/joe/2010february/iw3.php

Clark, R. E. (2005). Five common but questionable principles in multimedia learning. In R. E. Mayer (Ed.), *The Cambridge handbook of multimedia learning*. New York: Cambridge University Press. doi:10.1017/CBO9780511816819.007

Clewell, B. C., & Braddock, J. H. (2000). Influences on minority participation in mathematics, science, and engineering. In G. Campbell Jr, R. Denes, & C. Morrison (Eds.), *Access denied: Race, ethnicity, and the scientific enterprise* (pp. 89–136). New York, NY: Oxford University Press.

Cole, S., & Kosc, G. (2010). Quit surfing and start "clicking": One professor's effort to combat the problems of teaching the US survey in a large lecture hall. *The History Teacher*, *43*(3), 397–410.

Comm, J. (2006). *The AdSense Code: What Google Never Told You about Making Money with AdSense*. Garden City, NY: Morgan James Publishing, LLC.

Compeau, D. R., & Higgins, C. A. (1995). Computer self-efficacy: Development of measure and initial test. *Management Information Systems Quarterly*, *19*(2), 189–211. doi:10.2307/249688

Connor, E. (2009). Perceptions and uses of clicker technology. *Journal of Electronic Resources in Medical Libraries*, *6*(1), 19–32. doi:10.1080/15424060802705145

Courville, K. (2011, November). Educational technology: Effective leadership and current initiatives. *Paper presented at the2011 Louisiana Computer Using Educator's Conference*, New Orleans, LA.

Couse, L. J., & Chen, D. W. (2010). A tablet computer for young children? Exploring its viability for early childhood education. *Journal of Research on Technology in Education*, *43*(1), 75–98. doi:10.1080/15391523.2010.10782562

Craft, A. (2006). Fostering creativity with wisdom. *Cambridge Journal of Education*, *36*(3), 337–350. doi:10.1080/03057640600865835

Creswell, J. W. (2014). *Educational research: Planning, conducting and evaluating quantitative and qualitative research* (5th ed.). Upper Saddle River, NJ: Pearson Prentice Hall.

Cuban, L. (2000). Is spending money on technology worth it? *Education Week*, *19*(24), 42.

Cuban, L. (2001). *Oversold and underused: Computers in the classroom*. Cambridge, MA: Harvard University Press.

Culp, K. M., Honey, M., & Mandinach, E. (2005). A retrospective on twenty years of education technology policy. *Journal of Educational Computing Research*, *32*(3), 279–307. doi:10.2190/7W71-QVT2-PAP2-UDX7

Cyr, B., Horn, W., Miao, D., & Specter, M. (2014). *Security analysis of wearable fitness devices (Fitbit)*. Cambridge, MA: Massachusetts Institute of Technology.

da Cruz, L., Coley, B.F., Dorn, J., Merlini, F., Filley, E., Christopher, P., ... Dagnelie, G.(2013). The Argus II epiretinal prosthesis system allows letter and word reading and long-term function in patients with profound vision loss. Retrieved from http://bjo.bmj.com/content/early/2013/02/19/bjophthalmol-2012-301525.full

Dadwal, S., & Hassan, A. (2015). The augmented reality marketing: A merger of marketing and technology in tourism. In N. Ray (Ed.), *Emerging Innovative Marketing Strategies in the Tourism Industry* (pp. 78–96). Hershey, PA, USA: IGI Global. doi:10.4018/978-1-4666-8699-1.ch005

Daim, T. U., Letts, M., Krampits, M., Khamis, R., Dash, P., & Monalisa, M. (2012). IT infrastructure refresh planning for enterprises: A business process perspective. *Business Process Management Journal*, *17*(3), 1–16.

Dallimore, E. J., Hertenstein, J. H., & Platt, M. B. (2012). Impact of cold-calling on student voluntary participation. *Journal of Management Education*, *37*(3), 305–341. doi:10.1177/1052562912446067

Darwish, A., & Hassanien, E. (2011). Wearable and implantable wireless sensor network solutions for healthcare monitoring. *Sensors (Basel)*, *11*(6), 5561–5595. doi:10.3390/s110605561 PMID:22163914

Davies, C. (2009, July 1). Pocket tweet app turns your shirt into a Twitter bubble. *Slashgear.com*. Retrieved from http://www.slashgear.com/pocket-tweet-app-turns-your-shirt-into-a-twitter-bubble-0148404/

Davies, C. (2014). Google's Glass designer talks less-is-more in wearables. *Slashgear.com*. Retrieved from http://www.slashgear.com/googles-glass-designer-talks-less-is-more-in-wearables-16325496/

de Barros, J. A., Remold, J., da Silva, G. S., & Tagliati, J. (2004). Engajamento interativo no curso de Física I da UFJF. *Revista Brasileira de Ensino de Física*, *26*(1), 63–69.

de Bruin, E. D., Hartmann, A., Uebelhart, D., Murer, K., & Zijlstra, W. (2008). Wearable systems for monitoring mobility-related activities in older people: A systematic review. *Clinical Rehabilitation*, *22*(10-11), 878–895. doi:10.1177/0269215508090675 PMID:18955420

De Magalhaes, F. A., Vannozzi, G., Gatta, G., & Fantozzi, S. (2014). Wearable inertial sensors in swimming motion analysis: A systematic review. *Journal of Sports Sciences*, *33*(7), 732–745. doi:10.1080/02640414.2014.962574 PMID:25356682

De Wever, B., Van Keer, H., Schellens, T., & Valcke, M. (2009). Structuring asynchronous discussion groups: The impact of role assignment and self-assessment on students' levels of knowledge construction through social negotiation. *Journal of Computer Assisted Learning*, *25*(2), 177–188. doi:10.1111/j.1365-2729.2008.00292.x

Dede, C., Brown-L'Bahy, T., Ketelhut, D., & Whitehouse, P. (2004). Distance learning (virtual learning). In H. Bidgoli (Ed.), *The Internet encyclopedia* (pp. 549–560). New York: Wiley. doi:10.1002/047148296X.tie047

Deffeo, C. (2013). Wearable tech: 5 obstacles to going mass market. *Entrepreneur.com*. Retrieved from http://www.entrepreneur.com/article/229545

Deighton, J. A., & Kornfeld, L. (2007). *Digital interactivity: Unanticipated consequences for markets, marketing and consumers*. Boston: Harvard Business School.

Dellaserra, C. L., Gao, Y., & Ransdell, L. (2014). Use of integrated technology in team sports: a review of opportunities, challenges, and future directions for athletes. *Journal of Strength and Conditioning Research / National Strength & Conditioning Association, 28*(2), 556–73. Doi:10.1519/JSC.0b013e3182a952fb

De-Marcos, L., Hilera, J. R., Barchino, R., Jiménez, L., Martínez, J. J., Gutiérrez, J. A., & Otón, S. et al. (2010). An experiment for improving students performance in secondary and tertiary education by means of m-learning auto-assessment. *Computers & Education*, *55*(3), 1069–1079. doi:10.1016/j.compedu.2010.05.003

Denise, R. (2014). The brains behind the hoodie that texts. *Forbes.com*. Retrieved from http://www.forbes.com/sites/deniserestauri/2014/08/14/the-brains-behind-the-hoodie-that-texts-its-not-who-you-think/

Desai, A. N. (2015, June 4). Indian students develop device that converts sign language into speech and text—lending voice to the mute. *INQUISITR*. Retrieved from http://www.inquisitr.com/2144746/indian-students-develop-device-that-converts-sign-language-into-speech-and-text-lending-voice-to-the-mute/

Desinan, C. (1998). *Discutere la scuola. Ipotesi, contenuti e prospettive a confronto*. Milano, Italy: Franco Angeli.

Dewey, J. (1964). *Scuola e Società*. Firenze, Italy: La Nuova Italia.

Dhir, A. (2004). *The digital consumer technology handbook*. Oxford: Elsevier.

Diamond, J. (2002). Evolution, consequences and future of plant and animal domestication. *Nature*, *418*(6898), 700–707. doi:10.1038/nature01019 PMID:12167878

Dick, W., Carey, L., & Carey, J. O. (2003). *Courseware Authoring*. Boston, MA: Pearson Custom Publishing.

Dictionary, O. E. (2002). *Concise Oxford English Dictionary*. Oxford, UK: Oxford University Press.

Dochy, F., Segers, M., & Sluijsmans, D. (1999). The use of self-, peer- and co-assessment in higher education: A review. *Studies in Higher Education*, *24*(3), 331–350. doi:10.1080/03075079912331379935

Doherty, A. R., Hodges, S. E., King, A. C., Smeaton, A. F., Berry, E., Moulin, C. J., & Foster, C. et al. (2013). Wearable cameras in health: The state of the art and future possibilities. *American Journal of Preventive Medicine*, *44*(3), 320–323. doi:10.1016/j.amepre.2012.11.008 PMID:23415132

Dolan, B. (2014). Fitbit, Jawbone, Nike had 97 percent of fitness tracker retail sales in 2013. Retrieved from http://mobihealthnews.com/28825/fitbit-jawbone-nike-had-97-percent-of-fitness-tracker-retail-sales-in-2013/

Donthu, N., & Garcia, A. (1999). The Internet shopper. *Journal of Advertising Research*, *39*, 52–58.

Dooley, K. E. (2005). *Advanced methods in distance education: Applications and practices for educators, administrators and learners*. Hershey, PA, USA: IGI Global. doi:10.4018/978-1-59140-485-9

Doolittle, P., & Mariano, G. (2008). Working memory capacity and mobile multimedia learning environments: Individual differences in learning while mobile. *Journal of Educational Multimedia and Hypermedia*, *17*(4), 511–530.

Dündar, H., & Akçayır, M. (2014). Implementing tablet PCs in schools: Students' attitudes and opinions. *Computers in Human Behavior*, *32*(0), 40–46. doi:10.1016/j.chb.2013.11.020

ECAR Study of Undergraduate Students and Information Technology. (2014). *Educause Center for Analysis and Research*. Retrieved from http://net.educause.edu/ir/library/pdf/ss14/ERS1406.pdf

Education at a Glance 2014. (2014OCSE. OECD Indicators.

Education is the only solution: Remarks of U.S. Secretary of Education Arne Duncan at the U.S. Agency for International Development (USAID) Global Education Summit. (2013, August 6). *US Department of Education*.

Ehrenkranz, M. (2014) Wearable Tech for people with Disabilities: How a Tech Specialist is Modifying Google Glass to be More accessible. *Idigitaltimes.com*. Retrieved from http://www.idigitaltimes.com/wearable-tech-people-disabilities-how-tech-specialist-modifying-google-glass-be-more-accessible

Eisenberg, A. (2009). Inside these lenses, a digital dimension. *The New York Times.com*. Retrieved from http://www.nytimes.com/2009/04/26/business/26novel.html?_r=0

Emilliani, P. L., & Stephanidis, C. (2005). Universal access to ambient intelligence environments: Opportunities and challenges for people with disabilities. *IBM Systems Journal*, *44*(3), 605–619. doi:10.1147/sj.443.0605

Engineering: Short on fact? [Editorial]. (2012, February). *Quality Progress*, 45(2), 14-15. Retrieved from http://asq.org/quality-progress/2012/02/keeping-current.html

Erduran, S. & Villamanan, R. (2009). Cool Argument: Engineering Students´ Written Arguments about Thermodynamics in the Context of the Peltier Effect in Refrigeration. *Educacaión quimica* (pp. 119-125).

Eriksson, A. (2011). Google AdSense secrets revealed: All about residual income with AdSense. *Six-Bucks.com*. Retrieved from http://www.six-bucks.com/downloads/GoogleAdSenseSecretsRevealed.pdf

Essential conditions. (2009). *International Society for Technology in Education*. Retrieved from http://www.iste.org/docs/pdfs/netsessentialconditions.pdf?sfvrsn=2

European agency snapshot study. (2015). *IAB*. Retrieved from http://www.iabeurope.eu/files/5014/2470/1844/IAB_Agency_Snapshot_Survey_European_Results_FINAL_2.pdf

Evans, D. (2011). The Internet of Things: How the next evolution of the Internet is changing everything. *Cisco.com*. Retrieved from https://www.cisco.com/web/about/ac79/docs/innov/IoT_IBSG_0411FINAL.pdf

Evans, C. (2008). The effectiveness of m-learning in the form of podcast revision lectures in higher education. *Computers & Education*, *50*(2), 491–498. doi:10.1016/j.compedu.2007.09.016

Everson, K. (2015). Learning is all in the wrist. *Chief Learning Officer*, *14*(4), 18–21.

Eyal, S., Dagan, Y., & Baharav, A. (2012). Sleep in the Cloud: On How to Use Available Heart Rate Monitors to Track Sleep and Improve Quality of Life. In *Computing in Cardiology (CinC)* (pp. 329–332). IEEE. Retrieved from http://ieeexplore.ieee.org/stamp/stamp.jsp?tp=&arnumber=6420397

Fallik, O., Rosenfeld, S., & Eylon, B. S. (2013). School and out-of-school science: A model for bridging the gap. *Studies in Science Education*, *49*(1), 69–91. doi:10.1080/03057267.2013.822166

Fári, M. G., & Kralovánszky, U. P. (2006). The founding father of biotechnology: Károly (Karl) Ereky. *International Journal of Horticultural Science*, *12*(1), 9–12.

Fernández-López, Á., Rodríguez-Fórtiz, M. J., Rodríguez-Almendros, M. L., & Martínez-Segura, M. J. (2013). Mobile learning technology based on iOS devices to support students with special education needs. *Computers & Education*, *61*, 77–90. doi:10.1016/j.compedu.2012.09.014

Ferrer, F., Belvis, E., & Pamies, J. (2011). Tablet PCs, academic results and educational inequalities. *Computers & Education*, *56*(1), 280–288. doi:10.1016/j.compedu.2010.07.018

Ferrière, A. (1967). *Trasformiamo la scuola*. Firenze, Italy: La Nuova Italia.

Fibre-optic thread improves smart garments. (2015). *Wearable Technology Insights*. Retrieved from http://www.wearabletechnologyinsights.com/articles/7623/fibre-optic-thread-improves-smart-garments

Finkel, E. (2012). Getting the best ROI in technology: The latest, greatest tools won't buy you success without a plan. *District Administration*, *8*, 74.

FitzPatrick, K. A., Finn, K. E., & Campisi, J. (2011). Effect of personal response systems on student perception and academic performance in courses in a health sciences curriculum. *Advances in Physiology Education*, *35*(3), 280–289. doi:10.1152/advan.00036.2011 PMID:21908838

Flood, R. L. (1993). *Dealing with complexity: an introduction to the theory and application of systems science*. New York, NY: Springer Science & Business Media. doi:10.1007/978-1-4757-2235-2

Foote, C. (2015). Wear to learn!?!. *Internet@Schools, 22*(3), 12-13.

Forman, G. E. (2010). When 2-year-olds and 3-year-olds think like scientists. Early Childhood Research & Practice, 12 (2). Retrieved from http://ecrp.uiuc.edu/v12n2/forman.html

Fredrick, K. (2015). Dressing for the future: Wearable technology. *School Library Monthly*, *31*(4), 25–26.

Freire-Maia, N. (1995). *Gregor Mendel - Vida e Obra*. São Paulo, SP: TAQ.

Fulk, G. D., Combs, S., Danks, K. A., Nirider, C. D., Raja, B., & Reisman, D. S. (2014). Accuracy of 2 activity monitors in detecting steps in people with stroke and traumatic brain injury. *Physical Therapy*, *94*(2), 222–229. doi:10.2522/ptj.20120525 PMID:24052577

Fullan, M. (2013a). Great to excellent: Launching the next stage of Ontario's education agenda. *Department of Education*. Retrieved from http://www.edu.gov.on.ca/eng/document/reports/FullanReport_EN_07.pdf

Fullan, M. (2013b). *Stratosphere: Integrating technology, pedagogy, and change knowledge*. Canada: Pearson.

Fulton, K. (2012). Upside down and inside out: Flip your classroom to improve student learning. *Learning and Leading with Technology*, *39*(8), 12–17.

Gagne, R. M., & Brigs, L. J. (1979). *Principles of Instructional Design*. New York: Holt, Rinehart & Winston.

Galligan, L., Hobohm, C., & Loch, B. (2012). Tablet technology to facilitate improved interaction and communication with students studying mathematics at a distance. *Journal of Computers in Mathematics and Science Teaching*, *31*(4), 363–385.

Gardner, P. J., & Campagna, P. D. (2011). Pedometers as measurement tools and motivational devices: New insights for researchers and practitioners. *Health Promotion Practice*, *12*(1), 55–62. doi:10.1177/1524839909334623 PMID:19515864

Gartner Says in 2015, 50 Percent of People Considering Buying a Smart Wristband Will Choose a Smartwatch Instead. (2014). *Gartner*. Retrieved from http://www.gartner.com/newsroom/id/2913318

Gartner: 10 Business & Technology Trends for Education in 2015 [INFOGRAPHIC]. (2015). *Gartner*. Retrieved from https://mrmck.wordpress.com/2014/10/06/gartner-10-business-technology-trends-for-education-in-2015-infographic/

Gebru, M. T., Phelps, A. J., & Wulfsberg, G. (2012). Effect of clickers versus online homework on students' long-term retention of general chemistry course material. *Chemistry Education Research and Practice*, *13*(3), 325–329. doi:10.1039/C2RP20033C

Gedik, N., Hanci-Karademirci, A., Kursun, E., & Cagiltay, K. (2012). Key instructional design issues in a cellular phone-based mobile learning project. *Computers & Education*, *58*(4), 1149–1159. doi:10.1016/j.compedu.2011.12.002

Gee, J. (2011). Human action and social groups as the natural home of assessment: Thoughts on 21st century learning and assessment. In V.J. Shute, B.J. Becker (Eds.), Innovative Assessments for the 21st Century: Supporting Educational Needs. NY: Springer.

Gentry, T., Kriner, R., Sima, A., McDonough, J., & Wehman, P. (2015). Reducing the need for personal supports among workers with autism using an iPod touch as assistive technology: Delayed randomized control trial. *Journal of Autism and Developmental Disorders*, *45*(3), 669–684. doi:10.1007/s10803-014-2221-8 PMID:25212414

Gibson, D., Ostashewski, N., Flintoff, K., Grant, S., & Knight, E. (2013). Digital badges in education. *Education and Information Technologies*, 2013, 1–8.

Gillen, A. L., & Sherwin, F. J. III. (2008). Louis Pasteur's views on creation, evolution, and the genesis of germs. *Answers Research Journal*, *1*, 43–52.

Gochfeld, M., Volz, C., Burger, J., Jewett, S., Powers, C., & Friedlander, B. (2006). Developing a Health and Safety Plan for Hazardous Field Work in Remote Areas. *Journal of Occupational and Environmental Hygiene*, *3*(12), 671–683. doi:10.1080/15459620601009201 PMID:17050349

Goeddel, D. V., Kleid, D. G., Bolivar, F., Heyneker, H. L., Yansura, D. G., Crea, R., & Riggs, A. D. et al. (1979). Expression in Escherichia coli of chemically synthesized genes for human insulin. *Proceedings of the National Academy of Sciences of the United States of America*, *76*(1), 106–110. doi:10.1073/pnas.76.1.106 PMID:85300

Golden, A. (1978). A foliage penetration summary. Rome Air Development Center, Air Force Systems Command, Griffiss Air Force Base, New York.

Goldsmith, R. E., & Lafferty, B. A. (2002). Consumer response to web sites and their influence on advertising effectiveness. *Internet Research*, *12*(4), 318–328. doi:10.1108/10662240210438407

Gomes, W. (2011). Leadership in educational technology: Insights from the corporate world. *Journal of Leadership Studies*, *4*(4), 57–61. doi:10.1002/jls

Gonzalez, Z. D. (2014). Put a ring on it. *Wearable Technologies.com*. Retrieved from https://www.wearable-technologies.com/2014/04/put-a-ring-on-it/

Gourishetty, N. (2009). *Google AdSense*. [Master in Business Consulting Thesis]. Hochschule Furtwangen University, Schwarzwald, Germany.

Grant, I. S., & Phillips, W. R. (2008). Electromagnetism (2nd ed.). Manchester Physics, John Wiley & Sons.

Grant, S. (2013). 5 buckets for badge system design: You are here. *HASTAC*. Retrieved from http://www.hastac.org/blogs/slgrant/2013/10/23/5-buckets-badge-system-design-you-are-here

Grant, S. L. (2014). *What counts as learning: Open digital badges for new opportunities*. Irvine, CA: Digital Media and Learning Research Hub.

Greaves, T., & Hayes, J. (2008). America's digital schools 2008: The six trends to watch. *The Greaves Group, The Hayes Connection*. Retrieved from http://www.lcps.org/cms/lib4/VA01000195/Centricity/Domain/52/Americas_Digital_Schools_2008.pdf

Guillemette, R. A. (1995). The evaluation of usability in interactive information systems. In J. M. Carey (Ed.), *Human factors in information systems: Emerging theoretical bases* (pp. 207–221). Norwood, NJ: Ablex.

Guimaraes, T. (2012). Industry clockspeed's impact on business innovation success factors. *European Journal of Innovation Management*, *14*(3), 322–344. doi:10.1108/14601061111148825

Güler, Ç., Kılıç, E., & Çavuş, H. (2014). A comparison of difficulties in instructional design processes: Mobile vs. desktop. *Computers in Human Behavior*, *39*, 128–135. doi:10.1016/j.chb.2014.07.008

Gurrin, C., Hopfgartner, F., Hurst, W., Johansen, H., Lee, H., & O'Connor, N. (2014). MultiMedia Modeling. *Proceedings of the 20th Anniversary International Conference MMM '14* (Part 1). Berlin, GE: Springer

Hall, G. E. (2010). Technology's Achilles heel: Achieving high-quality implementation. *Journal of Research on Technology in Education*, *42*(3), 231–253. doi:10.1080/15391523.2010.10782550

Hall, T. L. (1996). *Utilizing multimedia toolbook 3.0*. Danvers, MA: Boud & Fraser.

Hargis, J. (2014). Eager adopters in education: Strategic plan ideas for integrating instructional technology. *Turkish Online Journal of Distance Education*, *15*(4), 14–22.

Hassan, A., & Shabani, N. (2015, May 26-29). eMarketing adoption in tourism and hospitality industry in London: Industry analysis and some narratives. *Proceedings of the 4th International Interdisciplinary Business-Economics Advancement Conference (IIBA)*, Las Vegas, USA. Retrieved from http://bit.ly/1BQqGnI

Hassan, A. (2012a). Key Components for an Effective Marketing Planning: A Conceptual Analysis. *International Journal of Management & Development Studies*, *2*(1), 68–70.

Hassan, A. (2012b). Rationalization of Business Planning Through the Current Dynamics of Tourism. *International Journal of Management & Development Studies*, *2*(1), 61–63.

Hassan, A. (2015). The customization of electronic word of mouth: An industry tailored application for tourism promotion. In S. Rathore & A. Panwar (Eds.), *Capturing, Analyzing and Managing Word-of-Mouth in the Digital Marketplace* (pp. 61–75). Hershey, PA, USA: IGI Global.

Hassan, A. (in press). Destination image formation: The function analysis of augmented reality application. In M. Khosrow-Pour (Ed.), *The Encyclopaedia of Information Science and Technology* (4th ed.). Hershey, PA, USA : IGI Global.

Hassan, A., & Ahmed, F. (2013b). The role of NGOs in preventing unemployment in Bangladesh. *Kuwait Chapter of Arabian Journal of Business and Management Review*, *2*(9), 1–12. doi:10.12816/0001227

Hassan, A., & Burns, P. (2014). Tourism policies of Bangladesh – A contextual analysis. *Tourism Planning & Development*, *11*(4), 463–466. doi:10.1080/21568316.2013.874366

Hassan, A., & Dadwal, S. (in press). Search engine marketing – An outlining of conceptualization and strategic application. In W. Ozuem & G. Bowen (Eds.), *Competitive Social Media Marketing Strategies*. Hershey, PA, USA : IGI Global.

Hassan, A., & Forhad, A. (2013a). The role of NGOs in the sustainable development of Bangladesh. *Present Environment & Sustainable Development*, *7*(2), 59–72.

Hassan, A., & Jung, T. (in press). Augmented reality as an emerging application in tourism education. In D. H. Choi, A. Dailey-Hebert, & J. S. Estes (Eds.), *Emerging Tools and Applications of Virtual Reality in Education*. Hershey, PA, USA: IGI Global. doi:10.4018/978-1-4666-9837-6.ch008

Hassan, A., & Rahman, M. (2015). Macromarketing perspective in promoting tourism: The case of the Buddhist Vihara at Paharpur. *Tourism Spectrum*, *1*(2), 13–19.

Hassan, A., & Ramkissoon, H. (in press). Augmented reality for visitor experiences. In J. N. Albrecht (Ed.), *Visitor Management*. Oxfordshire: CABI.

Hassan, A., & Sharma, A. (in press). Wildlife tourism: Technology adoption for marketing and conservation. In M. A. Khan & J. K. Fatima (Eds.), *Wilderness of Wildlife Tourism*. Waretown: Apple Academic Press, Inc.

Hatano, Y. (1997). Prevalence and Use of Pedometer. *Res J Walking*, *1*, 45–54.

Hawkins, D. (2014). Library services and Google Glass. *Libconf.com*. Retrieved from http://www.libconf.com/2014/04/09/library-services-google-glass/

Hax, A. C. (2010). *Reinventing your business strategy*. New York: Springer.

Heasley, S. (2012). IPod may ease transition for those with autism. *Disability Scoop*. Retrieved from http://www.disabilityscoop.com/2012/09/06/ipod-transition-autism/16398/

Heasley, S. (2014, September 18). Nearly 1 in 3 with disabilities in poverty. *Disabilityscoop*. Retrieved from http://www.disabilityscoop.com/2014/09/18/report-disabilities-poverty/19683/

Heinecke, W. F., & Blasi, L. (2004). *Methods of evaluating educational technology*. Greenwich, CT: Information Age.

Herbert, G. (2013). Regal's new closed caption glasses finally make movies enjoyable for deaf and hard-of-hearing. *Syracuse.com*. Retrieved from http://www.syracuse.com/entertainment/index.ssf/2013/06/closed_caption_glasses_movies_regal_theaters_deaf.html

Hernandez, E. (2014). Google Glass in the academic library. *CreativeLibraryPractice.org*. Retrieved from http://creativelibrarypractice.org/2014/03/25/google-glass-in-the-academic-library-the-exploration/

Herreid, C. F., & Schiller, N. A. (2013). Case studies and the flipped classroom. *Journal of College Science Teaching*, *42*(5), 62–66.

Higgitt, D., & Bullard, J. (2010). Assessing fieldwork risk for undergraduate projects. *Journal of Geography in Higher Education*, *23*(3).

Hill, D. (2012). Smart gloves turn sign language gestures into vocalized speech. *Singularity Hub*. Retrieved from http://singularityhub.com/2012/09/16/smart-gloves-turn-sign-language-gestures-into-vocalized-speech/

Hindhede, A. L. (2012). Negotiating hearing disability and hearing disabled identities. *Health*, *16*(2), 169–185. doi:10.1177/1363459311403946 PMID:21540252

Hmelo, C. E., Holton, D., & Kolodner, J. L. (2000). Designing to learn about complex systems. *Journal of the Learning Sciences*, *9*(3), 247–298. doi:10.1207/S15327809JLS0903_2

Hoebe, K., Janssen, E., & Beutler, B. (2004). The interface between innate and adaptive immunity. *Nature Immunology*, *5*(10), 971–974. doi:10.1038/ni1004-971 PMID:15454919

Hooke, R. (1961). Micrographia. London, UK.

Horizon Report – Library Edition. (2014). *New Media Center*. Retrieved from http://cdn.nmc.org/media/2014-nmc-horizon-report-library-EN.pdf

Hovsepian, D., Meardon, S., & Kernozek, T. W. (2014). Consistency and Agreement of Two Devices for Running Speed. *Athletic Training & Sports Health Care*, *6*(2), 67–72. doi:10.3928/19425864-20140306-02

Hsu, Y. C., Hung, J. L., & Ching, Y. H. (2013). Trends of educational technology research: More than a decade of international research in six SSCI-indexed refereed journals. *Educational Technology Research and Development*, *61*(4), 685–705. doi:10.1007/s11423-013-9290-9

Huba, M. E., & Freed, J. E. (2000). *Learner-Centered Assessment on College Campuses: Shifting the Focus from Teaching to Learning*. Boston, MA: Allyn and Bacon.

Humen, D. P., & Boughner, D. R. (1984). Evaluation of commercially available heart rate monitors. *Canadian Medical Association Journal*, *131*(6), 585–588. PMID:6478343

Husain, A. (2011). Problem-based learning: A current model of education. *Oman Medical Journal*, *26*, 295. doi:10.5001/omj.2011.74 PMID:22043442

Hwang, G. J., Yang, T. C., Tsai, C. C., & Yang, S. J. H. (2009). A context-aware ubiquitous learning environment for conducting complex science experiments. *Computers & Education*, *53*(2), 402–413. doi:10.1016/j.compedu.2009.02.016

Ikemoto, G. S., & Marsh, J. A. (2007). Chapter 5: Cutting through the "data-driven" mantra: Different conceptions of data-driven decision making. *Yearbook of the National Society for the Study of Education*, *106*(1), 105–131. doi:10.1111/j.1744-7984.2007.00099.x

Improving ed-tech purchasing: Identifying the key obstacles and potential solutions for the discovery and acquisition of personalized learning tools. (2014). *Digital Promise & Education Industry Association*. Retrieved from http://www.digitalpromise.org/page/-/dpdocuments/marketplace/Improving_Ed-Tech_Purchasing.pdf?nocdn=1

Introduction to wearable technology. (2014). *Wearable Devices*. Retrieved from http://www.wearabledevices.com/what-is-a-wearable-device/

Jackson, L. (2004). One-to-one computing: Lessons learned and pitfalls to avoid. *Education World*. Retrieved from http://www.educationworld.com/a_tech/tech/tech197.shtml

Jeng, Y.-L., Wu, T.-T., Huang, Y.-M., Tan, Q., & Yang, S. J. H. (2010). The add-on impact of mobile applications in learning strategies: A review study. *Journal of Educational Technology & Society*, *13*(3), 3–11.

Jenkinson, J. (2009). Measuring the effectiveness of educational technology: What are we attempting to measure? *Electronic Journal of e-Learning*, 7(3), 273–280.

Jeonghee, K., Hangue, P., Bruce, J., Rowles, D., Holbrook, J., Nardone, B., & Ghovanloo, M. et al. (2014). Qualitative assessment of Tongue Drive System by people with high-level spinal cord injury. *Journal of Rehabilitation Research and Development*, *51*(3), 451–465. PMID:25019667

Jewitt, C. (2012). An Introduction to Using Video for Research. *NCRM Working paper*. London

Johnson, D. (2015). The revolution will be wearable. *Industrial Safety & Hygiene News*, *49*(5), 1–30.

Johnson, D. L., & Maddux, C. D. (2008). Introduction: Effectiveness of information technology in education. *Computers in the Schools*, *24*(3/4), 1–6.

Johnson, L., Adams, S., Cummins, M., Estrada, V., Freeman, A., & Ludgate, H. (2013). *The NMC Horizon Report: 2013 Higher Education Edition*. New Media Consortium.

Jonassen, D. (2000). *Computers as mindtools for schools. Engaging critical thinking* (2nd ed.). Saddle River, NJ: Prentice-Hall.

Jordan, E., & Balmain, K. G. (1968). *Electromagnetic waves and radiating systems* (2nd ed.). Prentice-Hall.

Judson, E. (2002). Learning from past and present: Electronic response systems in college lecture halls. *Journal of Computers in Mathematics and Science Teaching*, *21*(2), 167–181.

Kadirire, J. (2009). Mobile Learning Demystified. In R. Guy (Ed.), *The Evolution of Mobile Teaching and Learning* (pp. 15–56). California, USA: Informing Science Press.

Kafai, Y. B., Peppler, K. A., Burke, Q., Moore, M., & Glosson, D. (2010, June). Fröbel's forgotten gift: textile construction kits as pathways into play, design and computation.*Proceedings of the 9th International Conference on Interaction Design and Children* (pp. 214-217). doi:10.1145/1810543.1810574

Kafai, Y., Fields, D., & Searle, K. (2013). Making connections across disciplines in high school e-textiles workshops. In L. Buechley, K. Peppler, M. Eisenberg, & Y. Kafai (Eds.), *Textile Messages: Dispatches from the World for e-Textiles and Education*. New York, NY: Peter Lang.

Kaivanto, E. K., Berg, M., Salonen, E., & de Maagt, P. (2011). Wearable circularly polarized antenna for personal satellite communication and navigation. *IEEE Transactions on Antennas and Propagation*, *59*(12), 4490–4496. doi:10.1109/TAP.2011.2165513

Karulf, E. (2008). Body area networks (BAN). Retrieved from http://www.cse.wustl.edu/~jain/cse574-08/ftp/ban.pdf

Katterfeldt, E., Dittert, N., & Schelhowe, H. (2009). EduWear: smart textiles as ways of relating computing technology to everyday life. In *Proceedings of the 8th International Conference on Interaction Design and Children* (IDC '09) (pp. 9-17). ACM, New York. DOI: http://doi.acm.org/10.1145/1551788.155179110.1145/1551788.1551791

Kay, R. H., & LeSage, A. (2009). A strategic assessment of audience response systems used in higher education. *Australasian Journal of Educational Technology*, *25*(2), 235–249.

Kay, R. H., & LeSage, A. (2009). Examining the benefits and challenges of using audience response systems: A review of the literature. *Computers & Education*, *53*(3), 819–827. doi:10.1016/j.compedu.2009.05.001

Keller, J. M. (1987). Development and use of the ARCS model of motivational design. *Journal of Instructional Development*, *10*(3), 2–10. doi:10.1007/BF02905780

Kelly, P. (2013). *Assessing the utility of wearable cameras in the measurement of walking and cycling* [Doctoral dissertation]. University of Oxford.

Kelly, P., Doherty, A., Berry, E., Hodges, S., Batterham, A. M., & Foster, C. (2011). Can we use digital life-log images to investigate active and sedentary travel behaviour? Results from a pilot study. *The International Journal of Behavioral Nutrition and Physical Activity*, *8*(1), 44. doi:10.1186/1479-5868-8-44 PMID:21599935

Kelly, P., Marshall, S. J., Badland, H., Kerr, J., Oliver, M., Doherty, A. R., & Foster, C. (2013). An ethical framework for automated, wearable cameras in health behavior research. *American Journal of Preventive Medicine*, *44*(3), 314–319. PMID:23415131

Kersten, P., & Ribbens, C. (2015). *Wearable games*. Retrieved from http://wearablegames.eu/

Kim, B. (2015). Keeping up with….gamification. *Association of College & Research Libraries*. Retrieved from http://www.ala.org/acrl/publications/keeping_up_with/gamification

Kim, P., Hagashi, T., Carillo, L., Gonzales, I., Makany, T., Lee, B., & Gàrate, A. (2010). Socioeconomic strata, mobile technology, and education: A comparative analysis. *Educational Technology Research and Development*, *59*(4), 465–486. doi:10.1007/s11423-010-9172-3

Kiriakidis, P. (2010). How Does Skype, as an Online Communication Software Tool, Contribute to K-12 Administrators' Level of Self-efficacy? *International Journal of Leadership in Educational Technology*, *1*(9), 1–29.

Klaiber, A. (2000). *The technology behind crusoe processors. Low-Power X86-Compatible Processors Implemented with Code Morphing Software*. Transmeta Technical Brief.

Klegeris, A., & Hurren, H. (2011). Impact of problem-based learning in a large classroom setting: Student perception and problem-solving skills. *Advances in Physiology Education*, *35*(4), 408–415. doi:10.1152/advan.00046.2011 PMID:22139779

Klopfer, E., & Sheldon, J. (2010). Augmenting your own reality: Student authoring of science-based augmented reality games. *New Directions for Youth Development*, *2010*(128), 85–94. doi:10.1002/yd.378 PMID:21240956

Klymkowsky, M. W., Underwood, S. M., & Garvin-Doxas, R. K. (2010). Biological Concepts Instrument (BCI): A diagnostic tool for revealing student thinking.

Kolikant, Y. B. D., Drane, D., & Calkins, S. (2010). "Clickers" as Catalysts for Transformation of Teachers. *College Teaching*, *58*(4), 127–135. doi:10.1080/87567551003774894

Komljenovica, D., Grovesb, W. A., & Kecojevic, V. J. (2008). Injuries in U.S. mining operations – A preliminary risk analysis. *Safety Science*, *46*(5), 792–801. doi:10.1016/j.ssci.2007.01.012

Krasilchik, M. (2000). Reformas e realidade: O caso do ensino das ciências. *São Paulo em Perspectiva*, *14*(1), 85–93. doi:10.1590/S0102-88392000000100010

Krueger, K. (2013). Forget ROI, the future of technology investment is all about value. *T.H.E. Journal*, *40*(6), 25–28.

Kuhlthau, C. C. (1991). Inside the search process: Information seeking from the user's perspective. *Journal of the American society for information science*, 42(5), 361-371. Retrieved from http://ptarpp2.uitm.edu.my/silibus/insidesearch2.pdf

Kukulska-Hulme, A. (2005). Mobile usability and user experience. In A. Kukulska-Hulme & J. Traxler (Eds.), *Mobile learning: A handbook for educators and trainers* (pp. 45–56). London: Routledge.

Kukulska-Hulme, A., Sharples, M., Milrad, M., Arnedillo-Sánchez, I., & Vavoula, G. (2009). Innovation in mobile learning: A European perspective. *International Journal of Mobile and Blended Learning*, *1*(1), 13–35. doi:10.4018/jmbl.2009010102

Kukulska-Hulme, A., & Traxler, J. (2007). Learning design with mobile and wireless technologies. In H. Beetham & R. Sharpe (Eds.), *Rethinking pedagogy for the digital age: Designing and delivering e-learning* (pp. 180–192). London, UK: Routledge.

Kumar, E., & Kohli, S. (2007). A strategic analysis of search engine advertising in web based-commerce. *Journal of Internet Banking and Commerce*, *12*(2), 1–13.

Kumar, S. (2015). Biosafety and Biosecurity Issues in Biotechnology Research. *Biosafety*, *4*(01), 153. doi:10.4172/2167-0331.1000e153

Kumar, V., Venkatesan, R., & Reinartz, W. J. (2006). Knowing what to sell when to whom. *Harvard Business Review*, May, 131–145. PMID:16515161

L'istruzione programmata. (1970AA.VV. Roma, Italy: Bulzoni.

Lachapelle, C. P., Phadnis, P., Jocz, J., & Cunningham, C. M. (2012). The impact of engineering curriculum units on students' interest in engineering and science. *Paper presented at theNARST Annual International Conference*, Indianapolis, IN, USA.

Lage, M. J., Platt, G. J., & Treglia, M. (2000). Inverting the classroom: A gateway to creating an inclusive learning environment. *The Journal of Economic Education*, *31*(1), 30–43. doi:10.1080/00220480009596759

Lagemann, E. C., & Shulman, L. S. (1999). *Issues in education research: problems and possibilities*. San Francisco: Jossey-Bass Publishers.

Lahlou, S. (2011). How can we capture the subject's perspective?: An evidence-based approach for the social scientist. *Social Sciences Information. Information Sur les Sciences Sociales*, *50*(4), 607–655. doi:10.1177/0539018411411033

Lalli, F., & Prunesti, A. (2013). *Wearable: Connect you with everything*. IQUII Social & Mobile Factory.

Lantz, M. E. (2010). The use of 'Clickers' in the classroom: Teaching innovation or merely an amusing novelty? *Computers in Human Behavior*, *26*(4), 556–561. doi:10.1016/j.chb.2010.02.014

Laudon, C. K., & Laudon, J. P. (2007). *Essentials of Business Information Systems*. Upper Saddle River, NJ: Pearson Education.

Laukkanen, R. M. T., & Virtanen, P. K. (1998). Heart rate monitors: State of the art. *Journal of Sports Sciences, 16*(Suppl. 1), 3–7. Doi:10.1080/026404198366920

Laurillard, D. (2012). *Insegnamento come scienza della progettazione. Costruire modelli pedagogici per apprendere con la tecnologia*. Milano, Italy: Franco Angeli.

Lauritzen, J., Muñoz, A., Luis Sevillano, J., & Civit, A. (2013). The usefulness of activity trackers in elderly with reduced mobility: A case study. In Studies in Health Technology and Informatics (Vol. 192, pp. 759–762). Doi:10.3233/978-1-61499-289-9-759

Lavery, H. (2012). Breaking down barriers with tablet technology for teaching. *Educause Review Online*.

Le Tellier, A. (2009). And you thought the Jawbone Headset was stylish. *LA Times.com*. Retrieved from http://latimesblogs.latimes.com/alltherage/2009/07/and-you-thought-the-jawbone-headset-was-stylish.html

Learner-centered psychological principles: A framework for school reform and redesign. (1997). Group of the Board of Educational Affairs. Washington, DC: American Psychological Association.

Learning science in informal environments: People, places, and pursuits. (2009National Research Council. Washington, DC: The National Academies Press.

Least developed countries: A UN classification. (2015). *World Bank.* Retrieved from http://data.worldbank.org/region/LDC

Lebeaux, R. (2013, October). Technology leadership quotes from the October 'CIO Decisions' e-zine. *TechTarget.* Retrieved from http://searchcio.techtarget.com/opinion/Technology-leadership-quotes-from-the-October-CIO-Decisions-e-zine

Lee, K., & Miller, K. E. (2003, December 1-3). A positive attitude-ad-brand relationship by customising banner advertisement design- An empirical study. *Proceedings of ANZMAC 2003 Conference* (pp. 25-33).

Lee, K., & Salman, R. (2012). The Design and Development of Mobile Collaborative Learning Application Using Android. *Journal of Information Technology and Application in Education,* 1(1), 1-8.

Lee, J. B., Mellifont, R. B., & Burkett, B. J. (2010). The use of a single inertial sensor to identify stride, step, and stance durations of running gait. *Journal of Science and Medicine in Sport, 13*(2), 270–273. doi:10.1016/j.jsams.2009.01.005 PMID:19574098

Lee, V. (2014). Learning Technologies and the Body: Integration and Implementation. In *Formal and Informal Learning Environments.* New York, NY: Routledge.

Lewis, W., Agarwal, R., & Sambamurthy, V. (2003). Sources of influence on beliefs about information technology use: An empirical study of knowledge workers. *Management Information Systems Quarterly, 27*(4), 657–678.

Li, T., Mei, T., Niu, X., & Ngo, C. (2010, April 26–30). PageSense: Style-wise web page advertising. *Proceedings of the World Wide Web,* Raleigh, North Carolina (pp. 1273-1276).

Lipman, V. (2014). 71% of 16-to-24-year-olds want wearable tech. Why don't I even want to wear a watch? *Forbes.com.* Retrieved from http://www.forbes.com/sites/victorlipman/2014/09/22/71-of-16-24s-want-wearable-tech-why-dont-i-even-want-to-wear-a-watch/

Liu, C., & Arnett, K. P. (2000). Exploring the factors associated with web site success in the context of electronic commerce. *Information & Management, 38*(1), 28–33. doi:10.1016/S0378-7206(00)00049-5

Livneh, H. (2012). On the origins of negative attitudes toward people with disabilities. In I. Marini & M. Stebricki (Eds.), *The psychological and social impact of illness and disability* (6th ed., pp. 13–14). New York, NY: Springer Publishing Company.

Long, T., Logan, J., & Waugh, M. (2014). Students' Perceptions of Pre-class Instructional Video in the Flipped Classroom Model: A Survey Study. *Proceedings of Society for Information Technology & Teacher Education International Conference.* VA: AACE.

Lovell, E., & Buechley, L. (2010, June). An e-sewing tutorial for DIY learning.*Proceedings of the 9th International Conference on Interaction Design and Children* (pp. 230-233). ACM. doi:10.1145/1810543.1810578

Lukowitz, P., Kirtein, T., & Troster, G. (2014). Wearable systems for health care applications. *Methods of Information in Medicine.* Retrieved from http://www.researchgate.net/publication/8480044_Wearable_systems_for_health_care_applications

Maas, T., & Lake, R. (2015). A blueprint for effective and adaptable school district procurement. *CRPE.* Retrieved from http://www.crpe.org/sites/default/files/crpe-report-effective-adaptable-school-district-procurement_0.pdf

Magolda, P. (2006). Critical fusion – Technology and equity in secondary education. *AACE Journal, 14*(3), 287–311.

Maidment, D. (2014). Advanced architectures and technologies for the development of wearable devices. *ARM*. Retrieved from http://www.arm.com/files/pdf/Advanced-Architectures-and-Technologies-for-the-Development-of-Wearable.pdf

Majority of the disabled facing discrimination, poverty. (2014, December 3). *Nation (Pakistan)*.

Maltese, A. V., & Tai, R. H. (2010). Eyeballs in the fridge: Sources of early interest in science. *International Journal of Science Education*, *32*(5), 669–685. doi:10.1080/09500690902792385

Maltese, A. V., & Tai, R. H. (2011). Pipeline persistence: Examining the association of educational experiences with earned degrees in STEM among US students. *Science Education*, *95*(5), 877–907. doi:10.1002/sce.20441

Mandinach, E. B., Honey, M., & Light, D. (2006, April). A theoretical framework for data-driven decision making. *Paper presented at theannual meeting of the American Educational Research Association*, San Francisco, CA.

Mandinach, E. B., & Honey, M. (2008). *Data-driven school improvement: Linking data and learning. Technology, Education Connections (TEC) Series*. New York, NY: Teachers College Press.

Mann, E. L., Mann, R. L., Strutz, M., Duncan, D., & Yoon, S. Y. (2011). Integrating engineering into K-6 curriculum: Developing talent in the STEM disciplines. *Journal of Advanced Academics*, *22*(4), 639–658. doi:10.1177/1932202X11415007

Mann, S. (1997). "Smart clothing": wearable multimedia computing and "personal imaging" to restore the technological balance between people and their environments.*Proceedings of the fourth ACM international conference on Multimedia* (pp. 163-174). ACM.

Mann, S. (1997). Wearable computing: A first step toward personal imaging. MIT, Cambridge, MA. *IEEE Computer Journal*, *30*(2), 24–28.

Mann, S., & Ferenbok, J. (2013). New Media and the Power Politics of Sousveillance in a Surveillance-Dominated World. *Surveillance & Society*, *11*(1/2), 18–34.

Maragliano, E. (1994). *Manuale di didattica multimediale*. Bari, Italy: Laterza.

Marchionini, G. (1995). *Information seeking in electronic environments*. University of Maryland. Retrieved from http://catdir.loc.gov/catdir/samples/cam031/94044629.pdf

Markoff, J. (2014). Device from Israeli start-up gives the visually impaired a way to read. *New York Times*, A5.

Martin, H. (2007). Stash your gadgets in these tech-ready togs. *LA Times.com*. Retrieved from http://www.latimes.com/travel/la-trw-roadtest24jun24-story.html

Martins, F. P. S., & Pereira, F. L. (2012). *O Ensino de Imunologia: Análise dos Principais Métodos e Recursos Didáticos Utilizados em Universidades Brasileiras*. Uberlândia: II Fórum Internacional Sobre Prática Docente Universitária.

Marzano, R., Pickering, D., & Pollock, J. (2001). *Classroom instruction that works*. Alexandria, VA: ASCD.

Mateescu, A., Rosenblat, A., & Boyd, D. (2015). Police Body-Worn Cameras (Working Paper). NYC: Data & Society Research Institute

Matsuda, Y., & Isomura, T. (2012). Finger Braille recognition system. In X. Ding (Ed.), *Advances in Character Recognition* (pp. 193–210). Rijeka, Croatia: Intech Open.

Matusov, E. (2007). Applying Bakhtin Scholarship on Discourse in Education: A Critical Review Essay. *Educational Theory*, *57*(2), 215–237. doi:10.1111/j.1741-5446.2007.00253.x

Mayer, R. E. (2009). *Multimedia learning* (2nd ed.). New York: Cambridge University Press. doi:10.1017/CBO9780511811678

Mayer, R. E., Stull, A., DeLeeuw, K., Almeroth, K., Bimber, B., Chun, D., & Zhang, H. et al. (2009). Clickers in college classrooms: Fostering learning with questioning methods in large lecture classes. *Contemporary Educational Psychology, 34*(1), 51–57. doi:10.1016/j.cedpsych.2008.04.002

Mazza, S., (2015). Insegnare ai tempi del cloud. eBook.

Mccall, C., & Mccall, W. V. (2012). Comparison of actigraphy with polysomnography and sleep logs in depressed insomniacs. *Journal of Sleep Research, 21*(1), 122–127. doi:10.1111/j.1365-2869.2011.00917.x PMID:21447050

McCombs, B., & Whisler, J. S. (1997). *The learner-centered classroom and school: Strategies for increasing student motivation and achievement.* San Francisco, CA: Jossey-Bass.

McElheran, K. S. (2012). *Do market leaders lead in business process innovation? The case(s) of e-business adoption.* Boston: Harvard Business School.

McGrath, E., Lowes, S., McKay, M., Sayres, M., & Lin, P. (2012). Robots underwater! Learning science, engineering, and 21st century skills: The evolution of curricula, professional development and research in formal and informal contexts. In B. S. Barker, G. Nugent, N. Grandgenett, & V. I. Adamchuk (Eds.), *Robots in K-12 education: A new technology for learning* (pp. 141–167). Hershey, PA, USA: IGI Global. doi:10.4018/978-1-4666-0182-6.ch007

McGuire, P.C., Elvira, J.G., Manfredi, J.A., Martinez, E.S., Ormo, J., Martinez, E.D., Ritter, H., Oesker, H., Haschke, R. & Ontrup, J., (2004). Field Geology with a wearable computer: First Results of the Cyborg Astrobiologist System.

Mcnamara, D. J., Gabbett, T. J., Chapman, P., Naughton, G., & Farhart, P. (2015). The Validity of Microsensors to Automatically Detect Bowling Events and Counts in Cricket Fast Bowlers. *International Journal of Sports Physiology and Performance, 10*, 71–75. Retrieved from10.1123/ijspp.2014-0062

Mearian, L. (2012). 'Body area networks' should free hospital bandwidth, untether patients (at-home use of disposable wireless monitors could be the biggest market of all. *Computerworld.* Retrieved from http://www.computerworld.com/article/2503882/healthcare-it/-body-area-networks--should-free-hospital-bandwidth--untether-patients.html

Mehalik, M. M., Doppelt, Y., & Schunn, C. D. (2008). Middle-school science through design-based learning versus scripted inquiry: Better overall science concept learning and equity gap reduction. *The Journal of Engineering Education, 97*(1), 71–85. doi:10.1002/j.2168-9830.2008.tb00955.x

Mei, T., Hua, X. S., Yang, L., & Li, S. (2007). VideoSense: towards effective online video advertising.*Proceedings of ACM MULTIMEDIA* (pp. 1075-1084).

Melanson, E. L., Knoll, J. R., Bell, M. L., Donahoo, W. T., Hill, J. O., Nysse, L. J., & Levine, J. A. et al. (2004). Commercially available pedometers: Considerations for accurate step counting. *Preventive Medicine, 39*(2), 361–368. doi:10.1016/j.ypmed.2004.01.032 PMID:15226047

Melchoir, A., & Cohen, F. (2005). *More than Robots: An evaluation of the FIRST robotics competition participant and institutional impacts.* Waltham, MA: Center of Youth and Communities, Heller School for Social Policy and Management, Brandeis University.

Mendel, G. (1996). Experiments in plant hybridization (1865). Verhandlungen des Naturforschenden Vereins Brünn.

Messick, S. (1994). The interplay of evidence and consequences in the validation of performance assessments. *Education Researcher, 23*(2), 13–23. doi:10.3102/0013189X023002013

Metz, R. (2014). Google Glass is dead: Long live smart glasses. *MIT Technology Review.* Retrieved from http://www.technologyreview.com/featuredstory/532691/google-glass-is-dead-long-live-smart-glasses/

Metz, R. (2015). Google Glass Is Dead; Long Live Smart Glasses. *MIT Technology Review*, 79.

Meyer, K. (2003). Face-to-face versus threaded discussions: The role of time and higher-order thinking. *Journal of Asynchronous Learning Networks*, *7*(3), 55–65.

Michael, J. (2007). What makes physiology hard for students to learn? Results of a faculty survey. *Advances in Physiology Education*, *31*(1), 34–40. doi:10.1152/advan.00057.2006 PMID:17327580

Michels, P. (2013, November 5). The high cost of iPad-mania in public schools. *The Texas Observer*. Retrieved from http://www.texasobserver.org/run-school-ipad-program/

Mihart, C. (2012). Impact of integrated marketing communication on consumer behaviour: Effects on consumer decision – Making process. *International Journal of Marketing Studies*, *4*(2), 121–129. doi:10.5539/ijms.v4n2p121

Milestones in Computer Development . (n. d.) Retrieved from http://www.cis.usouthal.edu/faculty/daigle/project1/timeline.htm

Militello, M., & Friend, J. (2013). *Principal 2.0: Technology and educational leadership*. Charlotte, NC: Information Age.

Miller, A. M. (2014). Wearable tech for people with disabilities. *U.S. News Health*. Retrieved from http://health.usnews.com/health-news/health-wellness/articles/2014/11/20/wearable-tech-for-people-with-disabilities

Miller, M. (2015). *The Internet of Things: How Smart TVs, Smart Cars, Smart Homes, and Smart Cities Are Changing the World*. Indianapolis, IN: Que Publishing.

Ministry of Education, FATIH Project. (2015). *MEB.gov*. Retrieved from http://www.meb.gov.tr/press-release/haber/8098/en/

Mohapatra, S. (2009). *Business process automation*. New Delhi: PHI Private Limited.

Molenda, M. (2003). In Search of the Elusive ADDIE Model. *Performance Improvement*, *42*(5), 34–36. doi:10.1002/pfi.4930420508

Moore, G. E. (1965). Cramming more components onto integrated circuits. *Electronics*, *38*(8), 114–117.

More than one million smart watches will be shipped in 2013. (2013). *ABIResearch*. Retrieved from https://www.abiresearch.com/press/more-than-one-million-smart-watches-will-be-shippe/

Moreira, M. A. (1999). Teorias de aprendizagem. São Paulo, SP: Editora pedagógica e universitária.

Morita, M. (2003). The mobile-based learning (MBL) in Japan. *Proceedings of the First Conference on Creating, Connecting and Collaborating through Computing* (C5'03) (pp. 128-129).

Morrison, J. R., Ross, S. M., Corcoran, R. P., & Reid, A. J. (2014, September 22). *Fostering market efficiency in ed-tech procurement*. Retrieved from http://b.3cdn.net/dpromise/4e2869abb217c9062f_z6m6v2v9u.pdf

Mortimer, E. F. & de Carvalho, A. M. P. (2013). Referenciais teóricos para análise do processo de ensino de ciências. *Cadernos de Pesquisa,* 96, 05-14.

Moss, R. (n. d.). Haptic technology: The next frontier in video games, wearables, virtual reality, and mobile electronics. *Gizmag.com*. Retrieved from http://www.gizmag.com/haptic-tech-vr-wearables-games-sightlence/35616/

Moss, K., & Crowley, M. (2011). Effective learning in science: The use of personal response systems with a wide range of audiences. *Computers & Education*, *56*(1), 36–43. doi:10.1016/j.compedu.2010.03.021

Mozilla Foundation and Peer 2 Peer University, in collaboration with The MacArthur Foundation (2012). *Open Badges for Lifelong Learning*. Retrieved from https://wiki.mozilla.org/images/b/b1/OpenBadges-Working-Paper_092011.pdf

Muhammad, M. A., & Hamida, E. B. (2014). Surveying wearable human assistive technology for life and safety critical applications: Standards, challenges and opportunities. Retrieved from http://www.mdpi.com/1424-8220/14/5/9153/htm

Murray, J. H. (1997). *Hamlet on the Holodeck*. Cambridge, MA: MIT Press.

Newman, J. (2015, January 12). Wearables go niche and narrow in the search for consumer relevance. *PC World*.

Nicholl, T. A., & Lou, K. (2012). A model for small-group problem-based learning in a large class facilitated by one instructor. *American Journal of Pharmaceutical Education*, *76*(6), 117. doi:10.5688/ajpe766117 PMID:22919093

Nielsen, J. (1990). Evaluating hypertext usability. In D. H. Jonassen & H. Mandl (Eds.), *Designing hypermedia for learning* (pp. 147–168). Berlin: Springer-Verlag. doi:10.1007/978-3-642-75945-1_9

Nielsen, J. (1993). *Usability engineering*. Boston: Academic Press.

Nielsen, J. (1997). Usability testing. In G. Salvendy (Ed.), *Handbook of Human Factors and Ergonomics* (2nd ed., pp. 1543–1568). John Wiley & Sons.

Niewolny, D. (2013) *How the internet of things is revolutionizing healthcare*. Retrieved from http://cache.freescale.com/files/corporate/doc/white_paper/IOTREVHEALCARWP.pdf

Noam, G. G., Biancarosa, G., & Dechausay, N. (2002). *Learning to bridge – bridging to learn: A model and action plan to increase engagement between schools and afterschool programs in Boston (Report commissioned by Boston's After-School for All Partnership Learning Goal Research)*. Cambridge, MA: Harvard Education Press.

Nortcliffe, A., & Middleton, A. (2013). The innovative use of personal smart devices by students to support their learning. In A. Wankel & P. Blessinger (Eds.), *Increasing student engagement and retention using mobile applications: Smartphones, Skype and texting technologies (cutting-edge technologies in higher education)*. Bingley, UK: Emerald Group Publishing Limited. doi:10.1108/S2044-9968(2013)000006D009

Norton, K. M. (2007). A brief history of prosthetics. *inMotion,17*(7). Retrieved from http://www.amputee-coalition.org/inmotion/nov_dec_07/history_prosthetics.html

Nugent, G., Barker, B., Grandgenett, N., & Welch, G. (2014). Robotics camps, clubs, and competitions: Results from a US robotics project. *Proceedings of 4th International Workshop Teaching Robotics, Teaching with Robotics &5th International Conference Robotics in Education Padova (Italy) July* (Vol. 18, No. 2014, pp. 11-18).

Nugent, G., Barker, B., & Grandgenett, N. (2012). The impact of educational robotics on student STEM learning, attitudes and workplace skills. In B. S. Barker, G. Nugent, N. Grandgenett, & V. I. Adamchuk (Eds.), *Robotics in K–12 education: A new technology for learning*. Hershey, PA, USA: IGI Global. doi:10.4018/978-1-4666-0182-6.ch009

Nugent, G., Barker, B., Grandgenett, N. F., & Adamchuk, V. (2010). Impact of robotics and geospatial technology interventions on youth STEM learning and attitudes. *Journal of Research on Technology in Education*, *42*(4), 391–408. doi:10.1080/15391523.2010.10782557

Nugent, G., Barker, B., Welch, G., Grandgenett, N., Wu, C., & Nelson, C. (2015). A model of factors contributing to STEM learning and career orientation. *International Journal of Science Education*, *37*(7), 1067–1088. doi:10.1080/09500693.2015.1017863

O'Brien, J. (2015, November 16). Breaking a different kind of language barrier: Sign language becomes sensor-based. *Mashable*.

O'Driscoll, G. (2014). Your Definitive Guide to Wearables. Dublin, IR: HomeMentors.

O'Toole, M. L., Douglas, P. S., & Hiller, W. D. B. (1998). Use of heart rate monitors by endurance athletes: Lessons from triathletes. *The Journal of Sports Medicine and Physical Fitness*. PMID:9830823

On this day in history. (2015). Historyorb. Retrieved from http://www.historyorb.com/

Online retail will drive innovation in interactive marketing. (2015). *Marketing Week*. Retrieved from http://bit.ly/1nbpjdM

Otani, A. (2013). Cornell music professor becomes Google Glass explorer.

Otrel-Cass, K., Cowie, B., & Maguire, M. (2010). Taking Video Cameras Into The Classroom. *Waikato Journal of Education, 15(2)*. doi:10.15663/wje.v15i2.117

Padden, C., & Humphries, T. (1988). *Deaf in America: Voices from a culture*. Cambridge, MA: Harvard University Press.

Palfrey, J., & Gasser, U. (2013). *Born digital: Understanding the first generation of digital natives*. Philadelphia, PA: Basic Books.

Parâmetros Curriculares Nacionais: Ciências Naturais. (1997). *Secretaria de Educação Fundamental*, Brasil.

Park, S., & Jayaraman, S. (2004). *E-health and quality of life: The role of wearable motherboard*. US National Library of Medicine, National Institute of Health: IOS Press.

Partnership for 21st Century Skills. (2011). *Common Core toolkit: A guide to aligning the Common Core Standards with the framework for 21st century skills*. Retrieved from http://www.p21.org/storage/documents/P21CommonCoreToolkit.pdf

Patassini, A. (2015). Google Drive e la didattica. Asiago, Italy: Coll@borare.

Pele, M., & Cimpeanu, C. (2012). *Biotechnology: An Introduction*. Boston, MA: Wit Press.

Pellerey, M., (1978). Compiti e insidie della tecnologia dell'educazione. Orientamenti pedagogici, 4, 7-13.

Peluso, D., (2012). The fast-paced iPad revolution: Can educators stay up to date and relevant about these ubiquitous devices? *British Journal of Educational Technology, 43(4)*, 125.-127.

Peppler, K., & Glosson, D. (2013). Stitching circuits: Learning about circuitry through e-textile materials. *Journal of Science Education and Technology*, *22*(5), 751–763. doi:10.1007/s10956-012-9428-2

Peppler, K., Gresalfi, M., Salen, K., & Santo, R. (2014). *Soft circuits: Crafting e-fashion with DIY electronics*. Cambridge, MA: MIT Press.

Peppler, K., Salen, K., Gresalfi, M., & Santo, R. (2014). *Short circuits: Crafting e-puppets with DIY electronics*. Cambridge, MA: MIT Press.

Peppler, K., Sharpe, L., & Glosson, D. (2013). E-Textiles and the new fundamentals of fine arts. In L. Buechley, K. Peppler, M. Eisenberg, & Y. Kafai (Eds.), *Textile Messages: Dispatches from the World for e-Textiles and Education*. New York, NY: Peter Lang.

Pereira, A., Miller, T., Huang, Y. M., Odell, D., & Rempel, D. (2013). Holding a tablet computer with one hand: Effect of tablet design features on biomechanics and subjective usability among users with small hands. *Ergonomics*, *56*(9), 1363–1375. doi:10.1080/00140139.2013.820844 PMID:23909815

Piaget, J. (1972). *The child's conception of the world*. Towota, NJ: Littlefield Adams.

Pichai, S. (2014). Sharing what's up our sleeve: Android coming to wearables. *Google Blog*. Retrieved from https://googleblog.blogspot.com/2014/03/sharing-whats-up-our-sleeve-android.html

Pickert, K. (2014, October 17). *How the iPad helped bring down the Los Angeles schools chief.* Retrieved from http://time.com/3514155/ipad-john-deasy-lausd-superintendent-resigns/

Pielot, M., & Boll, S. (2010). Tactile wayfinder: Comparison of tactile waypoint navigation with commercial pedestrian navigation systems. *Proceedings of the 8th International Conference on Pervasive Computing* (pp. 76-93).

Płomińska, N., Smagurauskaite, A., Platonova, E., & Gadzhuk, O. (2014). *Consumers go mobile in CEE- Mobile market overview.* Retrieved from http://www.iabeurope.eu/files/4414/1898/2815/Gemius_Consumers_go_mobile_in_CEE_2014_final.pdf

Polliack, A., & Scheinberg, S. (2006). A new technology for reducing shear and friction forces on the skin: Implications for blister care in the wilderness setting. *Wilderness & Environmental Medicine*, *17*(2), 109–119. doi:10.1580/PR30-05.1 PMID:16805147

Porta, M., & Rastelli, S. (2013). Lo studio dei tracciati oculari (eye-tracking) nella ricerca sul linguaggio In La ricerca sperimentale sul linguaggio: acquisizione, uso, perdita (pp. 165-197). Pavia University Press.

Postman, N. (1995). *The End of Education: redefining the value of school.* New York: Knopf.

Preece, J., Rogers, Y., Sharp, H., Benyon, D., Holland, S., & Carey, T. (1994). *Human-computer interaction, Workingham.* England: Addison-Wesley.

Prensky, M. (2008). The role of technology in teaching and the classroom. *Educational Technology*, 2008, 1–3.

Press alert: Global market for wearable technology worth $8.9bn in 2014. (2015, April 9). *Futuresource Consulting.* Retrieved from http://www.futuresource-consulting.com/2015-04-Wearable-Technology-7723.html

Privitera, D. (2015). Tourist valorisation of urban poverty: An empirical study on the web. *Urban Forum* (pp. 1-29). doi:10.1007/s12132-015-9259-3

Pulley, J. (2014). Exploring new frontiers with Google Glass. *Campus Technology Magazine*, *27*(11), 31–33.

Pultarova, T. (2015). Researchers propose body WiFi for the wearable technology age. *Engineering and Technology Magazine.* Retrieved from http://eandt.theiet.org/news/2015/apr/body-wifi.cfm

Qiu, K., Buechley, L., Baafi, E., & Dubow, W. (2013, June). A curriculum for teaching computer science through computational textiles.*Proceedings of the 12th International Conference on Interaction Design and Children* (pp. 20-27). ACM. doi:10.1145/2485760.2485787

Racoma, J. A. (2013). Wearable tech: A brief history and a look into the future. *Android Authority.* Retrieved from http://www.androidauthority.com/wearable-computing-history-238324/

Ranger, S. (2015). Before the Apple iWatch: A history of smart watches in pictures. *ZDNet.* Retrieved from [REMOVED HYPERLINK FIELD]http://www.zdnet.com/pictures/before-the-iwatch-a-history-of-smart-watches-in-pictures/8/

Rapporto annuale. (2011ISTAT. Roma, Italy: Centro Stampa e Riproduzione.

Rayport, J. F., & Bernard, J. J. (2004). Best face forward. *Harvard Business Review*, *82*(12), 47–52. PMID:15605565

Regni, R. (2004). *Geopedagogia. L'educazione tra globalizzazione, tecnologia e consumo.* Milano, Italy: Armando Editore.

Rego, M. C. (2000). *Vygotsky: uma perspectiva histórico cultural da educação.* Petrópolis, RJ: Editora Vozes Limitada.

Rehmat, A. P., & Bailey, J. M. (2014). Technology Integration in a Science Classroom: Preservice Teachers' Perceptions. *Journal of Science Education and Technology*, *23*(6), 744–755. doi:10.1007/s10956-014-9507-7

Reid, R. L. (1972). *Tecnologia dell'educazione. Lo sfondo e la necessità*. Brescia, Italy: La Scuola.

Reiser, R. A. (2001). A history of instructional design and technology: Part II: A history of instructional design. *Educational Technology Research and Development*, *49*(2), 57–67. doi:10.1007/BF02504928

Research Highlights in Technology and Teacher Education. (2011). *Ed/ITLib Digital Library*. Retrieved from http://storefront.acculink.com/aace

Ribeiro, J. (2015). *Educational technology decision-making: A case study on technology acquisition for 746,000 Ontario students* [Unpublished master's thesis]. Brock University, Ontario, Canada. Retrieved from https://dr.library.brocku.ca/handle/10464/6318

Ribeiro, J. (2014). Teacher technology adoption and the philosophy of fear. *Antistasis*, *4*(2), 30–33.

Richardson, J. C., & Ice, P. (2010). Investigating students´ level of critical thinking across instructional strategies in online discussions. *The Internet and Higher Education*, *13*(1-2), 52–59. doi:10.1016/j.iheduc.2009.10.009

Richardson, W. (2012). Gaming gains respect. *District Administration*, *48*(6), 50–51.

Richey, R. C., & Klein, J. D. (2014). *Design and development research: Methods, strategies, and issues*. New York, NY: Routledge. doi:10.1007/978-1-4614-3185-5_12

Rich, R. L. (2011). *A framework for synchronous web-based professional development [Thesis]*. Stockton, California.

Ripat, J., & Woodgate, R. (2011). The intersection of culture, disability and assistive technology. *Disability and Rehabilitation. Assistive Technology*, *6*(2), 87–96. doi:10.3109/17483107.2010.507859 PMID:20698763

Rising Above the Gathering Storm, Revisited: Rapidly Approaching Category 5 National Research Council. (2010). Washington, DC: The National Academies Press.

Riskowski, J., Todd, C., Wee, B., Dark, M., & Harbor, J. (2009). Exploring the effectiveness of an interdisciplinary water resources engineering module in an eighth grade science course. *International Journal of Engineering Education*, *25*(1), 181–195.

Rivoltella, P. C. (2014). Smart future. Didattica, media digitali ed inclusione. Milano, Italy: Franco Angeli.

Robak, M. J., & Ajmi, A. (2015). Google glass for the educator: A postmortem separating the reality from the hype and some thoughts for Google. *Computers in Libraries*, *3*, 12–15.

Robinshon, S. B. (1976). *Curricula scolastici come fondamento di ogni riforma*. Roma, Italy: Armando.

Rockinson-Szapkiw, A. J., Courduff, J., Carter, K., & Bennett, D. (2013). Electronic versus traditional print textbooks: A comparison study on the influence of university students' learning. *Computers & Education*, *63*, 259–266. doi:10.1016/j.compedu.2012.11.022

Roehl, A., Reddy, S. L., & Shannon, G. J. (2013). The Flipped Classroom: An Opportunity to Engage Millennial Students through Active Learning Strategies. *Journal of Family and Consumer Sciences*, *105*(2), 44–49. doi:10.14307/JFCS105.2.12

Romney, C. A. (2011). Tablet PC use in freshman mathematics classes promotes stem retention. *Proceedings of the 41st annual Frontiers in Education Conference* (pp. F1J-1 - F1J-7). doi:10.1109/FIE.2011.6142773

Rood, R. (2013). New closed-captioning glasses help deaf go out to the movies. *NPR.org*. Retrieved from http://www.npr.org/sections/alltechconsidered/2013/05/12/183218751/new-closed-captioning-glasses-help-deaf-go-out-to-the-movies

Ror, L. (2014). Moto 360: it's time. *Motorola.com.* Retrieved from http://motorola-blog.blogspot.com/2014/03/moto-360-its-time.html

Rosine, L. L. (1964). Advances in electronic circuit packaging.*Proceedings of the Fifth International Electronic Circuit Packaging Symposium, Boulder*, Colorado.

Roth, G. A., & Fee, E. (2011). The Invention of Braille. *American Journal of Public Health, 101*(3), 454. doi:10.2105/AJPH.2010.200865 PMID:21307377

Roullard, S. (2013). *The Internet of Things in the library.* Retrieved from http://libserra.com/the-internet-of-things-in-the-library/

Rowland, J. L., & Sasaki, J. E., Hickey, a, Mavilia, M., Tedesco, J., John, D., … Freedson, P. S. (2014). Validation of the Fitbit Wireless Activity Tracker(R) for Prediction of Energy Expenditure. *Journal of Medical Internet Research*, 149–155. doi:10.1123/jpah.2012-0495

Rubin, J. (1994). *Handbook of usability testing: How to plan, design, and conduct effective tests.* New York, NY: John Wiley & Sons.

Rusco, L. (2015). Wichita State professors invent accessory lens for Google Glass. *Witchita State News.* Retrieved from http://www.wichita.edu/thisis/stories/story.asp?si=2743

Rust, R. T., Lemon, K. N., & Zeithaml, V. A. (2004). Return on marketing: Using customer equity to focus marketing strategy. *Journal of Marketing*, *68*(1), 109–127. doi:10.1509/jmkg.68.1.109.24030

Ryan, E. B., Bajorek, S., Beaman, A., & Anas, A. P. (2005). "I Just Want You to Know That 'Them' is Me": Intergroup Perspectives on Communication and Disability. In *Intergroup Communication* (pp. 117–137). New York, NY: Peter Lang Publishing, Inc.

Sabhlok, R. (2013). Smart watches, Google Glass and the wearable technology chocolate box. *Forbes.com.* Retrieved from http://www.forbes.com/sites/rajsabhlok/2013/12/05/smart watches-google-glass-and-the-wearable-technology-chocolate-box/

Sadiku, M. N. O. (2007). *Elements of Electromagnetics* (4th ed., pp. 380–390). New York, USA: Oxford University Press.

Sana, F., Weston, T., & Cepeda, N. J. (2013). Laptop multitasking hinders classroom learning for both users and nearby peers. *Computers & Education*, *62*, 24–31. doi:10.1016/j.compedu.2012.10.003

Scardamalia, M., & Bereiter, C. (1993). Technologies for knowledge-building discourse. *Communications of the ACM*, *36*(5), 37–41. doi:10.1145/155049.155056

Scardamalia, M., & Bereiter, C. (1994). Computer support for knowledge-building communities. *Journal of the Learning Sciences*, *3*(3), 265–283. doi:10.1207/s15327809jls0303_3

Schach, S. R. (2005). *Objected-oriented and classical software engineering* (6th ed.). New York: McGraw-Hill.

Schellens, T., & Valcke, M. (2005). Collaborative learning in asynchronous discussion groups: What about the impact on cognitive processing? *Computers in Human Behavior*, *21*(6), 957–975. doi:10.1016/j.chb.2004.02.025

Scheltic, T. (2013). Glove Prototype Provides Mobile Communications Capabilities for the Blind-Deaf. *Wearable Tech News.* Retrieved from http://www.wearabletechworld.com/topics/wearable-tech/articles/343796-glove-prototype-provides-mobile-communications-capabilities-the-blind.htm

Scheuer, O., Loll, F., Pinkwart, N., & McLaren, B. M. (2010). Computer-supported argumentation: A review of the state of the art. *Computer-Supported Collaborative Learning*, *5*(5), 43–102. doi:10.1007/s11412-009-9080-x

Schmidt, J., Geith, C., Haklev, S., & Thierstein, J. (2009). Peer-to-peer recognition of learning in open education. *International Review of Research in Open and Distance Learning, 10*, 5. Retrieved from http://www.irrodl.org/index.php/irrodl/article/view/641

Schneps, M. H., Ruel, J., Sonnert, G., Dussault, M., Griffin, M., & Sadler, P. M. (2014). Conceptualizing astronomical scale: Virtual simulations on handheld tablet computers reverse misconceptions. *Computers & Education, 70*, 269–280. doi:10.1016/j.compedu.2013.09.001

Schur, L. A. (2003). Barriers or opportunities? The causes of contingent and part-time work among people with disabilities. *Industrial Relations, 42*(4), 589–622. doi:10.1111/1468-232X.00308

Schwabe, G., & Göth, C. (2005). Mobile learning with a mobile game: Design and motivational effects. *Journal of Computer Assisted Learning, 21*(3), 204–216. doi:10.1111/j.1365-2729.2005.00128.x

Scott, M. (2013). Google Glass: West Michigan online physics teacher takes students on virtual field trip to Switzerland. *Michigan Live.* Retrieved from http://www.mlive.com/news/grand-rapids/index.ssf/2013/05/google_glass_west_michigan_onl.html

Seels, B. B., & Richey, R. C. (1994). *Instructional technology: The definition and domains of the field Association for Educational Communications and Technology.* Washington, DC: Association for Educational Communications and Technology.

Sellen, A. J., Fogg, A., Aitken, M., Hodges, S., Rother, C., & Wood, K. (2007). Do life-logging technologies support memory for the past?: an experimental study using sensecam.*Proceedings of the SIGCHI conference on Human factors in computing systems* (pp. 81-90). ACM. doi:10.1145/1240624.1240636

Senior, M. (2013). Appetite for cheaper biologics spreads across the map. *Nature Biotechnology, 31*(12), 1059–1061. doi:10.1038/nbt1213-1059 PMID:24316622

Serra, A., Carboni, D., & Marotto, V. (2010). Indoor pedestrian navigation system using a modern smartphone.*Proceedings of the 12th International Conference on Human Computer Interaction with Mobile Devices and Services.* doi:10.1145/1851600.1851683

Servais, J. (2014). *Technological Determinism and Social Change: Communication in a Tech-Mad World.* Lanham, Maryland: Lexington Books.

Serving America's Future: Increasing College Readiness. (2012). American Association of State Colleges and Universities (AASCU), Task Force on College Readiness. Retrieved from http://www.aascu.org/CollegeReadiness/ExecutiveSummary/

Shackel, B. (1991). Usability-Context, framework, definition, design and evaluation. In B. Shackel & S. J. Richardson (Eds.), *Human factors for informatics usability* (pp. 21–37). Cambridge: Cambridge University.

Shapiro, S., MacInnis, D. J., & Heckler, S. E. (1997). The effects of incidental AdExposure on the formation of consideration sets. *The Journal of Consumer Research, 24*(1), 94–104. doi:10.1086/209496

Sharples, M., Milrad, M., Arnedillo, S. I., & Vavoula, G. (2009). Mobile learning: Small devices, big issues. In N. Balacheff, S. Ludvigsen, T. De Jong, A. Lasonder, & S. Barnes (Eds.), *Technology enhanced learning: Principles and products* (pp. 233–249). Heidelberg: Springer. doi:10.1007/978-1-4020-9827-7_14

Sheehy, A. (2015). Now that wi-fi is done it's time for wi-bo. *Generator Research.com.* Retrieved from: http://www.generatorresearch.com/tekcarta/analysis-insight/now-that-wi-fi-is-done-its-time-for-wi-bo-wireless-body/ July, 2015.

Sheehy, A. (2014). *8 Mind-blowing uses of wearable technology (Seriously…). Generator Research.*

Shum, H. Y., Liao, M., & Zhang, S. (Eds.), (2001, October 24-26). Advances in Multimedia Information Processing-Pcm 2001. *Proceedings of theSecond IEEE Pacific Rim Conference on Multimedia*, Beijing, China (*Vol. 2*). Springer Science & Business Media.

Shute, V., Rieber, L., & Van Eck (2012). Games…and…learning. In R. Reiser, & J. Dempsey (Eds.), *Trends and issues in instructional design and technology (3rd ed.)* Boston, MA: Pearson Education.

Silva, A. R., Pinho, S., Macedo, L. M., & Moulin, C. J. (2013). Benefits of sensecam review on neuropsychological test performance. *American Journal of Preventive Medicine*, *44*(3), 302–307. doi:10.1016/j.amepre.2012.11.005 PMID:23415129

Silverstein, A. M. (1982). History of immunology. Development of the concept of immunologic specificity, I. *Cellular Immunology*, *67*(2), 396–409. doi:10.1016/0008-8749(82)90231-3 PMID:6177425

Singer, R. W., & Perry, A. J. (2015). Wearables: The well-dressed privacy policy. *Intellectual Property & Technology Law Journal*, *17*(7), 24–27.

Siqueira-Batista, R., Gomes, A. P., Albuquerque, V. S., Madalon-Fraga, R., Aleksandrowicz, A. M. C., & Geller, M. (2009). Ensino de imunologia na educação médica: Lições de Akira Kurosawa. *Revista Brasileira de Educação Médica*, *33*(2), 186–190. doi:10.1590/S0100-55022009000200004

Skinner, B. F. (1976). *Il comportamento verbale*. Milano, Italy: Franco Angeli.

Smaldino, S. E., Lowther, D. L., & Russell, J. W. (2012). *Instructional technology and media for learning* (10th ed.). Boston, MA: Allyn & Bacon.

Smart phones are fading. Wearables are next. (2014). *CNNMoney*. Retrieved from http://money.cnn.com/2014/03/19/technology/mobile/wearable-devices/

Smith, D. (2009). The rise of the virtual trainer. *PDDnet.com*. Retrieved from http://www.pddnet.com/news/2009/07/rise-virtual-trainer

Smith, A. C., Stewart, R., Shields, P., Hayes-Klosteridis, J., Robinson, P., & Yuan, R. (2005). Introductory biology courses: A framework to support active learning in large enrollment introductory science courses. *Cell Biology Education*, *4*(2), 143–156. doi:10.1187/cbe.04-08-0048 PMID:15917873

Sommerville, I. (2015). *Software Engineering* (10th ed.). NY: Pearson.

Sorge, C. (2007). What happens: Relationship of age and gender with science attitudes from elementary to middle school. *Science Educator*, *16*, 33–37.

Soyka, M. B., van de Veen, W., Holzmann, D., Akdis, M., & Akdis, C. A. (2014). Scientific foundations of allergen-specific immunotherapy for allergic disease. *Chest*, *146*(5), 1347–1357. doi:10.1378/chest.14-0049 PMID:25367471

Specht, M., Ternier, S., & Greller, W. (2011). Mobile augmented reality for learning: A case study. *Journal for Research Center for Educational Technology*, *7*(1), 117–127.

Spector, J. M. (2013). Emerging Educational Technologies and Research Directions. *Journal of Educational Technology & Society*, *16*(2), 21–30.

Squire, K., & Jan, M. (2007). Mad city mystery: Developing scientific argumentation skills with a place-based augmented reality game on handheld computers. *Journal of Science Education and Technology*, *16*(1), 5–29. doi:10.1007/s10956-006-9037-z

Srisuwan, P., & Barnes, S. J. (2008). Predicting online channel use for an online and print magazine: A case study. *Internet Research, 18*(3), 266–285. doi:10.1108/10662240810883317

Stackpool, C. M. (2013). *The Accuracy of Various Activity Trackers in Estimating Steps Taken and Energy Expenditure* [Master's Thesis].

Starner, T. (1998). Real-time American Sign Language recognition using desk and wearable computer based video. *IEEE Transactions on Pattern Analysis and Machine Intelligence*, 20(12).

Starner, T. (2013). Google glass lead: How wearing tech on our bodies actually helps it get Out of our way. *Wired.com.* Retrieved from http://www.wired.com/2013/12/the-paradox-of-wearables-close-to-your-body-but-keeping-tech-far-away/

Steele, B., Belza, B., & Cain, K. (2003). Bodies in motion: monitoring daily activity and exercise with motion sensors in people with chronic pulmonary disease. *Journal of Rehabilitation Research and Development*. Retrieved from http://www.rehab.research.va.gov/jour/03/40/5sup2/pdf/steele.pdf

Stel, V. S., Smit, J. H., Pluijm, S. M. F., Visser, M., Deeg, D. J. H., & Lips, P. (2004). Comparison of the LASA Physical Activity Questionnaire with a 7-day diary and pedometer. *Journal of Clinical Epidemiology, 57*(3), 252–258. doi:10.1016/j.jclinepi.2003.07.008 PMID:15066685

Stemler, S. (2001). An overview of content analysis. *Practical Assessment, Research & Evaluation, 7*(17), 137–146.

Stewart, M. (2015). Through the Google GLASS. *Pn, 69*(2), 34-37.

Stowell, J. R., & Nelson, J. M. (2007). Benefits of electronic audience response systems on student participation, learning, and emotion. *Teaching of Psychology, 34*(4), 253–258. doi:10.1080/00986280701700391

Strauss, A., & Corbin, J. (2008). *Basics of qualitative research: Grounded theory procedures and techniques* (3rd ed.). Newbury Park, CA: Sage.

Ström, R., Vendel, M., & Bredican, J. (2014). Mobile marketing: A literature review on its value for consumers and retailers. *Journal of retailing and consumer services*, 21(6), 1001-1012.

Students, Computers and Learning. Making the Connection. (2015). *OECD*, Paris. doi:.10.1787/9789264239555-en

Stukus, D. R., Green, T., Montandon, S. V., & Wada, K. J. (2015). Deficits in allergy knowledge among physicians at academic medical centers. *Annals of Allergy, Asthma & Immunology, 115*, 51–55. PMID:26024806

Success stories. (2015). *Google*. Retrieved from http://gsmamobileeconomy.covaibalem.

Suh, M., Carroll, K., & Cassill, N. (2010). Critical Review on Smart Clothing Product Development. *Journal of Textile and Apparel, 6*(4), 1–18. Retrieved from http://ojs.cnr.ncsu.edu/index.php/JTATM/article/view/702

Sullivan, F. R. (2008). Robotics and science literacy: Thinking skills, science process skills and systems understanding. *Journal of Research in Science Teaching, 45*(3), 373–394. doi:10.1002/tea.20238

Sultan, N. (2015). Reflective thoughts on the potential and challenges of wearable technology for healthcare provision and medical education. *International Journal of Information Management, 35*(5), 521–526. doi:10.1016/j.ijinfomgt.2015.04.010

Sundeen, T. H., & Sundeen, D. M. (2013). Instructional technology for rural schools: Access and acquisition. *Rural Special Education Quarterly, 32*(2).

Sung, E., & Mayer, R. E. (2013). Online multimedia learning with mobile devices and desktop computers: An experimental test of Clark's methods-not-media hypothesis. *Computers in Human Behavior, 29*(3), 639–647. doi:10.1016/j.chb.2012.10.022

Support the IANR Extension division. (2013). *NUFoundation.* Retrieved from https://nufoundation.org/unl/institute-of-agriculture-and-natural-resources/colleges-and-areas/unl/extension-division

Takacs, J., Pollock, C. L., Guenther, J. R., Bahar, M., Napier, C., & Hunt, M. (2013). Validation of the Fitbit One activity monitor device during treadmill walking. *Journal of Science and Medicine in Sport, 17*(5), 496–500. doi:10.1016/j.jsams.2013.10.241 PMID:24268570

Tally, S. (2012). *Digital badges show students' skills along with degree.* Purdue University News.

Tanabe, S., Hirano, S., & Saitoh, E. (2013). Wearable Power-Assist Locomotor (WPAL) for supporting upright walking in persons with paraplegia. *NeuroRehabilitation, 33*(1), 99. doi:10.3233/NRE-130932 PMID:23949036

Tanenbaum, J. G., Williams, A. M., Desjardins, A., & Tanenbaum, K. (2013). Democratizing technology: pleasure, utility and expressiveness in DIY and maker practice.*Proc. of CHI'13* (pp. 2603-2612). doi:10.1145/2470654.2481360

Tao, X. E. (2001). *Smart Fibres, Fabrics and Clothing.* Smart Fibres, Fabrics and Clothing; doi:10.1533/9781855737600

Tassoni, E. C. M. (2000). *Afetividade e aprendizagem: a relação professor-aluno. Psicologia, análise e crítica da prática educacional.* Campinas: ANPED.

Tavor, T. (2011). Online advertising development and their economic effectiveness. *Australian Journal of Business and Management Research, 1*(6), 121–133.

Teachers know best: What educators want from digital instructional tools. (2014). *Bill & Melinda Gates Foundation.* Retrieved from http://collegeready.gatesfoundation.org/sites/default/files/Teachers%20Know%20Best_0.pdf

Ten Cate, O., & Durning, S. (2007). Peer teaching in medical education: Twelve reasons to move from theory to practice. *Medical Teacher, 29*(6), 591–599. doi:10.1080/01421590701606799 PMID:17922354

Tenreiro-Vieira, C. (2004). Formação em pensamento crítico de professores de ciências: Impacte nas práticas de sala de aula e no nível de pensamento crítico dos alunos. *Revista Electrónica de Enseñanza de las Ciencias, 3*(3), 228–256.

Terbizan, D. J., Dolezal, B. A., & Albano, C. (2002). Validity of Seven Commercially Available Heart Rate Monitors. *Measurement in Physical Education and Exercise Science, 6*(4), 243–247. doi:10.1207/S15327841MPEE0604_3

The 10 Skills Employers Most Want In 2015 Graduates. (2015). *Forbes.* Retrieved from http://www.forbes.com/sites/susanadams/2014/11/12/the-10-skills-employers-most-want-in-2015-graduates/

The definition of educational technology. (2004AECT, Association for Educational Communications and Technology. Washington, DC.

The European year for development: Education. (2015). *European Parliamentary Research Services.* Retrieved from http://epthinktank.eu/2015/02/27/the-european-year-for-development-education/

The Florida Agricultural and Mechanical University. (2015). *Welcome to the technology education program at Florida A&M University.* Retrieved from http://www.famu.edu/index.cfm?TechEd

The mobile economy 2015. (2015). *GSMA.* Retrieved from http://gsmamobileeconomy.covaibalem

The World Economic Forum & Boston Consultancy Group. (2015). *New vision for education unlocking the potential of technology.* Retrieved from http://www3.weforum.org/docs/WEFUSA_NewVisionforEducation_Report2015.pdf

There's finally something awesome to do with Google Glass. (2015). *CNNMoney.* Retrieved from http://money.cnn.com/2015/06/09/technology/google-glass-parkinsons/

Thompson, W. R. (2015). ACSM's Health & Fitness Journal. *ACSM's Health & Fitness Journal, 19*(6), 9–18. Retrieved from http://journals.lww.com/acsm-healthfitness/Fulltext/2015/11000/WORLDWIDE_SURVEY_OF_FITNESS_TRENDS_FOR_2016__10th.5.aspx

Thorpe, M., Holm, J., & van den Boer, G. (2014). *Discovering the decisions within your business processes using IBM blueworks live*. New York: International Business Machines Corporation.

Tobon Vallejo, D., Falk, T., & Maier, M. (2014). MS-QI:A Modulation Spectrum-Based ECG Quality Index for Telehealth Applications. In IEEE transactions on bio-medical engineering.

Toh, P. (2013). *The New Age of Consumer Wearables*. CreateSpace Independent Publishing Platform.

Tomkins-Lane, C. C., & Haig, A. J. (2012). A review of activity monitors as a new technology for objectifying function in lumbar spinal stenosis. *Journal of Back and Musculoskeletal Rehabilitation, 25*(3), 177–185. doi:10.3233/BMR-2012-0325 PMID:22935856

Topping, K. (1998). Peer Assessment between Students in Colleges and Universities. *Review of Educational Research, 68*(3), 249–276. doi:10.3102/00346543068003249

Topping, K. (2005). Trends in Peer Learning. *Educational Psychology, 25*(6), 631–645. doi:10.1080/01443410500345172

Tracey, M. (2009). Design and development research: A model validation case. *Educational Technology Research and Development, 57*(4), 553–571. doi:10.1007/s11423-007-9075-0

Tracy, K. W. (2012). Mobile application development experiences on Apple's iOS and Android OS. *Potentials, IEEE, 31*(4), 30–34. doi:10.1109/MPOT.2011.2182571

Traxler, J. (2005). Case studies: introduction and overview. In A. Kukulska-Hulme & J. Traxler (Eds.), *Mobile learning: A handbook for educators and trainers* (pp. 173–188). London: Routledge.

Trees, A. R., & Jackson, M. H. (2007). The learning environment in clicker classrooms: Student processes of learning and involvement in large university-level courses using student response systems. *Learning, Media and Technology, 32*(1), 21–40. doi:10.1080/17439880601141179

Tremblay, M. S., Esliger, D. W., Tremblay, A., & Colley, R. (2007). Incidental movement, lifestyle-embedded activity and sleep: New frontiers in physical activity assessment. *Canadian Journal of Public Health. Revue Canadienne de Santé Publique, 98*(Suppl. 2.). doi:10.1139/H07-130

Trost, S. G. (2001). Objective Measurement of Physical Activity in Youth: Current Issues, Future Directions. *Exercise and Sport Sciences Reviews, 29*(1), 32–36. doi:10.1097/00003677-200101000-00007 PMID:11210445

Trost, S. G., & O'Neil, M. (2014). Clinical use of objective measures of physical activity. *British Journal of Sports Medicine, 48*(3), 178–181. doi:10.1136/bjsports-2013-093173 PMID:24311601

Tsubono, Y., Tsuji, I., Fujita, K., Nakaya, N., Hozawa, A., & Ohkubo, T. … Hisamichi, S. (2002). Validation of walking questionnaire for population-based prospective studies in Japan: comparison with pedometer. Journal of Epidemiology / Japan Epidemiological Association, 12(4), 305–309. Doi:10.2188/jea.12.305

Tudor-Locke, C., Craig, C. L., Thyfault, J. P., & Spence, J. C. (2013). A step-defined sedentary lifestyle index : < 5000 steps / day (Review). *Applied Physiology, Nutrition, and Metabolism,* 38(2), 100–114. PubMed

Tudor-Locke, C., Pangrazi, R. P., Corbin, C. B., Rutherford, W. J., Vincent, S. D., Raustorp, A., & Cuddihy, T. F. et al. (2004). BMI-referenced standards for recommended pedometer-determined steps/day in children. *Preventive Medicine, 38*(6), 857–864. doi:10.1016/j.ypmed.2003.12.018 PMID:15193909

Tudor-Locke, C., Williams, J. E., Reis, J. P., & Pluto, D. (2002). Utility of pedometers for assessing physical activity: Convergent validity. *Sports Medicine (Auckland, N.Z.)*, *32*(12), 795–808. doi:10.2165/00007256-200232120-00004 PMID:12238942

Turban, E., McLean, E. R., & Wetherbe, J. C. (2008). *Information technology for management*. New Jersey: John Wiley and sons, Inc.

Twentyman, J. (2014). *A business magazine report based on wearable technology*. Retrieved from http://raconteur.net/technology/welcome-to-the-wearable-workplace

Ullrich, C., Shen, R., Tong, R., & Tan, X. (2010). A mobile live video learning system for large-scale learning-system design and evaluation. *IEEE Transactions on Learning Technologies*, *30*(1), 6–17. doi:10.1109/TLT.2009.54

Universe of education technology spend 2013. (2014, June 10). *Futuresource Consulting*. Retrieved from http://reports.futuresource-consulting.com/tabid/64/ItemId/177649/Default.aspx

Unwin, D. (1972). *Mezzi e metodi. Guida alle tecnologie educative*. Brescia, Italy: La Scuola.

Valtchev, S. S., Baikova, E. N. & Jorge, L. R. (2012). *Electromagnetic field as the wireless transporter of energy*. Facta Universitatis Ser. Electrical Engineering.

Van den Akker, J. (1999). Principles and methods of development research. In J. van den Akker, R. Branch, K. Gustafson, N. Nieveen, & T. Plomp (Eds.), *Design approaches and tools in education and training* (pp. 1–14). Dordrecht: Kluwer Academic Publishers. doi:10.1007/978-94-011-4255-7_1

Van den Akker, J., Gravemeijer, K., McKenney, S., & Nieveen, N. (2006). Introduction to educational design research. In J. Van den Akker, K. Gravemeijer, S. McKenney, & N. Nieveen (Eds.), *Educational design research* (pp. 3–7). London: Routledge.

Van Hilten, J. J., Hoogland, G., van der Velde, E. A., van Dijk, J. G., Kerkhof, G. A., & Roos, R. A. C. (1993). Quantitative Assessment of Parkinsonian Patients by Continuous Wrist Activity Monitoring. *Clinical Neuropharmacology*, 16(1), 36–45.

Van Kranenburg, R. (2011). *The Internet of Things*. Retrieved from http://www.researchgate.net/profile/Matt_Ratto/publication/228360933_The_Internet_of_Things/links/0912f513755ebd1e87000000.pdf

Van Laarhoven, T., Johnson, J. W., Van Laarhoven-Myers, T., Grider, K. L., & Grider, K. M. (2009). The effectiveness of using a video iPod as a prompting device in employment settings. *Journal of Behavioral Education*, *18*(2), 119–141. doi:10.1007/s10864-009-9077-6

Van Laerhoven, K., Borazio, M., Kilian, D., & Schiele, B. (2008). Sustained logging and discrimination of sleep postures with low-level, wrist-worn sensors. *Proceedings of the International Symposium on Wearable Computers ISWC* (pp. 69–76). Doi:http://doi.org/ doi:10.1109/ISWC.2008.4911588

Van Leeuwenhoek, A. (1722). *Arcana naturae detecta*. Langerak.

Veiga, I. P. A. (1991). *Repensando a didáctica*. Campinas, SP: Papirus Editora.

Verma, D. (2013). *Google finally reveals glass specifications. MyGlass app now live*. Retrieved from http://techlomedia.in/2013/04/google-glass-specification-revealed-15817/

Verma, A. S., Rastogi, S., Agrahari, S., & Singh, A. (2011). Biotechnology in the realm of history. *Journal of Pharmacy and Bioallied Sciences*, *3*(3), 321–323. doi:10.4103/0975-7406.84430 PMID:21966150

Villar, R., Beltrame, T., Ku, R., & Hughson, R. L. (2013). Validation of the Hexoskin wearable body metrics vest to predict activities of daily living. Toronto, Ontario, Canada: Canadian Society for Exercise Physiology Annual General Meeting. Retrieved from https://cdn.shopify.com/s/files/1/0284/7802/files/CSEP_Hexoskin_Poster_-_University_of_Waterloo.pdf?11148

Vincenzi, A. B. (2005). *Conoscere la classe e lo studente: costruire e usare questionari pedagogici*. Trento, Italy: Erickson.

Von Krogh, G., & Von Hippel, E. (2006). The promise of research on open source software. *Management Science*, *52*(7), 975–983. doi:10.1287/mnsc.1060.0560

Wagner, D., Laforge, P., & Cripps, D. (2013). Lecture Material Retention: a First Trial Report on Flipped Classroom Strategies in Electronic Systems Engineering at the University of Regina. *Paper presented at theCanadian Engineering Education Association Conference*, Montreal, Canada.

Wagstaff, J. (2015). The path to a wearable future lies in academia. *Reuters.com.* Retrieved from http://www.reuters.com/article/2015/05/04/us-tech-academia-idUSKBN0NP1YR20150504

Wallace, A. (2014). Social Learning Platforms and the Flipped Classroom. *International Journal of Information and Education Technology*, *4*(4), 293–296. doi:10.7763/IJIET.2014.V4.416

Ward, M. R., & Lee, M. J. (2000). Internet shopping, consumer search and product branding. *Journal of Product and Brand Management*, *9*(1), 6–20. doi:10.1108/10610420010316302

Wasik, B. (2013). Why wearable tech will be as big as the smart phone. *Wired.com.* Retrieved from http://www.wired.com/2013/12/wearable-computers/

Wassom, B. (2014). *Augmented Reality Law, Privacy, and Ethics: Law, Society, and Emerging AR Technologies.* Waltham, MA: Elsevier.

Watson, J. D., & Crick, F. H. (1953). Molecular structure of nucleic acids; a structure for deoxyribose nucleic acid. *Nature*, *171*(4356), 737–738. doi:10.1038/171737a0 PMID:13054692

Wearable computing - Challenges and opportunities for privacy protection. (2014). *Office of the Privacy Commissioner.* Retrieved from https://www.priv.gc.ca/information/research-recherche/2014/wc_201401_e.asp

Wearable computing FAQ. (1997). *MIT.* Retrieved from: https://www.media.mit.edu/wearables/lizzy/FAQ/FAQ.txt

Wearable technology dives into the pool with Xmetrics. (2014). *Sporttechie.* Retrieved from http://www.sporttechie.com/2014/11/10/wearable-technology-dives-pool-xmetrics/

Wearable technology to enter the Mining industry. (2014). *Mining.com.* Retrieved from http://www.mining.com/web/wearable-technology-to-enter-the-mining-industry/

Weibert, A., Marshall, A., Aal, K., Schubert, K., & Rode, J. (2014). Sewing interest in E-textiles: Analyzing making from a gendered perspective. In *Proceedings of the 2014 conference on Designing interactive systems* (DIS '14), New York, NY, USA (pp. 15-24). Doi: http://doi.acm.org/10.1145/2598510.260088610.1145/2598510.2600886

Weickgenant, J. (2009). Plenty of spinning, but more than just the D.J. *The New York Times.* Retrieved from http://intransit.blogs.nytimes.com/2009/07/15/plenty-of-spinning-but-more-than-just-the-dj/?_r=0

Wenger, E. (1998). *Communities of Practice - Learning, Meaning, and Identity.* New York: Cambridge University Press. doi:10.1017/CBO9780511803932

Wertsch, J. V. (2007). Mediation. In H. Daniels, M. Cole, & J. V. Wertsch (Eds.), *The Cambridge Companion to Vygotsky* (pp. 178–192). Cambridge: Cambridge University Press. doi:10.1017/CCOL0521831040.008

What are electromagnetic fields: Summary of health effects. (2015). World Health Organization.

Wieczorek, A., & Wright, M. (2012). History of agricultural biotechnology: How crop development has evolved. *Nature Education Knowledge*, *3*(10), 1–9.

Wisbey, B., Montgomery, P. G., Pyne, D. B., & Rattray, B. (2010). Quantifying movement demands of AFL football using GPS tracking. *Journal of Science and Medicine in Sport*, *13*(5), 531–536. doi:10.1016/j.jsams.2009.09.002 PMID:19897414

Wiznia, D., Korom, R., Marzuk, P., Safdieh, J., & Grafstein, B. (2012). PBL 2.0: Enhancing problem-based learning through increased student participation. *Medical Education Online*, *17*(0). doi:10.3402/meo.v17i0.17375 PMID:22778540

Woda, T. (n. d.). *Privacy and wearable technology* (White Paper). Retrieved from http://www.uknow.com/wp-content/uploads/2014/03/Location-Services_White_Paper.pdf

Woolman, T.A. (2013). Base metals prospecting in Kagera region, Tanzania using remote sensing and biogeochemistry analytics. *The Compass: Earth Science Journal of Sigma Gamma Epsilon*, *85*(4).

Woolman, T. A., & Yi, J. C. (2013). An Analysis of the Use of Predictive Modeling with Business Intelligence Systems for Exploration of Precious Metals Using Biogeochemical Data. *International Journal of Business Intelligence Research*, *4*(2), 39–53. doi:10.4018/jbir.2013040103

Wu, H., Lee, S. W., Chang, H., & Liang, J. (2013). Current status, opportunities, and challenges of augmented reality in education. *Computers & Education*, *62*, 41–49. doi:10.1016/j.compedu.2012.10.024

Wu, T., Dameff, C., & Tully, J. (2014). Integrating Google Glass into simulation-based training: Experiences and future directions. *Journal of Biomedical Graphics and Computing*, *4*(2), 49. doi:10.5430/jbgc.v4n2p49

Wu, W. H., Wu, Y. C. J., Chen, C. Y., Kao, H. Y., Lin, C. H., & Huang, S. H. (2012). Review of trends from mobile learning studies: A meta-analysis. *Computers & Education*, *59*(2), 817–827. doi:10.1016/j.compedu.2012.03.016

Yang, C. C. (2004). Exploring factors affecting the adoption of mobile commerce in Singapore. *Telematics and Informatics*, *22*(3), 257–277. doi:10.1016/j.tele.2004.11.003

Yilmaz, B., Asyali, M. H., Arikan, E., Yetkin, S., & Ozgen, F. (2010). Sleep stage and obstructive apneaic epoch classification using single-lead ECG. *Biomedical Engineering Online*, *9*(1), 39. doi:10.1186/1475-925X-9-39 PMID:20723232

Yoon, S. J., & Kim, J. H. (2001). The effects of perceived consumer characteristics on the choice and use of Internet ads. *Brand Management*, *8*(4/5), 346–364. doi:10.1057/palgrave.bm.2540033

Young, J. (2015). Wearable teaching? College to experiment with Apple Watch as a learning tool. *Chronicle.com*. Retrieved from http://chronicle.com/blogs/wiredcampus/wearable-teaching-college-to-experiment-with-apple-watch-as-learning-tool/56459?cid=wc&utm_source=wc&utm_medium=en

Young, J. G., Trudeau, M., Odell, D., Marinelli, K., & Dennerlein, J. T. (2012). Touch-screen tablet user configurations and case-supported tilt affect head and neck flexion angles. *Work (Reading, Mass.)*, *41*(1), 81–91. PMID:22246308

Yourstone, S. A., Kraye, H. S., & Albaum, G. (2008). Classroom Questioning with Immediate Electronic Response: Do Clickers Improve Learning? *Decision Sciences Journal of Innovative Education*, *6*(1), 75–88. doi:10.1111/j.1540-4609.2007.00166.x

Zaslavsky, C. (1982). *Tic-tac-toe: And other three-in-a-row games from ancient Egypt to the modern computer*. New York, NY: Crowell.

Zickuhr, K., & Smith, A. (2012). Digital differences. *Pew Research Center*. Retrieved from http://www.pewinternet.org/2012/04/13/digital-differences/

About the Contributors

Janet Holland, Ph.D., is an Associate Professor at Emporia State University in Emporia, Kansas, teaching graduate students in Instructional Design and Technology. She has served as president and conference chair for the Kansas Association for Educational and Communications Technology. Research and publication interests include human-computer interactions in the field of instructional design and technology, online learning, mobile learning, globalization, and wearable technologies. Dr. Holland has published many book chapters and journal articles and has presented internationally in Paris, London, Rome, and many other locations. She has won awards for both teaching and research at Emporia State University. As an instructional designer, new technologies are continually examined in an effort to inspire innovative teaching and learning practices.

* * *

Abdulrahman Alharby has received his Ph.D., B.Sc, and M.Sc. degrees in electrical engineering from King Saud University in 1993 and 2000 respectively. From 1994 to 1996, he was an electronic instructor in the Electronic Institute in Riyadh. From 1997 to 2001, he worked for R&D and Network Operation departments at the Saudi Telecom Company (STC). From 2006 to 2009, he was the Vice Dean of the College of Computer Science and Engineering at Taibah University. Presently, he is working as a Dean of the faculty of Computing and Information Technology at the Northern Border University, while also serving as an assistant Professor at The University of Dammam since 2010. His current research interests include network security, intrusion detections, cryptography, cloud computing security, and wearable devices technology. He has also reviewed many conferences' and journals' papers for many Institute of Electrical and Electronics Engineers (IEEE) conferences at the national and international level.

Lisbeth Amhag, Ph.D., is a Senior Lecturer in Pedagogy/e-Learning in the Faculty of Education and Society at Malmö University, Sweden. Since 2002, she has served as a university teacher, course leader and examiner/assessor to further the education of teachers while also teaching courses at the graduate and undergraduate levels in the discipline of Education on campus and at a distance. Furthermore, she has been active in postdoctoral research within the field of computer-supported collaborative learning and distance education. She has published several national and international peer-reviewed articles, book chapters and conference papers about computer-supported collaborative learning and formative peer assessment processes at a distance. Dr. Amhag is currently focusing on research and the further development of interactive webinars and web based examinations. She is also a member of the editorial board for The Journal of Educational Research in Sweden. Malmö University students motivated her to win a 2010 Teaching Award. Dr. Amhag has taught nursing, media, television journalism, and web editing courses.

Ibrahim Arpaci works as an Assistant Professor at Gaziosmanpasa University, in the Department of Computer Education and Instructional Technology. He holds a B.S. in Computer Education and Instructional Technology from Anadolu University - Eskisehir, Turkey, a M.S. and a Ph.D. in Information Systems both from the Middle East Technical University, Ankara, Turkey. His current research interests are in educational technology, information technology adoption, and global information technology management. His publications have appeared in such journals as the British Journal of Educational Technology, Computers in Human Behavior, and the Journal of Global Information Technology Management.

Bradley Barker, Associate Professor and Youth Development Specialist with Nebraska 4-H received his Ph.D. in Administration, Curriculum and Instruction in the area of Instructional Technology in 2002. Dr. Barker spent eight years with Nebraska Educational Telecommunications where he was an Interactive Media Producer. Dr. Barker has directed media productions for the CLASS project, the Nebraska Law Enforcement Training Center, and the Nebraska National Guard. Dr. Barker has been the Principal Investigator on two National Science Foundation Grants to develop the Nebraska 4-H Robotics and GPS/GIS program with the intent to scale-up the program to a national audience. Dr. Barker was also the PI on the National 4-H Robotics: Engineering for Today and Tomorrow curriculum development grant for National 4-H Council and CSREES. He is currently the PI on the NSF ITEST wearable technologies (WearTec) project. Dr. Barker's research interests include the development and evaluation of educational technology systems for STEM education in non-formal learning environments.

Claudia Marcia Borges Barreto, Masters and Ph.D. in Pathology, is a professor at the Universidade Federal Fluminense in Brazil. She has been teaching Immunology and Introduction to the teaching practices. Dr. Barreto develops research, teaching and extension projects with undergraduate and graduate students involving the design and pedagogical use of playful, artistic and technological tools aiming for their use at both primary and higher education.

Érika Bertozzi de Aquino Mattos has a Bachelor's degree in 6-12 Biology education and a Master's degree in Science and Biotechnology at Universidade Federal Fluminense in Brazil. Now she is undertaking her Ph.D. studies in pathology. Her research questions deal with the possibility of intervening in the regulation of the immunological system through functional foods. Concomitantly, she continues to develop teaching strategies to stimulate more active teaching at all levels of school from K-12 though graduate courses.

Mike Butler is an Associate Professor in the department of Health, Physical Education, and Recreation at Emporia State University. He received his BS and MS in Kinesiology from the California State University at Fullerton and his PhD in Human Motor Learning from Auburn University. Dr. Butler teaches biomechanics, motor learning, statistics, and technology at the undergraduate and graduate levels and helped create ESU's online masters degree in Health Physical Education and Recreation (HPER), the first accredited program of its kind in the United States. Dr. Butler has presented on technology at the state, regional, and national levels and has won awards for his teaching and work with technology.

Alessio Drivet obtained a degree in social sciences at the University of Trento in 1975. Since 1976, he has been a teacher of Mathematics, Probability and Statistics at the Commercial Institutes, retired from 2011. He has published books on mathematics and computer science at several publishing houses.

He has been engaged in training activities for the University of Turin and the Ministry of Education since 1987. Currently, he serves as a Contract Professor at the University of Turin.

Rob Gibson serves as the Director of Learning Technologies at Emporia State University, serving in that capacity for 8 years. Prior to accepting the position at Emporia State, Rob worked for the University of Colorado at Denver Health Sciences Center. Rob holds a degree in instructional technology and distance education, two Master's degrees, and three bachelor's degrees. His interests include emerging technologies, online learning, technologies for disabled students, and user interface design.

Neal Grandgenett is the Peter Kiewit Professor of Mathematics Education at the University of Nebraska at Omaha with research interests in the development and evaluation of technology-based learning environments in mathematics and science. He has authored over 80 articles and research papers related to the use of educational technology in the learning of mathematics and science. He is a current co-investigator on two NSF curriculum development projects, funded within the ITEST and DR-K12 programs. Dr. Grandgenett is a review editor for new curriculum applications in the Mathematics and Computer Education Journal, published internationally.

Azizul Hassan is a member of the Tourism Consultants Network of the UK Tourism Society, and is currently working towards a PhD at Cardiff Metropolitan University. His main areas of research are: technology-supported marketing in tourism; innovative marketing dynamics; destination branding in tourism; cultural heritage tourism; heritage interpretation; and sustainable management/marketing alternatives for cultural heritage industries. He is a regular reviewer of Tourism Analysis, the International Journal of Human Resource Management, the International Journal of Ecotourism, the eReview of Tourism Research (eRTR) and The International Interdisciplinary Business-Economics Advancement Journal.

Talitha Hudgins completed her Ph.D. at the University of Utah in Teaching and Learning with an emphasis in Educational Technology. Dr. Hudgins has taught technology to students and teachers at the K-12 and college levels for several years. She served as the Secondary Department chair for the School of Education for two consecutive terms and now serves as an Associate Dean of University College. Dr. Hudgins' research focuses on educational technology in teaching and learning, professional development, and pedagogical and content knowledge. She is passionate about researching how educational technology interweaves with teaching and learning for students and teachers.

Rosa Iaquinta has degrees in Theological Sciences (University of Naples), History (University of Calabria); Historical Sciences (University of Calabria); a Postgraduate Diploma (University of Calabria) in Leadership and Management in Education; Students' Special Needs, the social, economic and cultural disadvantage; Educational strategies for teaching; The teacher's function; Evaluation and learning sciences; a Ph.D. in Theory, Technology and History of Education (University of Macerata). Rosa is a teacher at the High Secondary School in Italy. Her research interests are oriented to the study of educational dynamics, pedagogy and school didactics: school education, curriculum studies, school learning; methodology of the didactic research: educational sciences, research methodologies, teamwork research; methodologies of the didactics education and evaluation: curriculum studies, learning studies, evaluation research; theories and practice of the writing didactics: writing research, Humanities, teacher education; analysis of the educational and professional practice: teacher education, research methodologies.

Tiziana Iaquinta is a permanent researcher and an auxiliary professor in general and social pedagogy at the University of Catanzaro Magna Graecia. Teacher of general pedagogy and education and community pedagogy. Teaching contact person at the Pole of Cosenza for the Degree in Health Professions. Member of the Orientation Commission at the University Magna Graecia. Startup Project Manager for the kindergarten at the Campus of the University of Catanzaro Magna Graecia. Her research activity is related to issues of general education and in particular focuses on the analysis of the methodology of workshops as an opportunity for active learning that supports a customized learning process. Recently, her research interests have widened in scope to include the pedagogy of pain and the educational issues about life ending.

Jenny Keshwani is an Assistant Professor of Biological Systems Engineering and a Science Literacy Specialist in the Institute of Agriculture and Natural Resources at the University of Nebraska-Lincoln. She is also the coordinator for the National Center for Agricultural Literacy. Dr. Keshwani teaches undergraduate courses in biomedical engineering and engineering properties of biological materials. Her main focus is promoting science and engineering education in both formal and informal settings through her research, extension, and outreach activities. One of her goals is to make science and engineering engaging and relevant by exposing people of all ages to the joys of exploration and problem solving while connecting science and engineering to their everyday lives. Dr. Keshwani is actively engaged in several cross-disciplinary regional and national efforts related to STEM education and outreach.

Ben Leduc-Mills is a researcher in the education department at SparkFun Electronics where he investigates strategies to infuse learning spaces with more hands-on activities. He received his Ph.D. in Computer Science from the University of Colorado, Boulder, where he worked in the Craft Technology Lab building physical interfaces aimed at getting children involved in designing objects for 3D printing. Prior to his arrival at Colorado, Ben spent time in New York City at Eyebeam, a premier art and technology center, and earned his Master's degree from NYU's Interactive Telecommunications Program (ITP). Ben has taught, presented, and published at numerous academic conferences, Maker Fairs, hacker spaces, universities, and other informal learning spaces, on building and teaching emerging technology to children and incorporating it into learning spaces of all shapes and sizes.

Paul E. Luebbers, Ph.D., CSCS*D, USATF-1, is an Associate Professor in the Health, Physical Education, and Recreation Department at Emporia State University in Emporia, Kansas. He received his doctorate from Virginia Commonwealth University, in Richmond, Virginia and taught in the Department of Kinesiology at Northern Illinois University prior to returning to his home state of Kansas. Paul has a Level 1 USA Track & Field certification and has coached track & field at middle school, high school, and collegiate levels. He has been an active member of the American College of Sports Medicine since 2002. Additionally, he has been an active member in the National Strength and Conditioning Association since 2002, when he first became certified as a Strength and Conditioning Specialist. Paul is currently serving on the Executive Council for the NSCA Nutrition, Metabolism, and Body Composition special interest group and is also the State Director for the Kansas Chapter of the NSCA. He has a wide range of research interests and has published in the areas of anaerobic power development and testing, resistance training, metabolism, and nutrition.

Isabelle Mazza Guimarães, BSc in biology, has a Master's degree and is now a Ph.D. graduate student at the Universidade Federal Fluminense in Niterói, Brazil, studying both applied immunology and student difficulties in learning immunology. She has acted as a student representative for the Graduate Program in Science and Biotechnology. Research and publication interests include Special Education (students with visual or hearing disabilities), wearable technologies and immunology. She has won a research award at the International Congress of Mucosal Immunology. As a future professor, she is concerned with providing effective education using new technologies and innovative processes.

Mehtab Mehdi received his Masters in Computer Application from UP Technical University (A State Govt. University of India) in 2006 and is pursuing a Ph.D. in Computer Science from Sri Venkateshwara University in India. He has over 9 years of teaching experience in a variety of Universities in both India and Saudi Arabia. He has written five international papers and one book based on the PL/SQL. His areas of research are in Information Security, Data mining, HCI and to some extent Automation Theory. He is working as a lecturer in the Dept. of Computing and Information Technology in Northern Border University, Rafha, Saudi Arabia.

Katrina R. Miller, Ed.D., is a tenured Professor in the Department of Counselor Education, at Emporia State University, Emporia, Kansas. She is coordinator of the Rehabilitation Counseling master's degree program, for which she teaches online and traditional courses. Miller is a certified rehabilitation counselor, a nationally certified counselor, and a licensed professional counselor. She has authored and was awarded a federal, five-year training grant to prepare counselors to assist persons with disabilities with pre-employment and employment training; and, she has published numerous papers on the communication concerns of deaf and hard of hearing individuals, including social attitudes towards cochlear implant technology, and the use of digital apps with consumers for employment skills development.

Carl Nelson is a professor in the Department of Mechanical and Materials Engineering at the University of Nebraska-Lincoln and a licensed professional mechanical engineer. He received MS and PhD degrees in mechanical engineering from Purdue University. He is an active researcher in the areas of mechanical design/analysis, robotics, medical devices, and STEM education.

Gwen Nugent is a Research Professor at the Nebraska Center for Research on Children, Youth, Families and Schools (CYFS) at the University of Nebraska-Lincoln and Schools (CYFS) at the University of Nebraska-Lincoln. Dr. Nugent coordinates implementation and research projects focusing on the development and delivery of instruction and training to improve student learning and teacher competencies. She has over 30 years experience in the design and evaluation of mediated instruction and training and is currently a principal investigator and evaluator for grants funded by the U.S. National Science Foundation, the Department of Education, and the Department of Agriculture.

Damara Goff Paris, EdD, NCC, CRC, LPC, completed a master's degree in Rehabilitation Counseling with the Deaf at Western Oregon University in Monmouth, Oregon, and a doctorate in Deaf Education and Deaf Studies at Lamar University at Beaumont, Texas. She is a Licensed Professional Counselor (LPC), certified rehabilitation counselor (CRC), and a nationally certified counselor (NCC). Upon receiving her master's degree, she began working as the director and lead counselor of the Connection Program, a mental health facility for individuals who are Deaf or Hard of Hearing (DHH) and have co-occurring disorders. Dr. Paris' administrative, counseling, teaching and research experiences have provided her with career opportunities in the private, non-profit, state and university fields. These positions have included directing a statewide telecommunications assistance program, overseeing a national marketing team with Sprint, and directing a doctoral program in Deaf Studies and Deaf Education. Her research interests are diverse and include indigenous, women's and disability studies, particularly the intersectionality of all three minorities. She also researches technology relevant to disabilities, and rehabilitation counseling issues. Dr. Paris has participated in or established research projects about Tribal Vocational Rehabilitation (TR), issues of the Deaf Professoriate, and anxiety of counselors-in-training. A qualitative researcher, she has published several articles regarding cross-cultural implications of being Native American and Deaf. Dr. Paris has also published six books and is working on a Gallaudet University Press (GUP) book pertaining to the criminal justice system and Deaf offenders.

Donatella Privitera is an Associate Professor at the University of Catania (Italy). Prior to becoming a fulltime academic, she had work experience in a multinational company in the marketing area. At present, she is teaching economic geography and its implications on tourism and regional development. She has presented papers at several conferences and published in international books. Research and publication interests include sustainable cities; green economy; e-tourism; food marketing; and transport geography. She has participated in a European Union (EU) project HERODOT –INTERREG IIIB Archimed 2000-2006 and is currently involved in cooperative national research projects. She is a reviewer for several national and international journals and a member of the editorial board of the International Journal of Sustainable Economies Management (IGI, USA), Economic Insight-Trends and Challenges (Petroleum-Gas University), and the Economics of Agriculture (Institute of Agricultural Economics, Serbia).

Jason Ribeiro received his Masters of Education degree in 2015 from Brock University and is currently pursuing a Ph.D. specializing in Leadership, Policy and Governance at the University of Calgary in Alberta, Canada. He has served as a K-12 teacher, student government representative and community leader. Jason has published scholarly articles on topics ranging from technology acquisition, teaching in online environments and teacher technology adoption. His work has been presented at over a dozen peer-reviewed, national and international academic conferences. Jason has been the recipient of such prestigious awards as the Brock University Spirit of Brock Medal, Jack M. Miller Excellence in Research Award and President's Surgite Award. His current research aims to understand how senior leadership teams can improve the impacts of their educational technology projects as they try to lead more complex, adaptive learning environments.

Alexander Goncalves da Silva has a Bachelor's degree in Biology at Universidade Federal Fluminense. At present, he is undertaking his Master's degree in Immunology and Inflammation at the Universidade Federal do Rio de Janeiro. Currently, he continues to study what influences Immunology teaching and learning processes.

Gerlinde Agate Platais Brasil Teixeira is a MD-Ph.D. and full professor at Universidade Federal Fluminense in Brazil. She has been teaching and researching Immunology for the last 30 years, where she has developed a mouse model to study food allergies and has published over 25 papers. In the last 15 years, she has also been involved in teaching training courses where she has developed over 10 teaching models. She has also participated in graduate courses as a professor and advisor in both teaching and Immunology research.

Daniel Watanabe, MS, is an Instructional Designer at Olympic College in Bremerton, Washington, working on a Title III grant from the US Department of Education. Mr. Watanabe taught graduate and undergraduate level courses in instructional design and technology as an adjunct faculty member and graduate teaching assistant at Emporia State University in Emporia, Kansas. Research and publication interests include mobile learning, augmented reality, and conceptual frameworks for online professional development courses. As an instructional designer and teacher, Mr. Watanabe continually seeks ways to relate to learners and make learning technologies accessible to learners as a tool to enhance content.

Thomas A. Woolman is the Senior Systems Architect and CEO at On Target Technologies. He has consulted in the business intelligence and enterprise performance management fields for over 20 years for industries ranging from investment and commercial banking, to minerals exploration, utilities, energy, pharmaceuticals, insurance, aerospace, manufacturing, airlines, defense and for the U.S. Government and municipalities. He has also consulted internationally in the field of economic geology, prospecting with biogeochemistry analytics and predictive modeling. He has a Master's degree in Physical Science (Earth Science concentration) and a graduate certificate in Geospatial Analysis from Emporia State University, a Master's degree in Business Intelligence and Analytics from Saint Joseph's University and obtained his B.S. in Information Systems from Drexel University. He is currently completing his MBA degree at Saint Joseph's University. His professional interests are in economic geology, biogeochemistry mineral exploration, data mining, predictive modeling, OLAP databases and enterprise performance management systems.

Index

www.ingramcontent.com/pod-product-compliance
Lightning Source LLC
LaVergne TN
LVHW062121300926
849161LV00010B/2516
9781522500698